Tomasz G. Smolinski, Mariofanna G. Milanova and Aboul-Ella Hassanien (Eds.)

Applications of Computational Intelligence in Biology

Studies in Computational Intelligence, Volume 122

Editor-in-chief
Prof. Janusz Kacprzyk
Systems Research Institute
Polish Academy of Sciences
ul. Newelska 6
01-447 Warsaw
Poland
E-mail: kacprzyk@ibspan.waw.pl

Tomasz G. Smolinski
Mariofanna G. Milanova
Aboul-Ella Hassanien
(Eds.)

Applications of Computational Intelligence in Biology

Current Trends and Open Problems

With 107 Figures and 43 Tables

 Springer

Dr. Tomasz G. Smolinski
Department of Biology
Emory University
1510 Clifton Road NE
Atlanta, Georgia 30322
USA
tomasz.smolinski@emory.edu

Professor Mariofanna G. Milanova
Department of Computer Science
University of Arkansas at Little Rock
2801 S. University Ave.
Little Rock, Arkansas 72204
USA
mgmilanova@ualr.edu

Professor Aboul-Ella Hassanien
Department of Quantitative Methods and
Information Systems
College of Business and Administration
Kuwait University
P.O. Box 5486
Safat, 13055
Kuwait
abo@cba.edu.kw

and

Department of Information Technology
Faculty of Computers and Information
Cairo University
5 Ahamed Zewal Street
Orman, Giza
Egypt
a.hassanien@fci-cu.edu.eg

ISBN 978-3-540-78533-0   e-ISBN 978-3-540-78534-7
DOI: 10.1007/978-3-540-78534-7

Studies in Computational Intelligence ISSN 1860-949X

Library of Congress Control Number: 2008922728

Cover design: Deblik, Berlin, Germany

Printed on acid-free paper

9 8 7 6 5 4 3 2 1

springer.com

To my mother
Aleksandra Smolińska

Tomasz G. Smolinski

To
my family

Mariofanna G. Milanova

To
my family

Aboul-Ella Hassanien

Preface

Computational Intelligence (CI) has been a tremendously active area of research for the past decade or so. There are many successful applications of CI in many subfields of biology, including bioinformatics, computational genomics, protein structure prediction, or neuronal systems modeling and analysis. However, there still are many open problems in biology that are in desperate need of advanced and efficient computational methodologies to deal with tremendous amounts of data that those problems are plagued by. Unfortunately, biology researchers are very often unaware of the abundance of computational techniques that they could put to use to help them analyze and understand the data underlying their research inquiries. On the other hand, computational intelligence practitioners are often unfamiliar with the particular problems that their new, state-of-the-art algorithms could be successfully applied for. The separation between the two worlds is partially caused by the use of *different languages* in these two spheres of science, but also by the relatively small number of publications devoted solely to the purpose of facilitating the exchange of new computational algorithms and methodologies on one hand, and the needs of the biology realm on the other.

The purpose of this book is to provide a medium for such an exchange of expertise and concerns. In order to achieve the goal, we have solicited contributions from both computational intelligence as well as biology researchers. We have collected contributions from the CI community describing powerful new methodologies that could, or currently are, utilized for biology-oriented applications. On the other hand, the book also contains chapters devoted to open problems in biology that are in need of strong computational techniques, so the CI community can find a brand new and potentially intriguing spectrum of applications.

We have divided the book into four major parts. Part I, *Techniques and Methodologies*, contains a selection of contributions that provide a detailed description of several theories and methods that could be (or to some extent already are) of great benefit to biologists dealing with problems of data analysis (including large-volume, imprecise, and incomplete data), search-space exploration, optimization, etc.

In Chapter 1, by *Andrew Hamilton-Wright and Daniel W. Stashuk*, present a statistically based pattern discovery tool that produces a rule-based description of complex data through the set of its statistically significant associations. Although many biologists are familiar with statistical data analysis methods and use them quite extensively to quantify and summarize their data, they relatively rarely employ more complex statistically motivated techniques to thoroughly mine their databases in search of concise patterns and/or classification rules. This chapter provides an introduction to such types of analysis, supported by a discussion of the performance of the presented technique on a series of biologically relevant data distributions.

Chapter 2, by *Lech Polkowski and Piotr Artiemjew*, describes another very powerful methodology designed to deal with analysis of inexact and/or incomplete data. By utilizing the presented theory of Rough Sets (RS), biologists would be able to cope with uncertainty due to ambiguity of classification caused by incompleteness of knowledge, so prevalent in many real-life databases.

Chapters 3, by *Thomas McTavish and Diego Restrepo, and 4, by Antonio López Jaimes and Carlos A. Coello Coello*, treat on selected aspects of the field of Evolutionary Computation (EC). EC, itself being biologically-inspired (by the principles of evolution, natural selection, etc.), is by no means being used exclusively in biological problems and has found many applications across many disciplines. Very potent and flexible approaches based on Genetic Algorithms (GAs), Evolution Strategies (ESs), Multi-Objective Evolutionary Algorithms (MOEAs), have been very successfully applied to many problems of function optimization, parameter space exploration, etc. The first of the two chapters presents an introduction to Evolutionary Algorithms (EAs) in general and their applicability to various biological problems, focusing on EAs' use as an optimization technique for fitting parameters to a biological model. The second chapters concentrates on MOEAs (as applied to problems where multi-criteria optimization is necessary) and discusses some of their potential uses in biology.

Part II of the book, *Current Trends*, presents a selection of chapters describing specific existing and on-going applications of Computational Intelligence in biology.

In Chapter 5, *Wit Jakuczun*, introduces a technique based on a hybridization of wavelets and Support Vector Machines (SVMs), called Local Classifiers, for analysis and classification of signals. The author also presents an application of the method to a real-life problem of analysis of local field potentials recorded within barrel cortex of an awake rat.

Ryszard Tadeusiewicz, in Chapter 6, presents a systematic study of an application of Artificial Neural Networks (ANNs) for evaluation of biological activity of chemical compounds. Even though ANNs are quite commonly utilized for such prediction tasks, little is known about the specific parameters of a network that should be employed for a given dataset. In this chapter, the author compares several different ANN architectures and provides a detailed discussion of benefits and weaknesses of each one of them.

In the study described in Chapter 7, by *Bai-Tao Zhou and Joong-Hwan Baek*, the authors use machine vision to detect distinctive behavioral phenotypes of thread-shape microscopic organism, such as *Caenorhabditis elegans* (*C.elegans*). The first part of this chapter introduces an animal auto-tracking and imaging system capable of following an individual animal for extended periods of time by saving a time-indexed image sequences representing its locomotion and body postures. Then the authors present a series of image processing procedures for gradually shrinking the thread-shape representation into a Bend Angle Series expression (BASe), which later is the foundation of n-order-difference calculation for static and locomotion pattern extraction. Finally, for mining distinctive behaviors, the Hierarchical Density Shaving (HDS) clustering method is applied for compacting, ranking and identifying unique static and locomotion patterns, which combined represent distinctive behavioral phenotypes for a specific species.

In Chapter 8, *Dah-Jye Lee et al.* introduce a simple and accurate real-time contour matching technique specifically for applications involving fish species recognition and migration monitoring. The authors describe FishID, a prototype vision system that employs a software implementation of their newly developed contour matching algorithms. They also discuss the challenges involved in the design of this system, both hardware and software, and present results from a field test of the system at Prosser Dam in Prosser, Washington, USA.

Chapter 9, by *Dawn R. Magness, Falk Huettmann, and John M. Morton*, propose to utilize the technique called Random Forests to predicted species distribution maps that would serve as a metric for ecological inventory and monitoring programs. The authors use Random Forests a highly accurate bagging classification algorithm to build multi-species avian distribution models using data collected as part of the Kenai National Wildlife Refuge Long-term Ecological Monitoring Program (LTEMP). The chapter provides some background on the application of this method to real-life data and a discussion of its value for evaluating climate change impacts on species distributions.

Concluding Part II is Chapter 10, by *John T. Langton, Elizabeth A. Gifford, and Timothy J. Hickey*, in which the authors present a number of visualization methods and tools for investigating large, multidimensional data sets, which are very common in virtually every biology lab nowadays. The chapter focuses on approaches that have been used to analyze a model neuron simulation database but could be applied to other domains.

Part III, *Open Problems*, contains a collection of chapters describing biological problems, for which, although some computational methods have been utilized to deal with them, there is still a need for efficient CI techniques to be applied.

Chapter 11, by *Laila A. Nahum and Sergio L. Pereira*, calls for an integrated approach between molecular evolution and Computational Intelligence to be applied to analysis of molecular sequences to determine gene function and phylogenetic relationships of organisms. The authors point out areas in the field of phylogenomics (*phylogenetics* and *genomics*) that require further development, such as computational tools and methods to manipulate large and diverse data sets, which may ultimately lead to a better system-based understanding of biological processes in different environments.

In Chapter 12, *Vladik Kreinovich and Max Shpak*, discuss the computational aspects of aggregability for linear and non-linear systems, which directly relate to aggregability in biological systems such as population genetic systems and multi-species (multi-variable) systems in ecology. More specifically, the authors investigate the problem of conditional aggregability (i.e., aggregability restricted to modular states) and aggregation of variables in biologically relevant quadratic dynamical systems.

In the study described in Chapter 13, by *Ying Xie et al.*, the authors designed and implemented a prototype of a conceptual biology research support platform that consists of a set of interrelated information extraction, mining, reasoning, and visualizing technologies to automatically generate several types of biomedical hypotheses and to facilitate researchers in validating generated hypotheses. Such a platform could utilize vast amounts of published biomedical data, via interacting with certain search engines such as PubMed, to enhance and speed up biomedical research. The chapter presents a detailed description of the proposed approach as well as a discussion of future research and development directions and needs.

Chapter 14, by *Cengiz Günay et al.*, concludes this part of the book. This chapter constitutes mini-proceedings of the Workshop on Physiology Databases and Analysis Software that was a part of the Annual Computational Neuroscience Meeting CNS*2007 that took place in Toronto, Canada in July of 2007 and comprises of several selected contributions provided by the participants. In Section 14.2, Thomas M. Morse discusses the current uses and potential applications of CI for electrophysiological databases (EPDBs). Sections 14.3 by Padraig Gleeson et al., 14.4 by Horatiu Voicu, 14.5 by Cengiz Günay, and 14.6 by Peter Andrews et al., describe some currently available data-exchange and analysis platforms and implementations. Finally, Sections 14.7 by Gloster Aaron and 14.8 by Jean-Marc Fellous present some interesting open problems in electrophysiology with examples of analysis techniques, including CI-motivated approaches.

We have decided to devote the last section of the book, Part IV, entirely to *Cognitive Biology*, as it symbolizes a perfect common ground for the intersection of Computational Intelligence and biology. On one hand, the CI community has already tremendously benefited from the insights derived from research in this sub-field of biology (e.g., development of multi-agent systems has been an extensively pursued area of CI for several years now), but on the other, there is a seemingly endless sea of opportunities for applications of CI methodologies to further study cognition.

In Chapter 15, *Stan Franklin and Michael H. Ferkin* advocate studying animal cognition by means of computational control architectures based on biologically and psychologically inspired, broad, integrative, hybrid models of cognition. The authors introduce the LIDA (Learning Intelligent Distribution Agent) model. By using this model, animal experiments can be replicated in artificial environments by means of virtual software agents controlled by such architectures. The study described in this chapter explores the possibility of such experiments using a virtual or a robotic vole to replicate, and to predict, the behavior of live voles, thus applying computational intelligence to cognitive ethology.

Chapter 16, by *David Windridge and Josef Kittler*, is a survey of the fundamental constraints upon self-updating representation in cognitive agents of natural and artificial origin. The authors argue that perception-action frameworks provide an appropriate basis for the development of an empirically meaningful criterion for validating perceptual categories. In this scenario, hypotheses about the agent's world are defined in terms of environmental affordances (characterized in terms of the agent's active capabilities). Accordingly, the grounding of such a priori 'bootstrap' representational hypotheses is ensured via the process of natural selection in biological agents capable of autonomous cognitive-updating.

The editors are very grateful to the authors of the contributions included in this volume and to the referees for their tremendous service by critically reviewing the chapters. We would especially like to thank Prof. Janusz Kacprzyk, Editor-in-chief of the series *"Studies in Computational*

Intelligence," and Dr. Thomas Ditzinger, Senior In-house Editor of Springer Verlag, Germany, for their help, editorial assistance, and excellent cooperation to produce this important scientific work. We sincerely hope that this collection of contributions from biologists and CI practitioners alike will prove useful to researchers in both those fields and that it will facilitate a productive dialog between the two spheres of science and result in fruitful collaborations and scientific advancements on both sides.

Atlanta, Georgia, USA
Little Rock, Arkansas, USA
Cairo, Egypt
December, 2007

Tomasz G. Smolinski
Mariofanna G. Milanova
Aboul-Ella Hassanien
(editors)

Contents

Part III Open Problems

Part IV Cognitive Biology

List of Contributors

Gloster Aaron
Department of Biology,
Wesleyan University
Middletown, CT 06459, USA
gaaron@wesleyan.edu

Peter Andrews
Cold Spring Harbor Laboratory
Cold Spring Harbor, NY 11724, USA
paa@drpa.us

James K. Archibald
Department of Electrical and
Computer Engineering,
Brigham Young University
Provo, UT 84602, USA
jka@byu.edu

Piotr Artiemjew
Department of Mathematics and
Computer Science,
University of Warmia and Mazury
Żołnierska 14, Olsztyn, Poland
artem@matman.uwm.edu.pl

Joong-Hwan Baek
School of Electronics, Telecommuni-
cation, and Computer Engineering,
Korea Aerospace University
Koyang City, South Korea
jhbaek@kau.ac.kr

Hemant Bokil
Cold Spring Harbor Laboratory
Cold Spring Harbor, NY 11724, USA
bokil@cshl.edu

Carlos A. Coello Coello
CINVESTAV-IPN
Evolutionary Computation Group,
Departamento de Computación,
Av. IPN No. 2508
Col. San Pedro Zacatenco
México, D.F. 07360, Mexico
ccoello@cs.cinvestav.mx

Sharon Crook
Department of Mathematics and
Statistics, Arizona State University
Tempe, Arizona, USA
Sharon.Crook@asu.edu

Aaron W. Dennis
Department of Electrical and
Computer Engineering,
Brigham Young University
Provo, UT 84602, USA
aaron_dennis@byu.net

Jean-Marc Fellous
Department of Psychology,
University of Arizona
Tucson, AZ 85721, USA
fellous@email.arizona.edu

Michael H. Ferkin
Department of Biology,
The University of Memphis
Memphis, TN 38152, USA
mhferkin@memphis.edu

Stan Franklin
Institute of Intelligent Systems,
FedEx Institute of Technology,
The University of Memphis
Memphis, TN 38152, USA
franklin@memphis.edu

Elizabeth A. Gifford
Michtom School of Computer
Science, Brandeis University
Waltham, MA 02254, USA
egifford@cs.brandeis.edu

Padraig Gleeson
Department of Physiology,
University College London
London, UK
p.gleeson@ucl.ac.uk

Cengiz Günay
Department of Biology,
Emory University
Atlanta, Georgia 30322, USA
cgunay@emory.edu

Andrew Hamilton-Wright
School of Rehabilitation Therapy,
Queen's University
Computing and Information Science,
University of Guelph
Mathematics and Computer Science,
Mount Allison University
Kingston, ON K7K 1T3, Canada
andrewhw@ieee.org

Timothy J. Hickey
Michtom School of Computer
Science, Brandeis University
Waltham, MA 02254, USA
tim@cs.brandeis.edu

Falk Huettmann
University of Alaska, EWHALE Lab,
Institute of Arctic Biology,
Department of Biology & Wildlife
Fairbanks, AK 99775, USA
fffh@uaf.edu

Wit Jakuczun
Nencki Institute of
Experimental Biology
Pasteura 3 St.,
02-093 Warsaw, Poland
WLOG Solutions
Harfowa 1A/25 St.,
02-389 Warsaw, Poland
W.Jakuczun@wlogsolutions.com

Antonio López Jaimes
CINVESTAV-IPN
Evolutionary Computation Group,
Departamento de Computación,
Av. IPN No. 2508
Col. San Pedro Zacatenco
México, D.F. 07360, Mexico
tonio.jaimes@gmail.com

Jayasimha Katukuri
Center for Advanced
Computer Studies,
University of Louisiana at Lafayette
Lafayette, LA 70503, USA
jrk8907@cacs.louisiana.edu

Josef Kittler
School of Electronics and Physical
Sciences, University of Surrey
Guildford, Surrey, GU2 7XH,
United Kingdom
J.Kittler@surrey.ac.uk

David Kleinfeld
Department of Physics,
University of California, San Diego
La Jolla, CA 92093, USA
dk@physics.ucsd.edu

Vladik Kreinovich
University of Texas at El Paso
El Paso, TX 79968, USA
vladik@utep.edu

John T. Langton
Michtom School of Computer
Science, Brandeis University
Waltham, MA 02254, USA
psyc@cs.brandeis.edu

Dah-Jye Lee
Department of Electrical and
Computer Engineering,
Brigham Young University
Provo, UT 84602, USA
djlee@ee.byu.edu

Catherine Loader
Department of Statistics,
University of Auckland
Auckland 1142, New Zealand
c.loader@stat.auckland.ac.nz

William W. Lytton
Department of Physiology,
Department of Pharmacology and
Neurology,
State University of New York -
Downstate
Brooklyn, NY 11203, USA
billl@neurosim.downstate.edu

Dawn R. Magness
University of Alaska, EWHALE Lab,
Department of Biology & Wildlife
Fairbanks, AK 99775, USA
Kenai National Wildlife Refuge,
U.S. Fish and Wildlife Service
Soldotna, AK 99669, USA
dawn.magness@uaf.edu

Hiren Maniar
Cold Spring Harbor Laboratory
Cold Spring Harbor, NY 11724, USA
maniar@cshl.edu

Thomas McTavish
Computational Bioscience Program,
University of Colorado
Health Sciences Center
P.O. Box 6511, Mail Stop 8303
Aurora, CO 80045, USA
Thomas.McTavish@uchsc.edu

Samar Mehta
School of Medicine, State University
of New York - Downstate
Brooklyn, New York 11203, USA
sbmehta@gmail.com

Partha P. Mitra
Cold Spring Harbor Laboratory
Cold Spring Harbor, NY 11724, USA
mitra@cshl.edu

Thomas M. Morse
Department of Neurobiology,
Yale University
New Haven, CT 06510, USA
tom.morse@yale.edu

John M. Morton
Kenai National Wildlife Refuge,
U.S. Fish and Wildlife Service
Soldotna, AK 99669, USA
john_m_morton@fws.gov

Laila A. Nahum
Bay Paul Center,
Marine Biological Laboratory
Woods Hole, MA 02543, USA
laila@nahum.com.br

Sergio L. Pereira
Department of Natural History,
Royal Ontario Museum
Toronto, ON M5S 2C6, Canada
sergio.pereira@utoronto.ca

Lech Polkowski
Polish-Japanese Institute of
Information Technology
Koszykowa 86,
02008 Warszawa, Poland
polkow@pjwstk.edu.pl

Tony Presti
Araicom Research, LLC.
1355 Peachtree Rd NE
Atlanta, GA, USA
tony.presti@araicom.com

Vijay V. Raghavan
Center for Advanced
Computer Studies,
University of Louisiana at Lafayette
Lafayette, LA 70503, USA
Raghavan@cacs.louisiana.edu

Diego Restrepo
Department of Cell and
Developmental Biology,
University of Colorado
Health Sciences Center
RC-1 South, Room 11119
PO Box 6511, Mail Stop 8108
Aurora, CO 80045, USA
Diego.Restrepo@uchsc.edu

Robert B. Schoenberger
Symmetron, LLC a div. of
ManTech International Corp.
Fairfax, VA 22033, USA
Robert.Schoenberger@mantech-ist.com

Dennis K. Shiozawa
Department of Biology,
Brigham Young University
Provo, UT 84602, USA
dennis_shiozawa@byu.edu

Max Shpak
University of Texas at El Paso
El Paso, TX 79968, USA
mshpak@utep.edu

Angus Silver
Department of Physiology,
University College London
London, UK
a.silver@ucl.ac.uk

Tomasz G. Smolinski
Department of Biology,
Emory University
Atlanta, Georgia 30322, USA
tsmolin@emory.edu

Daniel W. Stashuk
Systems Design Engineering,
University of Waterloo
Davis Centre, DC-2613
200 University Ave. W
Waterloo, ON N2L 3G1, Canada
stashuk@pami.uwaterloo.ca

Volker Steuber
Department of Physiology,
University College London
London, UK
School of Computer Science,
University of Hertfordshire
Hatfield, Herts, UK
V.Steuber@ucl.ac.uk

Ryszard Tadeusiewicz
AGH University of Science and
Technology
30 Mickiewicza Ave.,
30-059 Krakow, Poland
rtad@agh.edu.pl

David Thomson
Department of Mathematics and
Statistics, Queen's University
Kingston, ON, Canada
djt@mast.queensu.ca

Horatiu Voicu
Department of Neurobiology and
Anatomy, University of Texas Health
Science Center
Houston, TX 77030, USA
horatiu@voicu.us

David Windridge
School of Electronics and
Physical Sciences,
University of Surrey
Guildford, Surrey, GU2 7XH,
United Kingdom
D.Windridge@surrey.ac.uk

Ying Xie
Department of Computer Science
and Information Systems,
Kennesaw State University
Kennesaw, GA 30144, USA
yxie2@kennesaw.edu

Bai-Tao Zhou
School of Electronics, Telecommuni-
cation, and Computer Engineering,
Korea Aerospace University
Koyang City, South Korea
zhou@kau.ac.kr

List of Referees

Arvin Agah
Department of Electrical Engineering
and Computer Science
The University of Kansas
Lawrence, Kansas 66045, USA
agah@ku.edu

James Archibald
Department of Electrical and
Computer Engineering,
Brigham Young University
Provo, Utah 84602
jka@ee.byu.edu

Joong-Hwan Baek
School of Electronics, Telecommuni-
cation, and Computer Engineering,
Korea Aerospace University
Koyang City, South Korea
jhbaek@kau.ac.kr

Vitoantonio Bevilacqua
Polytechnic of Bari
70125 Bari, Italy
bevilacqua@poliba.it

Grzegorz M. Boratyn
National Center for
Biotechnology Information
U.S. National Library of Medicine
Bethesda, MD 20894, USA
boratyng@ncbi.nlm.nih.gov

Dongqing Chen
Computer Vision & Image Processing
(CVIP) Laboratory
Department of Electrical and
Computer Engineering
University of Louisville
Louisville, KY, 40292, USA
dqchen@cvip.louisville.edu

Carlos A. Coello Coello
CINVESTAV-IPN
Evolutionary Computation Group,
Departamento de Computación,
Av. IPN No. 2508
Col. San Pedro Zacatenco
México, D.F. 07360, Mexico
ccoello@cs.cinvestav.mx

Zhihua Cui
State Key Laboratory for Manufac-
turing Systems Engineering,
Xi'an Jiaotong University
Xi'an, Shaanxi, 710049, China
cuizhihua@gmail.com

Christine Decaestecker
Laboratory of Image Synthesis and
Analysis,
Université Libre de Bruxelles
Brussels, Belgium
cdecaes@ulb.ac.be

Anca Doloc-Mihu
Center for Advanced
Computer Studies,
University of Louisiana at Lafayette
Lafayette, LA 70503, USA
adoloc12345@yahoo.com

Aly A. Farag
Computer Vision & Image Processing
(CVIP) Laboratory
Department of Electrical & Com-
puter Engineering
University of Louisville
Louisville, KY, 40292, USA
farag@cvip.louisville.edu

Gary Fogel
Natural Selection, Inc.
San Diego, CA 92121, USA
gfogel@natural-selection.com

Adam E. Gaweda
Kidney Disease Program
University of Louisville
Louisville, KY 40202, USA
adam.gaweda@louisville.edu

Samik Ghosh
Biological Networking
Research Group,
Department of Computer Science
and Engineering,
The University Of Texas at Arlington
Arlington, TX 76019, USA
fsghosh@cse.uta.edu

Cengiz Günay
Department of Biology,
Emory University
Atlanta, Georgia 30322, USA
cgunay@emory.edu

Benjamin Haibe-Kains
Machine Learning Group,
Université Libre de Bruxelles
Brussels, Belgium
Functional Genomic Unit,
Institut Jules Bordet
Brussels, Belgium
bhaibeka@ulb.ac.be

Andrew Hamilton-Wright
School of Rehabilitation Therapy,
Queen's University
Computing and Information Science,
University of Guelph
Mathematics and Computer Science,
Mount Allison University
Kingston, ON K7K 1T3, Canada
andrewhw@ieee.org

Barbara Hammer
Computer Science Department,
Technical University Clausthal
Clausthal-Zellerfeld, Germany
hammer@in.tu-clausthal.de

Ray R. Hashemi
Department of Computer Science,
Armstrong Atlantic State University
Savannah, GA 31419, USA
ray.hashemi@yahoo.com

Aboul Ella Hassanien
Information Technology Department,
FCI, Cairo University
Orman, Giza, Egypt
Information System Department,
CBA, Kuwait University
Safat, 13055, Kuwait
a.hassanien@fci-cu.edu.eg

Ilkka Havukkala
Knowledge Engineering and Discov-
ery Research Institute, KEDRI,
Auckland University of Technology
Auckland 1142, New Zealand
ilkka.havukkala@aut.ac.nz

Timothy J. Hickey
Michtom School of Computer
Science, Brandeis University
Waltham, MA 02254, USA
tim@cs.brandeis.edu

Falk Huettmann
University of Alaska, EWHALE Lab,
Institute of Arctic Biology,
Department of Biology & Wildlife
Fairbanks, AK 99775, USA
fffh@uaf.edu

Wit Jakuczun
Nencki Institute of
Experimental Biology
Pasteura 3 St.,
02-093 Warsaw, Poland
WLOG Solutions
Harfowa 1A/25 St.,
02-389 Warsaw, Poland
W.Jakuczun@wlogsolutions.com

Fatma Gürel Kazancı
Department of Biology,
Emory University
Atlanta, Georgia 30322, USA
fgurelk@emory.edu

John T. Langton
Michtom School of Computer
Science, Brandeis University
Waltham, MA 02254, USA
psyc@cs.brandeis.edu

Dah-Jye Lee
Department of Electrical and
Computer Engineering,
Brigham Young University
Provo, UT 84602, USA
djlee@ee.byu.edu

Heitor Silvério Lopes
Bioinformatics Laboratory
Federal University of Technology –
Paraná
80230-901 Curitiba – Brazil
hslopes@utfpr.edu.br

Thomas McTavish
Computational Bioscience Program,
University of Colorado
Health Sciences Center
P.O. Box 6511, Mail Stop 8303
Aurora, CO 80045, USA
Thomas.McTavish@uchsc.edu

Filippo Menolascina
Polytechnic of Bari
70125 Bari, Italy
National Cancer Institute 'Giovanni
Paolo II'
70126 Bari, Italy
f.menolascina@ieee.org

Mariofanna G. Milanova
Computer Science Department,
University of Arkansas at Little Rock
Little Rock, Arkansas 72204, USA
mgmilanova@ualr.edu

Radhakrishnan Nagarajan
University of Arkansas for Medical
Sciences,
Little Rock, AR 72205, USA
nagarajanradhakrish@uams.edu

Vijayaraj Nagarajan
Department of Biological Sciences,
The University of Southern
Mississippi
Hattiesburg, MS 39406, USA
vijayaraj.nagarajan@usm.edu

Samuel Neymotin
Dept. Biomedical Engineering,
SUNY Downstate Medical Center
Brooklyn, NY 11203, USA
samn@neurosim.downstate.edu

F. G. (Pat) Patterson, Jr.
Institute of Intelligent Systems,
FedEx Institute of Technology,
The University of Memphis
Memphis, TN 38152, USA
drfgp2@gmail.com

Lech Polkowski
Polish-Japanese Institute of
Information Technology
Koszykowa 86,
02008 Warszawa, Poland
polkow@pjwstk.edu.pl

Astrid A. Prinz
Department of Biology,
Emory University
Atlanta, Georgia 30322, USA
aprinz@emory.edu

Rafał Scherer
Department of Computer Engineering, Częstochowa University of
Technology
36 Armii Krajowej Ave.,
42-200 Częstochowa, Poland
rafal.scherer@gmail.com

Frank-Michael Schleif
Medical Department,
University Leipzig
Leipzig, Germany
schleif@informatik.uni-leipzig.de

Jahangheer Shaik
Department of Electrical and
Computer Engineering,
The University of Memphis
Memphis, TN 38152, USA
jshaik@memphis.edu

Tomasz G. Smolinski
Department of Biology,
Emory University
Atlanta, Georgia 30322, USA
tsmolin@emory.edu

Ryszard Tadeusiewicz
AGH University of Science and
Technology
30 Mickiewicza Ave.,
30-059 Krakow, Poland
rtad@agh.edu.pl

David Windridge
School of Electronics and
Physical Sciences,
University of Surrey
Guildford, Surrey, GU2 7XH,
United Kingdom
D.Windridge@surrey.ac.uk

Ying Xie
Department of Computer Science
and Information Systems,
Kennesaw State University
Kennesaw, GA 30144, USA
yxie2@kennesaw.edu

Jun Zhang
Department of Computer Science,
SUN Yat-sen University
Guangzhou, 510275, China
junzhang@ieee.org

Bai-Tao Zhou
School of Electronics, Telecommuni-
cation, and Computer Engineering,
Korea Aerospace University
Koyang City, South Korea
zhou@kau.ac.kr

Part I

Techniques and Methodologies

1

Statistically Based Pattern Discovery Techniques for Biological Data Analysis

Andrew Hamilton-Wright[1,3,3] and Daniel W. Stashuk[4]

[1] School of Rehabilitation Therapy, Queen's University, Canada
[2] Computing and Information Science, University of Guelph, Canada
[3] Mathematics and Computer Science, Mount Allison University, Canada
   andrewhw@ieee.org
[4] Systems Design Engineering, University of Waterloo, Canada
   stashuk@pami.uwaterloo.ca

Summary. A statistically based pattern discovery tool is presented that produces a rule-based description of complex data through the set of its statistically significant associations. The rules resulting from this analysis capture all the patterns observable within a data set for which a statistically sound rationale is available. The validity of such patterns recommends their use in cases where the rationale underlying a decision must be understood. High-risk decision making systems, a milieu familiar to many biologically-related problem domains, is the likely area of application for this technique. An analysis of the performance of this technique on a series of biologically relevant data distributions is presented, and the relative merits and weaknesses of this technique are discussed.

1.1 Introduction

Biological data is being collected into ever-larger databases, requiring strong analytic tools to help discover the important relationships encoded therein. Performing such discovery using statistically based analysis provides a simple and comprehensive understanding of the relationships discovered. There are, however, few general-purpose tools available that will uncover, with statistical rigour, the intricate and high-order relationships present in complex biological data and present the results in a form useful for classification or categorization of new data samples.

This chapter describes a classification tool that chooses labels based on rules derived only from *statistically significant* associations discovered in analysis of training data, providing a rule set that is insensitive to random noise in the training data, and is *transparent*: that is, they can be easily interpreted, explained, rationalized, confirmed, and made explicit, such that they can be considered verifiable acquired knowledge. In contrast, many other techniques do not concern themselves with statistical rigour, finding rules through the

A. Hamilton-Wright and D.W. Stashuk: *Statistically Based Pattern Discovery Techniques for Biological Data Analysis*, Studies in Computational Intelligence (SCI) **122**, 3–31 (2008)
www.springerlink.com   © Springer-Verlag Berlin Heidelberg 2008

reduction of observed error; this leads both to an inability to generalize from training data and to being unable to understand how classifications are made.

Application of this tool to the classification of biological data provides a means of supporting clear and transparent decision making in high-risk domains. When computational support for high-risk decision making is provided, the transparency, explainability and statistical rigour of the decisions made has comparable importance to the overall accuracy of the system. In particular, transparency provides a means of incorporating multiple sub-decisions into an informed overall characterization while allowing all levels of the decision process to be understood by a human decision maker or stakeholder. The provision of such clarity rests on the quality of the patterns found, as well as on their ability to make characterizations.

The gathering of biological data is frequently quite expensive, both in terms of the time required for acquisition and in terms of the materials and equipment needed. It is therefore important that researchers be able to explore the data inter-relationships using tools that will locate informative patterns as well as provide good measures of the quality of the patterns found.

Broadly defined, algorithms that extract rules from training data through observation are referred to as "association classifiers" [1–12] and the extraction of such rules is termed "association mining" [13, 14]. Initially associated with an algorithm called "Apriori" used for mining databases [1], the term association classifier refers to an algorithm in which relationships discovered through analysis of training data are codified as rule-based patterns for use in classification. As our rules will make no assumptions regarding the relationships between features in the input data, we produce "generalized association rules" as described in [2], in contrast to the "pure" rules of [1].

Association classifiers are distinct from the more general class of supervised classification algorithms in that association classifiers deal explicitly with rule-based logic, while many other algorithms deal in metrics measuring a fit to a multi-dimensional error surface. To a party interested in the inference behind a particular classification, the rule-based approach provides a degree of transparency that is not available from distance-based inference.

The term "association mining" therefore encompasses many approaches, including such techniques as:

- tree generation algorithms [15, 16];
- extension matrix based techniques [17–21];
- fuzzy rule generation techniques, including :
    - fuzzy systems configured through evolutionary algorithms [22–25];
    - other neuro-fuzzy systems used for rule production [26–38];
- contingency table approaches, frequently based on rough sets [39–42]; and
- schemes using statistical clustering techniques to create input membership functions, combined with the generation of contingency tables [43–45].

All of these systems function by inspecting a training data set and producing a set of "rules" to be used in classification. The mechanism and quality of the rules produced may vary, however all these algorithms learn in a supervised fashion, and all produce a rule-set that in some way captures a model of the underlying data.

Previously presented as "Pattern Discovery" [9–12, 46–51], we present here a technique that combines the attributes of an association classifier with a technique that extracts patterns based on a test of statistical validity. As the name "Pattern Discovery" has come to be synonymous in the literature with the field of association mining itself, we will refer here to the algorithm detailed in [49] as "Statistically Based Pattern Discovery", or SBPD.

The remainder of this chapter will explore the design, utility and application of the SBPD algorithm. Section 1.2 describes the aims, scope and overview of the algorithm, including the generation of a measure to estimate classifier confidence in Section 1.2.4. Section 1.3 evaluates the performance of the SBPD algorithm in relation to other popular classifiers on synthetic data, while Section 1.3.5 discusses the types of biological data problems upon which it has been applied, and those upon which it is likely to provide insight.

1.2 Statistically Based Pattern Discovery Algorithm

The underlying ideas of the SBPD algorithm were originally developed by Wang and Wong [9–12], where its application was limited to discrete data domains only. It has since been extended to function with continuous valued and mixed-mode data [46, 48, 49], and has been applied by the authors to problems of interest in electrophysiology [50, 51], as well as other real-world biological data problems [49]. An overview of the major concepts of the SBPD algorithm follows; readers looking to implement the algorithm are directed to [12] and [46, 49].

The SBPD algorithm functions by observing the occurrence of "polythetic events" in a training data set. A polythetic event is an event defined by the occurrence of a combination of feature values. In an application in medical informatics, the record for a patient of a given age, gender, and disease state would constitute the observation of a polythetic event; in a gene sequencing problem this would be the observation of a specific combination of base pairs.

The objective of the SBPD algorithm is to find relations between specific values of features of a training data set. In this regard, SBPD is *event based* in contrast to *random-variable based* methods. A given value of a feature is considered a primary event. Co-occurrence of a specific ensemble of primary events is considered a high order event. High order events that occur a significant number of times are considered to be patterns that reflect the nature of the data studied and provide knowledge useful for inference and reasoning.

Discovered patterns can be ranked by the statistical confidence of their veracity and clearly explainable. This transparency in the data mining results of the SBPD algorithm allows classifiers based on this technique to effectively contribute to complex and high-risk decision support systems.

1.2.1 Event Based Data

The event based methodology differs from the random variables of statistical studies in examining relationships between specific values in a data set that is inherently discrete. In order to apply event based methods to continuous data values, a discretization or quantization must be applied to all data, resulting in a coarser approximation of the continuous data values.

Quantization may be accomplished with sufficient resolution and appropriate distribution of boundaries such that important relationships between feature values are adequately represented in the discretized space, and can then be discovered by the SBPD algorithm [47]. The probabilistic and statistical basis of the SBPD algorithm defines a relationship between the number of features and the number of training examples, and limits the quantization resolution (or number of quantization intervals) that can be practically used. Therefore, the "cost" of quantization on the performance of SBPD classifiers is of interest when compared with standard continuous-valued data classification methods such as artificial neural networks based on the back-propagation of training errors [52–56], and Bayesian classifiers. These will be explored in later sections of this chapter, once the basic SBPD algorithm has been introduced.

The strengths of the SBPD technique may be summarized as follows:

- there is no need to have an *a priori* knowledge of the distribution of the data; in particular, there is no assumption that the available data follows a Gaussian Normal distribution as is the case with many Bayesian techniques [57–60];
- the SBPD technique avoids the problems of over-fitting that are common with error-minimization techniques such as error back-propagation artificial neural networks;
- the simplicity of inference in the SBPD technique gives a straightforward explanation of the support for any given outcome; this factor allows it to be used in critical decision making paradigms, such as clinical applications [48]; and finally,
- the accuracy of classification of the SBPD technique is comparable to other commonly used classifiers.

1.2.2 Discovering Patterns in Training Data

Consider a set of training data presented as an array of N rows, each of length $M{+}1$. Each row, or input vector, contains M input feature values and a single class label, $Y{=}y_k$. The value for each feature is drawn from a set of discrete or quantized values specific to that feature; the class labels are drawn from a set of K possible class labels. Every input vector can thus be considered to be a single $M{+}1$-order event relating its input values to a label. Each vector element $\mathbf{x}_j$ will contain a value for either a single input feature, or a label. Elements corresponding to an input feature ($j \in 1 \cdots M$), will correspond either to one of $\boldsymbol{\nu}_j$ discrete observed values from the set of possible values describing feature j, or may be identified as a (discrete) quantization interval identifier. Each possible combination of m primary events selected from *within* a vector can be considered a sub-event of order m, $m \in \mathbb{I}$, $1 \le m \le M+1$.

Primary (or first order) events are represented as $\mathbf{x}_l^1$, while an event of order m is represented as $\mathbf{x}_l^m$, with l indicating a particular sub-event within the list of all sub-events of order m occurring in a particular input event $\mathbf{x}$.

Events of interest with respect to classification must be of order 2 or greater (*i.e.*, polythetic) and be an association of at least one input feature value (a primary event) and a specific class label.

SBPD analysis begins by counting the number of occurrences of all observed events among the N vectors forming the set of training data. Statistically significant events (or "patterns") within this set are then discovered through analysis of the adjusted residual of each event.

Adjusted Residual

If $o_{\mathbf{x}_l^m}$ is the observed number of occurrences of event $\mathbf{x}_l^m$ in a training data set, and $e_{\mathbf{x}_l^m}$ is the expected number of occurrences of the same event under an assumed model of uniform random chance, then we can calculate $e_{\mathbf{x}_l^m}$ using

$$e_{\mathbf{x}_l^m} = N \prod_{\substack{i=1 \\ \mathbf{x}_{li}^1 \in \mathbf{x}_l^m}}^{m} \left(\frac{o_{\mathbf{x}_{li}^1}}{N} \right) \tag{1.1}$$

where $o_{\mathbf{x}_{li}^1}$ is the number of occurrences of each primary event $\mathbf{x}_{li}^1 \in \mathbf{x}_l^m$ and N is the total number of observations made (*i.e.*, the number of rows of training data). We can calculate an unbiased estimate called the standardized residual [61] of these values using

$$z_{\mathbf{x}_l^m} = \frac{o_{\mathbf{x}_l^m} - e_{\mathbf{x}_l^m}}{\sqrt{e_{\mathbf{x}_l^m}}}, \tag{1.2}$$

and, if $v_{\mathbf{x}_l^m}$ is the maximum likelihood estimate of the variance of $z_{\mathbf{x}_l^m}$, then

$$r_{\mathbf{x}_l^m} = \frac{z_{\mathbf{x}_l^m}}{\sqrt{v_{\mathbf{x}_l^m}}} \tag{1.3}$$

will produce the adjusted residual: a Normal distribution with zero mean and unit standard deviation *if* the values of $\mathbf{x}_l^m$ are drawn from a uniform random distribution [61, 62].

As given in [9], $v_{\mathbf{x}_l^m}$ may be easily calculated using

$$v_{\mathbf{x}_l^m} = var\left(\frac{o_{\mathbf{x}_l^m} - e_{\mathbf{x}_l^m}}{\sqrt{e_{\mathbf{x}_l^m}}} \right) = 1 - \prod_{\substack{i=1 \\ \mathbf{x}_{li}^1 \in \mathbf{x}_l^m}}^{m} \left(\frac{o_{\mathbf{x}_{li}^1}}{N} \right), \tag{1.4}$$

providing a means of calculating all of these values simply by counting the occurrences of each event type.

The formulation of (1.3) allows the values of $r_{\mathbf{x}_l^m}$ to be scored to evaluate the null hypothesis that the event associations represented by $\mathbf{x}_l^m$ *occur randomly and independently.* As $r_{\mathbf{x}_l^m}$ follows a Normal distribution if this is true, we choose a threshold of $h_0 = 1.96$ as values of $|r_{\mathbf{x}_l^m}| > 1.96$ indicate that, to a 95% confidence level, the observed value is unlikely to be generated by the assumed distribution. Events whose $|r_{\mathbf{x}_l^m}| > 1.96$ are deemed statistically significant, and are considered "patterns".

Significance is calculated in absolute terms because combinations of events that occur significantly less frequently than would be expected by the null hypothesis (patterns with a negative $r_{\mathbf{x}_l^m}$) are just as significant and potentially discriminative as those that occur more frequently. Such a pattern indicates that the given features rarely or never associate; a situation that has interesting implications when attempting to apply class labels [63, 64].

The set of all such patterns discovered comprise the set of *all statistically valid associations* relating any feature values. Once the collection of statistically valid patterns is produced, those patterns that contain a value for the label column may be used as rules. Each such rule indicates that if the input portion of the rule is matched, then the pattern provides support for using the associated label value in classification. There may, however, be multiple matching patterns with differing order and potentially differing associated labels. Techniques to combine these patterns into a single suggested labelling are described in Section 1.2.3.

The Meaning of Statistically Significant Associations

It is entirely possible that a significant association may be found that is *not* deterministic for the label value. The existence of a discovered pattern indicates that observations of the values making up the pattern occur with some structure: this structure pertains to a model of the data distribution, however the distribution may contain overlapping classes. If the classes overlap, the discovered patterns will simultaneously support multiple different possible classifications.

Fig. 1.1 presents a set of observations describing a data set with two input features (x_1, x_2) and a single output Y, which corresponds to a class label, $Y \in \{A, B, C\}$. Portion 1.1(a) gives a graphical description of the distribution of data values, placed into a grid by the (discrete) values of (x_1, x_2). The data space is shown as a grid of cells containing polythetic events defined by input values $(\mathbf{x})$ and a class label. The number of instances of a letter in a given cell corresponds to the number of observed occurrences of that event. Fig. 1.1(b) shows a tabulation of the number of occurrences for each class, as well as an indication of whether any significant rules were found for the given event described. For the moment, we will ignore the fact that this set of data is far too small to have confidence in the patterns found. This topic will be discussed

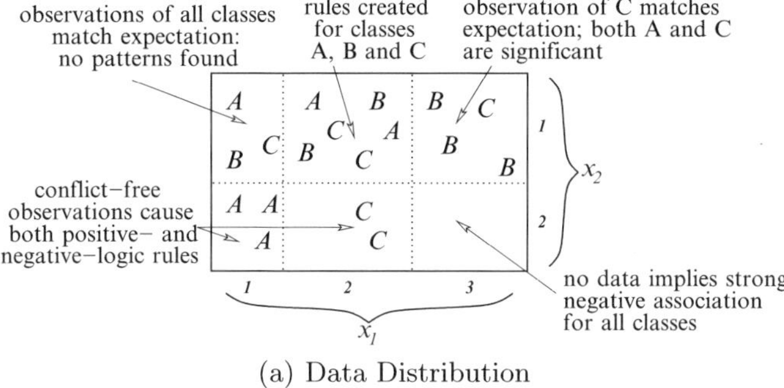

(a) Data Distribution

$x_1\ x_2 | A\ B\ C | Significance?$

x_1	x_2	A	B	C	Significance?
1	1	1	1	1	no
1	2	3	0	0	yes
2	1	2	2	2	yes
2	2	0	0	2	yes
3	1	0	3	1	yes
3	2	0	0	0	yes
Total		6	6	6	

(b) Tabulated Observations

x	Y	**x**	Y	**x**	Y
$(1,1) \mapsto$	$\emptyset$	$(2,1) \mapsto$	ABC	$(3,1) \nmapsto$	A
				$(3,1) \mapsto$	B
$(1,2) \mapsto$	A	$(2,2) \mapsto$	C		
$(1,2) \nmapsto$	BC	$(2,2) \nmapsto$	AB	$(3,2) \nmapsto$	ABC

(c) Rules Found from Patterns

Fig. 1.1. Hypothesis Testing Using the Adjusted Residual

in Section 1.2.6 when considering expectation and the need for reasonable training set sizes. Fig. 1.1(c) shows the patterns found through examination of this data. These patterns provide a model characterizing the distribution of the class labels: the patterns found are descriptive of the distribution, but may not be discriminative for classification.

This example consists of 6 cells (defining events) containing 6 observations, resulting in an expectation of $e_{\mathbf{x}_l^m}=1$ observation per cell. When we examine cell $\mathbf{x}=(1,1)$ of Fig. 1.1(a) and discover a single occurrence, our characterization is that no significant patterns are present. Event $\mathbf{x}=(1,1)$ is therefore shown in Fig. 1.1(c) as implying nothing ($\emptyset$).

By examining Fig. 1.1, one can see that the construction of patterns is performed independently for each class. A rule is created for event $\mathbf{x}=(1,2)$ indicating that the label A is supported by this event; independently, negative-logic rules are constructed indicating that the labels B and C are refuted. Together, this set of rules unequivocally associates this event with class A. Conversely, for event $\mathbf{x}=(2,1)$ there is an above-expectation observed occurrence of every label: this causes independent construction of rules supporting classification using all the label values A, B and C.

Events containing only one label, such as cells $\mathbf{x}=(1,2)$ and $\mathbf{x}=(2,2)$, form simple, unambiguous rules, and capture portions of the data space for which there is no conflict.

In contrast, the cell at $\mathbf{x}=(2,1)$ contains by far the majority of the observed data, fully $\frac{1}{3}$ of all observations, but there is no clear label to assign to this event as the number of occurrences of each class are equal. Considering the data from a modelling point of view, it is clear that such an event is statistically quite interesting, and a representation of these patterns provides a model of the distribution and inter-relationships of the data space, independent of the task of generating classifications for these events. We therefore construct patterns for *all three* classes for event $\mathbf{x}=(2,1)$, indicating that there is strong support for each class. This allows our model to distinguish such an event from one with "no information" (*i.e.*, expected "random" occurrence) such as $\mathbf{x}=(1,1)$, and also from negative information, such as that of cell $\mathbf{x}=(3,2)$.

The patterns found by using the adjusted residual may therefore be considered as a functional model of the distribution of data values in the observed space. A corollary here is that the application of these patterns to classification needs to take into account the fact that there may be patterns asserting support for multiple differing classifications for an event; the nature of this knowledge may not support discrimination between the various classes. Consider $\mathbf{x}=(2,1)$ and $\mathbf{x}=(3,1)$: in the first case, the patterns will indicate multiple possible classifications (*i.e.*, all of A, B and C), while in the second case complementary patterns will assert support for a likely B, and refute the extremely unlikely A. No pattern is produced for class C in event $\mathbf{x}=(3,1)$ as it contains the number of observations that were predicted by the model.

If a new input value corresponding to $\mathbf{x}=(2,1)$ is observed, our list of patterns tells us that this value is likely to occur (its occurrence is an expected event), but there is *no information* to support any particular classification: all observed labels are equally likely. In this case, an input vector may be left unclassified: this will be detectably different than cases where labels can be assigned. This behaviour provides a strength not commonly seen in classifiers: a possibility of a graceful degradation to "no decision" in scenarios when insufficient information is available to make a robust decision.

In contrast, an observation of a new input value corresponding to $\mathbf{x}=(3,2)$ may be described as *completely unexpected*, indicating we have a rare and indeed unprecedented phenomenon: we again have no information on a likely classification, but now because we have never seen any similar event.

Further, it is possible for rules to exist with conflicting labels that have significantly different number of observations. For instance in a larger data set (not shown here), it is possible that a cell contains a larger than expectation number of occurrences of both classes A and B, with the frequency of A being double that of B. In such a case, both A and B are recorded as (conflicting) rules, resulting in a method of recording the relative frequencies of occurrence as well as a likely label (A). We can discriminate between the differing degrees of support for classification using a rule weight, as discussed below.

It is clear that the use of the adjusted residual will provide a set of patterns that capture all of the information present in the data space. Such a set of patterns forms a complete model for a given level of statistical significance in which all knowledge regarding data relationships is captured.

1.2.3 Classification Using Statistically Based Rules

Once a set of patterns has been obtained, these patterns may be used as rules to fill in a missing label column (or, for that matter, any other missing value). The importance of the rules varies, depending on how well they capture unambiguous information regarding a potential labelling; to measure this importance a weighting value is supplied for each rule. Two different weighting measures have been proposed, and each will be discussed here.

The initial proposal by Wang and Wong [9–12] called "weight of evidence," provides rule weights based on the information content of each pattern. The second approach, described in later work by the authors of this chapter [48, 49], provides a $[-1 \ldots 1]$ bounded distance measure.

Weight of Evidence Associated With Patterns

In the original presentation of SBPD, Wang [12] suggests the use of a statistic called "weight of evidence" (or WOE) to measure the discriminative power of a pattern.

If $(Y{=}y_k)$ represents the label portion of some pattern $\mathbf{x}_l^m$, the remaining portion (consisting of the input feature values) is referred to as $\overset{\star}{\mathbf{x}}_l^m$. The mutual information between these two components can be calculated using [9]

$$I(Y{=}y_k : \overset{\star}{\mathbf{x}}_l^m) = \ln \frac{Pr(Y{=}y_k | \overset{\star}{\mathbf{x}}_l^m)}{Pr(Y{=}y_k)}. \tag{1.5}$$

A WOE relative to a particular labelling $y_k \in Y$ can be calculated as

$$\text{WOE}\left(\frac{Y{=}y_k}{Y{\neq}y_k} \,\middle|\, \overset{\star}{\mathbf{x}}_l^m\right) = I(Y{=}y_k : \overset{\star}{\mathbf{x}}_l^m) - I(Y{\neq}y_k : \overset{\star}{\mathbf{x}}_l^m),$$

or

$$\text{WOE} = \ln \frac{Pr(\overset{\star}{\mathbf{x}}_l^m, Y{=}y_k)\,Pr(Y{\neq}y_k)}{Pr(Y{=}y_k)\,Pr(\overset{\star}{\mathbf{x}}_l^m, Y{\neq}y_k)}. \tag{1.6}$$

WOE thereby provides a measure of how discriminative $\overset{\star}{\mathbf{x}}_l^m$ is in relation to a label y_k. The WOE value may be thought of as a measure of the relative probability of the co-occurrence of $\overset{\star}{\mathbf{x}}_l^m$ and y_k (*i.e.*, the "odds" of labelling correctly). The domain of WOE values is $[-\infty \cdots \infty]$, where $-\infty$ indicates those patterns that never occur in training data with the specific class label y_k; ∞ indicates patterns that only occur with the specific class label y_k. These infinite valued WOE patterns are the most descriptive relationships found in the training data set.

Occurrence Weighted Patterns

An alternative weighting scheme for discovered patterns, providing a means of weighting the patterns without resorting to infinite values, is to use the relative number of occurrences of $\mathbf{x}_l^{\overset{*}{m}}$ [47–51].

In this scheme, pattern weights are created using the combined co-occurrence of events, calculated through the number of occurrences:

$$
\mathcal{W}_\rangle =
\begin{cases}
\dfrac{o_{\mathbf{x}_l^m}}{o_{\mathbf{x}_l^{\overset{*}{m}}}} & \text{if } r_{\mathbf{x}_l^m} > 0 \\[2em]
\dfrac{o_{\mathbf{x}_l^m} - e_{\mathbf{x}_l^m}}{e_{\mathbf{x}_l^m}} & \text{if } r_{\mathbf{x}_l^m} \le 0
\end{cases}
\tag{1.7}
$$

As this weighting creates "assertions" supporting or refuting a classification based purely on the observed occurrence of sub-events, we will term this "occurrence" weighting. Each assertion will be a value in a $[-1 \ldots 1]$ bounded space indicating the range from complete refutation of the associated class label (-1), through to complete support (1).

Combination of Patterns

Two different strategies of rule combinations have been used to calculate the total support $\mathcal{A}_k$ for each of K observed classes:

"Independent" Rule Firing: Support $(\mathcal{A}_k)$ for each y_k (possible class label) is evaluated in turn by considering the highest-order pattern with the greatest adjusted residual from the set of all patterns occurring in an input data vector to be classified, accumulating the weight of this pattern in support of the associated label. All features of the input data vector matching this pattern are then excluded from further consideration for this y_k, and the next-highest order occurring pattern is considered. This continues until no patterns match the remaining input data vector, or all the features in the input data vector are excluded. This "independent" method of accumulating pattern weights estimates the accumulation of the probabilities of independent random variables, maintaining feature independence.

"All" Rule Firing: all matching rules are used to calculate $\mathcal{A}_k$, allowing the information presented by the features to be unequal, depending on the structure of the rules matched. The theoretical drawback of this scheme is the possibility that a feature that is used in many rules may come to determine the classification outcome, effectively causing features captured by few patterns to have less importance in classification. The choice of rule weighting therefore comes into play as the weights should capture the importance of the rule as a collection of feature values, rendering independent consideration of each features less important.

Once a value of $\mathcal{A}_k$ is produced for each possible labelling, this allows the selection of

$$\kappa = \underset{k}{\operatorname{argmax}}^{K} \mathcal{A}_k \qquad (1.8)$$

to produce κ, the index for which the maximum assertion is found. The corresponding class name then forms the suggested label value.

The negative weights associated with negative rules allow the set of $\mathcal{A}_k$ values to correspond to degree of assertion in support or refutation of a given labelling where positive rules provide positive support and negative rules provide refutation. A degree of relative support for each class is therefore possible by comparing the values of $\mathcal{A}_k$ across the K possible classes.

The suggested label values are therefore produced by simply evaluating statistical confidence estimations through (1.3) and information or occurrence probability measures associated with observed rules. This provides a level of transparency to the system that allows its use in critical decision support milieux, where the suggested outcome is to be incorporated by a human user into a larger decision making context.

To fully support such applications, a means of producing an estimated decision confidence measure must accompany the suggested labelling.

1.2.4 Confidence Estimation of SBPD Suggestions

Every decision support systems must include a confidence measure to indicate the predicted probability of error. Such a confidence measure provides a means to alert the user to the distinction between classifications made with relatively strong or relatively weak support. A confidence measure is therefore critical in systems intended for use in high-risk or critical decision making applications.

If such a confidence measure is present, the clarity of the system is improved, and decision-making users will be more comfortable with the limits of the system as its measured ability to make confident decisions degrades. Several techniques may be applied to produce confidence measures based on this algorithm, some of which have been explored in depth in [65–67]. The simplest and strongest of the measures evaluated so far is based on observation of the relative values of the inter-class assertions, $\mathcal{A}_k$, by defining the highest and second-highest class assertions made as τ and τ_2. The difference between these values $(\tau - \tau_2)$ measures the conflicting support, termed δ.

When using the occurrence based weightings of (1.7), the τ and τ_2 values are bounded $[-1 \ldots 1]$, resulting in a range of $[0 \ldots 2]$ for δ; this in turn provides a bounded range of (δ, τ) pairs. As discussed in [67], an evaluation of the relative probability of observing a given (δ, τ) pair provides a confidence value that accurately describes the probability of error in the suggested outcome. The relative probability may be constructed using any form of histogram technique, such as Parzen windows [60, pp. 164].

The SBPD system thereby provides a simple and transparent means to obtain suggested labels based on observation of training data values. Further,

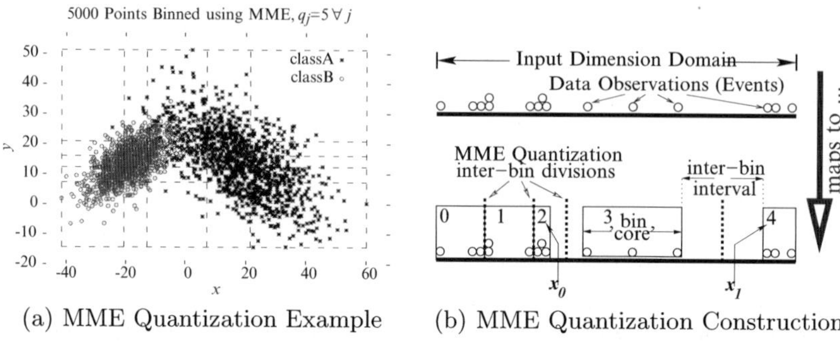

(a) MME Quantization Example   (b) MME Quantization Construction

Fig. 1.2. MME Quantization

both the inference underlying the suggestion, as well as the construction of a suggestion confidence indicating the likely probability of error, are produced in a simple and transparent way. As all parts of the system are based on the relative occurrence of observed events, the statistical support is easy to understand, describe, and ultimately, easy to incorporate into a larger decision making context.

Together with a confidence measure, this transparency provides the necessary conditions for the use of the SBPD system in decision support. The following section will therefore explore the performance of the SBPD system.

1.2.5 Pattern Analysis on Data with Continuous Values

In most real-world problems, data is continuous-valued and must be quantized to be used by the SBPD algorithm. Based on training data and a marginal maximum entropy partitioning (MME) scheme [68, 69], an optimal mapping of continuous-valued data into bounded "bins" is determined for each feature.

The MME algorithm divides observed points among q quantization "bins" for each feature, placing the bin boundaries so that an equal number of points fall into each "bin." Such a partition is shown in Fig. 1.2(a).

This is achieved over the set of observed values by:
- sorting all the values for a given feature j, $j \in 1 \cdots M$;
- dividing the sorted list into q_j "bins" of $\frac{N}{q_j}$ values each;
- calculating a minimum cover or "core" of each bin covering all data points;
- creating quantization intervals that cover gaps between the calculated "cores" of adjacent bins by extending the bin interval to the midpoint of the gaps.

This strategy results in a set of quantization intervals seamlessly covering the entire range of observations, based on bins defined by the $\frac{N}{q_j}$ points in each bin.

As is visible in Fig. 1.2(a), this causes the area covered by each quantization interval to be inversely related to the density of the points within them.

Regions of the data space with a high density of points are represented more precisely, using a greater number of intervals, while sparsely covered regions are represented with large quantization intervals. It is clear in Fig. 1.2(a) that the intervals become more closely spaced in the centre of the figure measured along the y-axis: this is due to the fact that the density of both class distributions is highest in the central part of the figure. On the x-axis, one sees that the two dense centres of distributions A and B are each assigned to a smaller quantization interval, while the region between the distribution means is grouped together into one larger, albeit less dense, shared interval. This shared interval will be one in which conflicting rules are produced.

Fig. 1.2(a) also demonstrates that the bin boundaries are derived by the location of the data points themselves; this is most clear when examining the edges of the figure, where the dashed lines indicating the quantization bin boundaries exactly intersect the outermost points of the range in x and y.

MME quantization has the effect of regularizing the information content of the data space such that the expectation, and thus the information, of each bin is equal [69]. For purposes of discovering patterns, this optimization of the information content provides the greatest discriminative power possible without resorting to a class-dependent quantization scheme. In such a scheme, the quantization grid would be constructed in a class-aware fashion, attempting to orient discretization bin bounds along inter-class decision boundaries [70–72].

Once quantization bins are constructed, input features are then assigned to the bin whose interval contains their value. Feature values assigned to the same bin represent the same primary event. This process is illustrated in Fig. 1.2(b), which shows that continuous-valued observations falling on a number line (top row) are used to construct quantization bins covering the observations. The midpoints of any gaps between bins are used as the inter-bin boundaries. In this example, there are $q_j=5$ quantization intervals for each $j \in M$.

1.2.6 Expectation Limits and Data Quantization

The quality of SBPD classifications depends upon two major factors: the number of training records available, and the quantization resolutions q_j that govern the number of MME bins feature values are divided among.

Given a fixed number of training records, it is important to choose q_j to ensure that enough records are present to allow a statistically sound decision to be made regarding the discovery of each possible high order pattern. As the estimate of the occurrence of a high order event is simply the product of the occurrence of the independent primary events (which MME attempts to keep equal across all bins), we can calculate an estimate of this value by simply calculating a product relating the number of rows of training data available (N), the quantization resolutions used for MME (q_j) and the number of features used to represent the class distribution (M) that defines the highest order observable event. This relation is

$$\mathcal{E}_{\mathbf{x}_l^m} = N \prod_{i=1}^{m} \frac{1}{q_j}, \qquad j = \mathrm{column}(i, \mathbf{x}_l^m) \tag{1.9}$$

in which the function "$\mathrm{column}(i, \mathbf{x}_l^m)$" selects the column index of primary event i within the polythetic event $\mathbf{x}_l^m$.

In (1.9), both an increase in any q_j and an increase in m (the order of the event) decreases $\mathcal{E}_{\mathbf{x}_l^m}$. This is a consequence of the well-known "curse of dimensionality" [73] which affects any multi-variate inference technique.

As a precaution against recording spurious patterns when $\mathcal{E}_{\mathbf{x}_l^m}$ is low, the SBPD classifier does not consider as patterns any events for which there are fewer than 5 expected occurrences. This is done in order to prevent the discovery of high-order patterns caused by the occurrence of only one or two instances of a high-order event (possibly generated by chance). Such a pattern, if accepted, will have a very high weight, as the chances of observing the same randomly-generated pattern with a different label during training is quite small. The existence of such patterns will diminish classifier performance, as the use of a highly weighted but spurious pattern would corrupt the evaluation of any remaining features in the induction of the correct class label value.

The inclusion of such patterns would cause a type of "over-fitting:" the construction of patterns observed only in the training data, but not representative of its underlying distribution [60, pp. 464]. Over-fitting is essentially due to the nature of the noise in the training data, and is common in many classifier paradigms, in particular artificial neural networks [34, 52–56] where this behaviour is particularly problematic because of a lack of transparency in the learning behaviour.

The significant patterns found represent captured knowledge, and will therefore be generalizable. The cost is that the true degree of fit is limited both by the choice of h_0 and by our requirement of an expectation of 5 occurrences; each of these choices admit the possibility that statistically insignificant, yet correct, rules may be discarded. The choice of these thresholds ideally should reflect the amount of noise present in the data domain; as this is not an easily determinable factor, relatively conservative values are suggested here.

The amount of training data available places constraints on the extent to which input data can be quantized and can limit the performance of SBPD classifiers. This is a significant constraint when used in biological data analysis, as the "curse of dimensionality" may cause an unsolvable problem due to a limited number of samples available from high-dimensional data. In such a case, it is advisable to attempt a feature reduction or feature selection before applying any type of classifier. The reader is referred to [74, 75] for an overview of the major techniques.

1.3 Experimental Validation Results and Discussion

To exercise the various classifiers, synthetic data distributions were generated. These data distributions have been chosen in order to provide data that is representative of that encountered in real-world biological problems.

Though synthetically created, the data studied here has relevance to that found in biological measurements. In particular, as it is difficult to know in advance the true distribution of biological data sets, a selection of data sets with known interesting structure will be more informative than a long list of real-world problems where the structure is unknown.

Many statistical tools are available to deal with data that has a Normal distribution, and these have been applied successfully to biological problems whose data has a strong central tendency (spectrographic densities, binary measures, *etc.*). If data is truly Normally distributed, it is optimally treated probabilistically with the NDDF (Normal Density Discriminant Function) classifier described in [60, pp. 26–41].

Biological data generally does not follow this distribution however; measures of abundance such as population measures, cellular counts, and any multiplicative measure are more likely characterized as logarithmically distributed, and further, any grouping measure containing multiple subgroups is likely to exhibit bi- or multi-modal behaviour.

We will therefore examine log-Normal and bimodal distributions, as well as a "spiral" distribution that demonstrates an extremely difficult data distribution space, in order to demonstrate the behaviour of the classifier in these circumstances. Additionally, we will examine the effects of errors in the labelling of training data as this problem is particularly salient to biological data studies, where it is frequently difficult to obtain unbiased gold-standard data of high quality, and label errors in training data are expected.

1.3.1 Synthetic Class Distribution Generation

We will use a simple probabilistic model for data generation, beginning with randomly generated Normal distributions and then colouring them in various ways to produce the desired data sets.

Four feature Normal data distributions were generated with differing covariance matrices and major orientations. Complete details on data generation are available in [47, 49]. Clouds of data were thereby created that intersect non-orthogonally and which have differing variances and covariances in each dimension, and in each class. The four classes were separated into different quadrants in Euclidean space by projecting the mean vector of each class away from the origin by separation factors of:

$$s_i \in \mathbf{S}, \quad \mathbf{S} = \left\{ \frac{1}{8}, \frac{1}{4}, \frac{1}{2}, 1, 2, 3, 4 \right\} \tag{1.10}$$

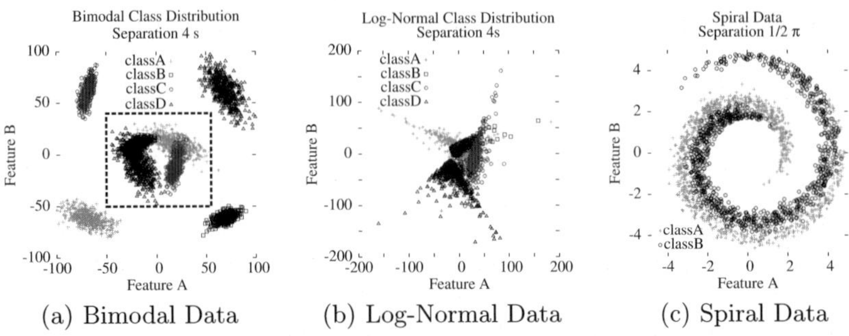

Fig. 1.3. Sample Data Distributions

and combining s_i with a measure of the overall class covariance. The s parameter thereby controls the difficulty of the problem due to overlap. At $s=\frac{1}{8}$, the data distributions overlap a great deal, but separate by $s=4$. Sample data with $s=4$ is shown in Fig. 1.3(a), in which the four distributions within the dotted box were generated using this method.

Log-Normal class distributions were produced by taking each point generated in a $\mathcal{N}(0,1)$ distribution and calculating

$$\mathbf{u}' = e^{\mathbf{u}\zeta}, \qquad \mathbf{u} \in \mathcal{N}(0,1) \tag{1.11}$$

to transform each point to be log-Normally distributed, controlled by the shape parameter ζ. Each point in the resulting log-Normal distribution is then coloured by applying a rotation and covariance. Separate experiments were done with $\zeta \in \{1.0, 1.5, 2.0\}$. The skewness values of the resulting class distributions were on average 4.58 for $\zeta=1.0$, 13.80 for $\zeta=1.5$ and 27.33 for $\zeta=2.0$. Sample data with $\zeta=1.5$ is shown in Fig. 1.3(b).

Covaried bimodal class distributions were created by starting with the s_i Normal distributions and adding a second set of points for each class. The second cluster location is set by translating the mean away from the origin in a diametrically opposite direction by $4\sqrt{v_{\max}}$, where $v_{\max}$ is the maximum variance value specified in the defining covariance matrix. Thus, along with a cluster of points centred at (s, s), a second cluster would be placed at $(-4s\sqrt{v_{\max}}, -4s\sqrt{v_{\max}})$. Each cluster contained half the total number of points.

This algorithm was repeated for all sets, generating a layout of pairs of clusters around the origin, shown in Fig. 1.3(a) for $s=4$, which displays the first two dimensions of the four-feature data and shows clearly that no straight line can be drawn across the field to separate any one class from any other.

Spiral class distributions were produced by using a $\mathcal{N}(0,1)$ distribution to control the scatter, in radians, from points along a spiral function. A second class was generated by choosing a similar set of points and separation was introduced by rotating the entire set around the origin by a specified amount in units of π radians. Further details on generation can be found in [47, 49]. A sample pair of distributions with a separation of $\frac{1}{2}\pi$ is shown for 2-dimensional data in Fig. 1.3(c).

The topology of this data distribution is far more extreme than those expected in biological data analysis; this data set is provided as it is one for which the quantization assumptions of the SBPD system are particularly problematic. The performance on this data distribution will therefore provide insight into how SBPD deals with particularly difficult distributions, and by extension what a "worst case" performance will be.

Training Error

For each class distribution studied the effect of training error $\mathcal{T}_{\text{err}}$ on the performance of the classifiers was examined. In each summary table a column indicating $\mathcal{T}_{\text{err}}=0.1$ is included for $N=1000$ points; this is the same data set used for the standard $N=1000$ test, however 10% of the records for each class have had their true label value replaced with a value chosen randomly from the other class labels.

This emulates a "gold-standard" data set in which there is actually a significant amount (*i.e.*, 10%) of error in the expert-supplied labels. While it is expected that a true gold-standard data set will not contain such a high fraction of errors in supplied labels, this experiment is provided to show the response of the various classifiers as the quality of the labelling in the training data becomes more suspect.

1.3.2 Comparative Evaluation

Error Back-Propagation Classifier Construction

A simple error back-propagation artificial network classifier (BP) was constructed using the algorithms provided in [55, 56]. The learning rate and momentum for the classifiers were fixed across all BP configurations using the typical values of 0.0125 and 0.5, respectively. Training for each experiment was run for a maximum of 10^5 epochs, until the overall error dropped below 2.5×10^{-3}, or until the derivative of the error dropped below 1.25×10^{-3}. Several choices for the number of hidden nodes were studied, the best results were found with $H=\{10, 20\}$.

Normal Density Discriminant Function Classifier Construction

The NDDF classifier [60, pp. 41] applies a whitening transform to each input record, and calculates a normalized distance to each mean. The smallest calculated distance is the optimal classification, assuming all classes are in fact Normal distributions. See [60, Chapter 2] further information.

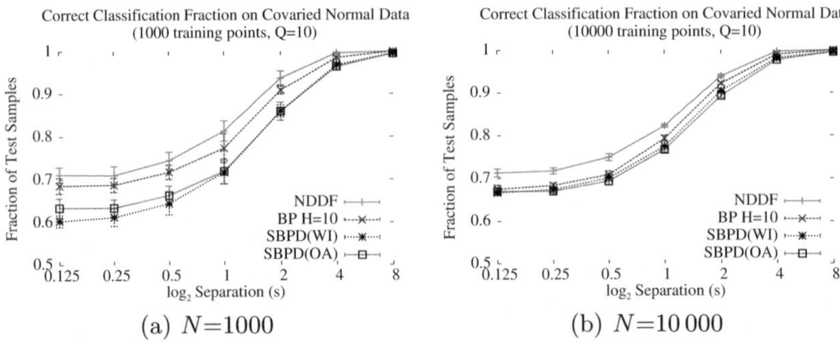

(a) N=1000 $\qquad$ (b) N=10 000

Fig. 1.4. Covaried Normal Data Results

SBPD Configuration

For simplicity, we will consider per-feature q_j as constant for all M input features; this will be notated by the quantization resolution value Q. Several different quantization values were studied (Q={5, 10, 20}). Two configurations of SBPD will be analyzed: "WOE" weightings with "Independent" rule firings and "Occurrence" weightings with "All" rule firings. These will be identified in the results as SBPD(WI) and SBPD(OA), respectively.

Weighted Performance Measure

To compare performance between classifiers, the statistic

$$\mathcal{P} = \frac{\sum_{i=0}^{|\mathbf{S}|} \frac{p_i^{\text{class}}}{s_i}}{\sum_{i=0}^{|\mathbf{S}|} \frac{1}{s_i}} \tag{1.12}$$

was used, where s_i is a separation value, $|\mathbf{S}|$ indicates the number of separations tested and p^{class} (the performance at a single separation) is the product of: {the fraction of records classified}×{the fraction of correct classifications}. This generates a single (scalar) value representing the overall performance of the classifier weighed across all separations and averaged across 10 trials with jackknife cross-validation as described in [47, 48]. Performance at lower separations is given more significance, as these problem are harder.

1.3.3 Evaluation Results

Covaried Normal class distribution performance results calculated using (1.12) are shown in the left half of Table 1.1, along with standard deviation values. Fig. 1.4(a) displays the results over all separations tested for 1000 and Fig. 1.4(b) for 10 000 training examples. These data are plotted using $\log_2$ along

Table 1.1. Covaried and Log-Normal Data Performance Summary

Classifier	Covaried Data			Log-Normal $N=10^3$		
	$N=10^3$	$N=10^3$ $\mathcal{T}_{\mathrm{err}}=0.1$	$N=10^4$	$s=1$	$s=1.5$	$s=2$
SBPD(WI) $Q=5$	0.66 ± 0.03	0.65 ± 0.02	0.69 ± 0.01	0.92 ± 0.02	0.82 ± 0.02	0.74 ± 0.02
SBPD(WI) $Q=10$	0.63 ± 0.02	0.63 ± 0.02	0.69 ± 0.01	0.92 ± 0.01	0.82 ± 0.02	0.75 ± 0.02
SBPD(WI) $Q=20$	0.61 ± 0.02	0.60 ± 0.02	0.66 ± 0.01	0.92 ± 0.01	0.79 ± 0.02	0.70 ± 0.03
SBPD(OA) $Q=5$	0.66 ± 0.03	0.65 ± 0.02	0.68 ± 0.01	0.90 ± 0.02	0.81 ± 0.02	0.74 ± 0.02
SBPD(OA) $Q=10$	0.66 ± 0.02	0.65 ± 0.02	0.69 ± 0.01	0.90 ± 0.01	0.81 ± 0.02	0.76 ± 0.03
SBPD(OA) $Q=20$	0.60 ± 0.02	0.59 ± 0.02	0.68 ± 0.01	0.88 ± 0.02	0.78 ± 0.02	0.69 ± 0.03
BP $H=10$	0.71 ± 0.02	0.70 ± 0.02	0.70 ± 0.01	0.96 ± 0.01	0.90 ± 0.02	0.81 ± 0.03
BP $H=20$	0.72 ± 0.02	0.69 ± 0.01	0.72 ± 0.01	0.96 ± 0.01	0.92 ± 0.01	0.85 ± 0.02
NDDF	0.74 ± 0.02	0.73 ± 0.02	0.74 ± 0.01	0.89 ± 0.01	0.61 ± 0.04	0.39 ± 0.05

the x axis to provide greater visual separation. Error bars indicate standard deviation across jackknife trials.

In this figure, BP is shown only with $H=10$ hidden nodes as this was the highest performance configuration tested. The covaried class distributions are quite rich in high-order information, and were easily separated by the NDDF classifier, which can be seen providing the optimal upper bound in each of Figs. 1.4(a) and 1.4(b).

The performance of the SBPD classifier was somewhat lower than that of either BP or NDDF, and only a small difference between the SBPD(OA) and SBPD(WI) performance is visible in Fig. 1.4(a), with the SBPD(OA) performance being noticeably higher at lower separations. Once a separation of $1s$ is reached, there is no longer a difference between the different SBPD classifier implementations. Fig. 1.4(b) shows no real difference between the SBPD or BP classifiers, though the NDDF classifier still shows that a marked difference exists in all these cases between the recorded performance and optimality.

These analyses are supported by an examination of Table 1.1, which reports similar results for SBPD(OA) versus SBPD(WI), but both being somewhat weaker than that of BP classifier. Considering the $\mathcal{T}_{\mathrm{err}}=0.1$ column in Table 1.1, we can see that the SBPD classifiers can tolerate moderate training data error. It is apparent that, for these class distributions, quantization at $Q=10$ exhibits the strongest performance for the SBPD based classifiers; there is a notable performance decrease when comparing any of the tests using $Q=10$ with those using $Q=5$ or $Q=20$.

Log-Normal class distribution results in Fig. 1.5(a) show that the performance of the SBPD classifier remained high as the data distribution deviated from Normal, while the NDDF classifier performance dropped remarkably as the skewness of the data distributions increased. This is summarized across all classifiers in the right portion of Table 1.1, in which it can be seen that the

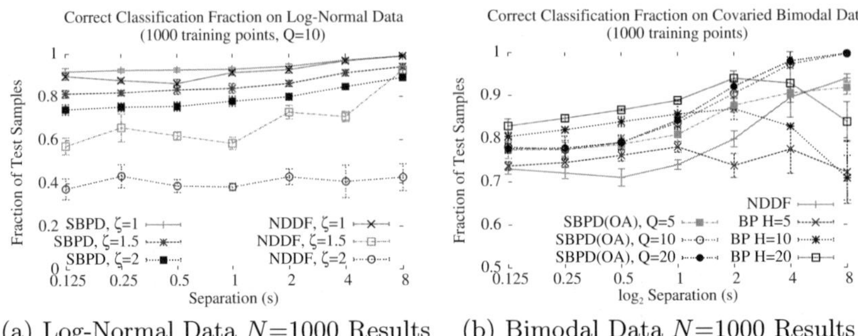

(a) Log-Normal Data N=1000 Results   (b) Bimodal Data N=1000 Results

Fig. 1.5. Log-Normal and Bimodal Data Results

Table 1.2. Weighted Performance: Bimodal and Spiral Class Distributions

Classifier	Bimodal $N=10^3$	Bimodal $N=10^3$ $\mathcal{T}_{\mathrm{err}}=0.1$	$N=10^4$	Spiral $N=10^3$	Spiral $N=10^3$ $\mathcal{T}_{\mathrm{err}}=0.1$	$N=10^4$
SBPD(WI) $Q=5$	0.78±0.02	0.78±0.02	0.82±0.01	0.65±0.03	0.64±0.03	0.67±0.01
SBPD(WI) $Q=10$	0.79±0.01	0.79±0.02	0.83±0.00	0.70±0.03	0.68±0.02	0.74±0.01
SBPD(WI) $Q=20$	0.79±0.02	0.78±0.02	0.82±0.00	0.63±0.02	0.61±0.02	0.74±0.01
SBPD(OA) $Q=5$	0.80±0.01	0.80±0.01	0.81±0.01	0.65±0.03	0.64±0.03	0.64±0.01
SBPD(OA) $Q=10$	0.82±0.01	0.81±0.01	0.84±0.01	0.70±0.03	0.69±0.02	0.73±0.01
SBPD(OA) $Q=20$	0.78±0.02	0.78±0.01	0.83±0.00	0.43±0.03	0.37±0.04	0.73±0.01
BP $H=10$	0.82±0.01	0.81±0.03	0.82±0.01	0.75±0.03	0.74±0.03	0.79±0.02
BP $H=20$	0.85±0.01	0.82±0.02	0.85±0.01	0.77±0.03	0.73±0.02	0.81±0.01
NDDF	0.73±0.01	0.67±0.01	0.73±0.01	0.49±0.03	0.50±0.02	0.50±0.01

BP classifier responds to changes in skewness with a stability comparable to that of the SBPD classifiers.

Covaried bimodal class distributions results are shown across all separations for N=10 000 in Fig. 1.5(b), and are summarized for all bimodal class distributions studied in Table 1.2.

As there is no single hyper-plane which can linearly divide any two classes for the bimodal and spiral data distributions, NDDF is no longer really useful as a classifier, and the comparison between SBPD and BP classifiers becomes much more important.

The SBPD classifiers with N=1000, Q=20 and Q=30 again showed poor performance, however with lower Q or greater N, the SBPD classifier performance matched that of the other classifiers. Note that the BP classifier performance suddenly decreases at high separation for these class distributions; this is due to the training algorithm locating, in this problem, by "local optima", or a solution that optimizes only a small portion of the data space. NDDF performance seems high in Table 1.2, but can be seen to be poor at high separations in Fig. 1.5(b); the high score placed on low-separation values by (1.12) favours NDDF here when the separation is too low for the multiple modes to be discerned. Label errors in training data, as seen in Table 1.2, have little effect on the performance of the SBPD classifiers, as shown in the $\mathcal{T}_{\mathrm{err}}=0.1$ column.

Spiral class distribution performance values are summarized in Table 1.2. We see in these results that the performance of the SBPD classifiers remain close to that of BP classifiers, and that $Q=10$ provides good performance in all of the cases examined. The performance of $Q=20$ is high when $N=10\,000$, however when $N=1000$ the performance is much lower. Here too, the $\mathcal{T}_{\mathrm{err}}=0.1$ column shows a stability against training data error with the SBPD classifiers being less affected than the BP classifiers.

1.3.4 Evaluation Analysis

The results of these evaluations clearly show that the SBPD algorithm is an effective data mining tool. Furthermore, these results demonstrate that SBPD classifiers can be effective components within a higher level decision support system. These results also show that the SBPD classifiers are sensitive to having sufficient training data, though their requirements do not exceed those of other popular classifiers, such as BP. The SBPD and BP classifiers are trained in a similar way; a set of training data is examined and the essential relationships are extracted. These relationships or patterns are then used in classification to provide labels for new test data. Both SBPD and BP classifiers can be applied to linearly and non-linearly separable class distributions, however each structure the decision space quite differently.

The SBPD classifiers construct a contingency table from discretized training values forming a hyper-cell division of the input space and makes classification decisions using a nonlinear-weighted information-theory or occurrence based estimation of the most likely class calculated using the patterns occurring in the input data vector. BP classifiers make a decision by performing a regression on multiple weighted hyper-planes within the subdivided space.

Both SBPD and BP classifiers benefit from large amounts of representative, labelled training data and have configuration parameters that affect their classification performance. When SBPD is applied to continuous-valued data, the number of intervals (*i.e.*, the resolution) used to quantize the data relative to the number of features and the amount of training data available is important. The number of hidden nodes and the learning rate and momentum are important factors for BP. Therefore, the performance of various configurations of these two classification schemes was compared so that some insight into the impact of these factors on the practical use of SBPD and BP classifiers could be obtained.

The results for the covaried data distributions demonstrate that when the value for Q (the number of quantization bins) rises, the performance may fall as shown in Table 1.1. This behaviour is a result of the need for sufficient amounts of data to reach the expectation limit (1.9) within the SBPD classifier to reliably discover high-order patterns. Without the presence of high-order patterns, performance suffers.

We also see that the performance of the SBPD classifiers is reasonably close to the optimal performance of an NDDF classifier for these simple, linearly separable class distributions, implying that a SBPD classifier can reach optimal classification performance when the quantization adequately represents the underlying data set and sufficient N allows optimal pattern discovery (*i.e.*, all existing high order patterns are found).

With a high value for Q and a small training data set size, an insufficient number of observations will occur for the highest order events to be confidently observed and discovered as patterns. As a result, events which may form patterns in the underlying data set are not discovered during training, in turn providing a poorer pattern space in which to make decisions during classification. This is the process responsible for the low performance at $Q{=}20$ with $N{=}100$ or 1000. Increasing the size of the training data set will overcome this as the number of occurrences of each high order event will rise; this is not a satisfactory solution in all cases however, as often sufficient training data is simply not available. In such cases, where reliable training data is difficult to produce, lowering the value of Q to produce a coarser division of the feature space may provide a more viable alternative. Considering the accuracy of the classification when $Q{=}10$ in Table 1.1 we see that choosing a lower value of Q does not necessarily penalize the performance of the SBPD classifier; instead, for these class distributions, the strong ability to generalize arising from the patterns discovered allows correct classification decisions to be made while discretizing the features at a lower resolution.

When examining the log-Normal data in Fig. 1.5(a) and Table 1.1 it is apparent that the performance of the SBPD classifier is independent of the distribution of the underlying random elements within the data, while the assumption of a Normal distribution made by the NDDF classifier penalizes its performance. As mentioned earlier the skewness of the log-Normal distributions increase along with the shape parameter ζ. Fig. 1.5(a) indicates that as the distributions become more skewed, the performance of the NDDF classifier drops, until by $\zeta{=}2$, the NDDF classifier is essentially guessing.

In contrast, the SBPD classifier performance is only slightly affected as skewness increases, even though in the tails of these distributions there is now insufficient data available for SBPD to be able to create acceptable patterns to characterize this space. The performance of the SBPD classifier is very stable compared with that of the NDDF classifier in this case. This demonstrates that SBPD classifier performance is not strongly tied to the inherent shape of the class distribution, nor to the distribution of the noise present during measurement. In particular, assumptions that any of the class distributions be Normal are not required.

The bimodal class distributions in Table 1.2 and Fig. 1.5(b) are not linearly separable, but still contain a high degree of internal structure, as the covariance of each mode matched the covariance in the unimodal case. All the non-linear classifiers found this problem relatively easy, out-performing the NDDF classifier from the outset.

Notable again was the deviation in the performance of the SBPD classifiers as Q changes; again $Q=10$ was the optimal value shown because of the balancing between discretization resolution and statistically sufficient expectation. In particular the performance of $Q=20$ was noticeably lower as there were not enough training samples to support this level of quantization as high order patterns were not discovered. These class distributions are clearly divided, although in a non-linear way, and the class divisions follow the orthogonal orientation of the feature quantization space, so SBPD performance was quite similar for $Q=5$ and $Q=10$.

For the spiral data, the effect of N can be clearly seen. Noting the performance of the SBPD classifiers with $Q=20$ when $N=1000$, it can be seen that the SBPD classifiers do not have a large enough training set to characterize this complex data. Once a training set of sufficient size is available, or if the quantization value is kept reasonably low, the SBPD classifier performance rises to rival that of the BP classifier, which performs admirably in this case.

Table 1.2, when $N=1000$, demonstrates both the strengths and weaknesses of the SBPD classifier:

- as the shape of the underlying distributions is curved, the resolution of the discretization bins is desired to be high, thus $Q=5$ has poor performance;
- the number of training examples is not sufficient to support 20 quantization intervals so performance is very low for $Q=20$;
- using a Q value high enough to capture as much resolution in the data as possible without defeating the threshold placed on expectation (*i.e.*, $Q=10$), allows a performance comparable to BP to be achieved.

At high Q and low N, the SBPD(OA) classifier performs abysmally, as there are not enough training examples to discover any patterns when separation is low. This leads to a large number of unassigned values, and biases the performance statistic to low values.

When examining the results for each of the jackknife sets, it was found that at separation 0.125, no rules were produced at all for the $N=100$ case, only 2 rules were produced at separation 0.5, and the largest number of rules produced was 8, all of order 2. In contrast, at $N=1000$ up to 92 rules were produced to characterize this data.

The low number of rules produced indicates that very few of the quantization intervals are covered by patterns. Data falling outside of these cells will remain unclassified, as no knowledge pertinent to classification in these regions has been captured. In cases where classifications do occur, they are performed based on few and conflicting rules.

With a low number of training samples, the SBPD training produces only very few, low-order rules as it attempts to capture the spiral data distribution in an orthogonally based quantization grid. The evaluation of these conflicting rules produces very low total class weights, showing that there is very little information in the rule base for this underspecified and very difficult problem.

The minimal effect of the training error on the performance of the SBPD classifiers is due to the statistical rigour used in defining a pattern. This allows erroneously labelled training examples to be ignored, providing they do not occur a statistically valid (and therefore unlikely) number of times. Training errors therefore will not affect the patterns discovered, nor subsequent classifications made.

This is perhaps the most significant result shown in terms of the application within the sphere of biological data analysis, as it implies that the effect of errors within the labelling of a "gold-standard" data set are limited, and that a certain degree of error will not change the overall outcome. The effect of these errors will be to lower the rule weightings, and therefore the decision confidence, of a final suggested labelling. Only if the errors have a significant bias will a faulty knowledge base be expected.

Overall Performance Analysis

For a given value of Q, as N is increased, the relative performance of the SBPD classifiers improves, relative to the BP classifiers. This performance improvement is a function of the ability of the SBPD algorithm to discover higher-order patterns with the requisite statistical confidence (*i.e.*, $\mathcal{E}_{\mathbf{x}_l^m} \geq 5$). The discovery of these higher order patterns allow more accurate classification decisions to be made.

The interplay between N and Q is such that if Q is too low, increasing N will have little effect. Conversely, if Q is to be set to a high value, a large N will be required before any high order patterns are available. For the continuous-valued class distributions studied, it seems that $Q{=}10$ provides a reasonable compromise between sufficient quantization resolution and the ability to discern high order patterns without the need for large training sets which are unlikely to be available in practice. Correspondingly, data sets of size no larger than $N{=}1000$ are sufficient to support characterization through SBPD patterns and achieve high performance during classification.

When examining the effect of differing data distributions, it was shown that the SBPD algorithm is largely insensitive to variations in data topology, and is not reliant on assumptions such as Normality. This is expected, as like BP, the underlying concepts of the SBPD algorithm avoid any specific assumptions about data distribution topology. The only assumption made by the SBPD algorithm is the null hypothesis that unrelated events are independent and uniformly distributed.

The SBPD classifiers performed well for all of the class distributions studied, even though in general, they are disadvantaged by the fact that the orthogonality and interval distribution of their discretization space is created without regard to class boundaries. The decision surfaces of the NDDF classifiers can therefore out-perform those of SBPD classifiers by creating an optimal hyper-plane when the data is linearly separable. The BP classifier can similarly create non-orthogonal planes to represent the class distributions, in both linearly separable data and in the general cases. In this regard, improved SBPD classifier performance could be expected if class-dependent discretization schemes were used.

1.3.5 Real Data Applications in the Biological Sciences

The features of the SBPD system, including both its performance and transparency, allow its application to varied fields of data. The authors are applying this technique within clinical decision support tools [50, 51], in which characterizations of disease in electrophysiological muscular data are obtained.

Further biological applications evaluated by the authors [49] include validation of the system using the Wisconsin Breast Cancer and Heart Disease databases. These data are available respectively under the keys `breast-cancer-Wisconsin` and `heart-disease` in the University of California, Irvine "Machine Learning" data repository [76].

Comparative results for these data sets were strong when measured against the Nefclass-J classifier [34]. For the breast cancer database an average performance of 0.97 ± 0.02 correct classifications was obtained via SBPD versus 0.91 ± 0.02 for Nefclass; for the heart disease database statistically comparable classification was obtained in both cases (SBPD 0.83 ± 0.38, Nefclass-J 0.82 ± 0.30). A full discussion of these experiments is available in [49].

What is clear from all of the experimental evidence is that the SBPD system provides comparable performance to techniques such as those derived from artificial neural networks, while avoiding the over-fitting problems and delivering transparent results through the selection of patterns based on a filter to ensure statistical validity.

1.4 Conclusions

The SBPD algorithm presented here provides a means of examining a set of training data and capturing the knowledge therein to produce a statistically sound model of the contained relationships as a set of patterns. The patterns found can be used to assign labels on new data resulting in an association classifier whose performance is comparable to, but in general slightly lower than, other commonly used, high performance classifiers such as the popular error back-propagation artificial neural networks when tested on a variety of data class distributions, of continuous- or mixed-mode data.

The explainability of the patterns found and clear methodology of obtaining their weights makes this technique useful when high-risk decisions are being made, such as systems where human decision makers are being assisted in complex problems. In such applications, the transparency and explainability of the inference provided may outweigh the raw performance available from a more opaque classification system.

Further, the SBPD classifier is easily configured, simply requiring sufficient quantization resolution to adequately represent the important aspects of the class distributions while maintaining the expectation bound $\mathcal{E}_{\mathbf{x}_l^m}$; the general desire is to increase the number of quantization intervals until the expected number of occurrences in a hyper-cell is near statistical reason. Statistical uncertainty can be easily avoided with only a cursory examination of the dimensionality of the class distributions and knowledge of the number of training examples available.

References

1. Agrawal R, Srikant R (1994) Fast Algorithms for Mining Association Rules, In: Proc. 20th Int. Conf. Very Large Data Bases, 487–499. Morgan Kaufmann, Santiago, Chile
2. Srikant R, Agrawal R (1997) Mining generalized association rules, Fut Gen Comp Sys 13(2–3):161–180
3. Ma X, Wang W, Sun Y (2003) Associative Classifier Modeling Method Based on Rough Set Theory and Factor Analysis Technology, In: Proc. Int. Conf. Sys., Man and Cyb., vol. 3, 2412–2417. SMC '03
4. Do TD, Hui SC, Fong AC (2005) Prediction Confidence for Associative Classification, In: Proc. 4th Int. Conf Mach. Learn. Cybern., vol. 4, 1993–1997. IEEE, Guangzhou
5. Sun Y, Wong AKC, Wang Y (2006) An overview of associative classfiers. Tech. rep., University of Waterloo, Canada
6. Sun Y, Wang Y, Wong AKC (2006) Boosting an associative classfier, IEEE Trans Knowledge Data Eng 18(7):988–992
7. Wang K, Zhou S, He Y (2000) Growing Decision Trees On Support-Less Association Rules, In: Proc. 6th ACM SIGKDD Int. Conf. on Know. Disc. Data Mining, 265–269. SIGKDD'00, Boston
8. Veloso A, Meira Jr W, Zaki MJ (2006) Lazy Associative Classification, In: Proc. 6th Int. Conf. Data Mining, 645–654. ICDM'06, Hong Kong
9. Wong AKC, Wang Y (1997) High-order pattern discovery from discrete-valued data, IEEE Trans Knowledge Data Eng 9(6):877–893
10. Wang Y, Wong AKC (2003) From association to classification: Inference using weight of evidence, IEEE Trans Knowledge Data Eng 15(3):764–767
11. Wong AKC, Wang Y (2003) Pattern discovery: A data driven approach to decision support, IEEE Trans Syst, Man, Cybern C 33(1):114–124
12. Wang Y (1997) High Order Pattern Discovery and Analysis of Discrete-Valued Data Sets. Ph.D. Thesis, Systems Design Engineering, University of Waterloo
13. Alhammady H, Ramamohanarao K (2006) Using emerging patterns to construct weighted decision trees, IEEE Trans Knowledge Data Eng 18(7):865–876
14. Yin X, Han J (2003) CPAR: Classification based on Predictive Association Rules, In: 3rd SIAM Int. Conf. Data Mining. SDM'03, San Francisco
15. Quinlan JR (1993) C4.5 : Programs for Machine Learning. Morgan Kaufman

16. Quinlan JR (1996) Learning first-order definitions of functions, Journal of Artificial Intelligence Research 5:139–161
17. Yager RR, Filev DP (1996) Relational partitioning of fuzzy rules, Fuzzy Sets & Sys 80:57–69
18. Chong A, Gedon TD, Wong KW, Koczy LT (2001) A Histogram-Based Rule Extraction Technique for Fuzzy Systems, In: Fuzzy Systems FUZZ-IEEE'01, 638–641. FUZZ-IEEE '01, Melbourne, Australia
19. Hong TP, Lee CY (1996) Induction of fuzzy rules and membership function from training examples, Fuzzy Sets & Sys 84:33–47
20. Wang XZ, Wang YD, Xu XF, Ling WD, Yeung DS (2001) A new approach to fuzzy rule generation: Fuzzy extension matrix, Fuzzy Sets & Sys 123:291–306
21. Xing H, Huang SH, Shi J (2003) Rapid development of knowledge-based systems via integrated knowledge acquisition, Artificial Intelligence for Engineering Design, Analysis and Manufacturing 17:221–234
22. Spiegel D, Sudkamp T (2003) Sparse data in the evolutionary generation of fuzzy models, Fuzzy Sets & Sys 138:363–379
23. Hoffmann F (2004) Combined boosting and evolutionary algorithms for learning of fuzzy classification rules, Fuzzy Sets & Sys 141:47–58
24. Ishibuchi H, Nozaki K, Yamamoto N, Tanaka H (1994) Construction of fuzzy classification systems with rectangular fuzzy rules using genetic algorithms, Fuzzy Sets & Sys 65(2/3):237–253
25. Cordón O, Herrera F, Hoffmann F, Magkalena L (2001) Genetic Fuzzy Systems: Evolutionary Tuning and Learning of Fuzzy Knowledge Bases, chap. 11, 375–382. World Scientific, Singapore
26. Ghosh A, Pal NR, Pal SK (1993) Self-organization for object extraction using a multilayer neural network and fuzziness measures, IEEE Trans Fuzzy Syst 1(1):54–68
27. Mitra S, Hayashi Y (2000) Neuro-fuzzy rule generation: Survey in soft computing framework, IEEE Trans Neural Networks 11(3):748–768
28. Labbi A, Gauthier E (1997) Combining fuzzy knowledge and data for neuro-fuzzy modeling, J of Intell Sys 6(4)
29. Pal SK, Mitra S (1999) Neuro-Fuzzy Pattern Recognition : Methods in Soft Computing. Wiley-Interscience
30. Pedrycz W (1995) Fuzzy Sets Engineering. CRC Press
31. Kruse R, Gebhardt JE, Klawonn F (1994) Foundations of Fuzzy Systems. John Wiley & Sons, New York
32. Nauck D, Klawonn F, Kruse R (1997) Foundations of Neuro-Fuzzy Systems. John Wiley & Sons, New York
33. Höppner F, Klawonn F, Kruse R, Runkler TA (1999) Fuzzy Cluster Analysis. Chichester, England
34. Nauck D, Kruse R (1996) Designing Neuro-Fuzzy Systems Through Backpropagation, 203–228. Kluwer Academic Publishers, Boston, Dordrecht, London
35. Nauck D, Kruse R (1997) A neuro-fuzzy method to learn fuzzy classification rules from data, Fuzzy Sets & Sys 89(3):277–288
36. Gabrys B (2004) Learning hybrid neuro-fuzzy classifier models from data: To combine or not to combine?, Fuzzy Sets & Sys 147(1):39–56
37. Song Q, Kasabov NK (2005) NFI: A neuro-fuzzy inference method for transductive reasoning, IEEE Trans Fuzzy Syst 13(6):799–808
38. Wang JS, Lee CSG (2002) Self-adaptive neuro-fuzzy inference systems for classification applications, IEEE Trans Fuzzy Syst 10(6):790–802

39. Shen Q, Chouchoulas A (2002) A rough-fuzzy approach for generating classification rules, Pat Rec 35:2425–2438
40. Tsumoto S (2002) Statistical Evidence for Rough Set Analysis, In: FUZZ-IEEE '02 [77], 757–762. Published in [78]
41. Ziarko W (2002) Acquisition of Hierarchy-Structured Probabalistic Decision Tables and Rules from Data, In: FUZZ-IEEE '02 [77], 779–784. Published in [78]
42. Bean CL, Kambhampati C, Rajasekharan S (2002) A Rough Set Solution to a Fuzzy Set Problem, In: FUZZ-IEEE '02 [77], 18–23. Published in [78]
43. Kukolj D (2002) Design of adaptive Takagi-Sugeno-Kang fuzzy models, Applied Soft Computing 2:89–103
44. Chen L, Tokuda N, Zhang X, He Y (2001) A new scheme for an automatic generation of multi-variable fuzzy systems, Fuzzy Sets & Sys 120:323–329
45. Chen MY (2002) Establishing Interpretable Fuzzy Models from Numeric Data, In: Proceedings of the 4th World Congress on Intelligent Control and Automation, vol. 3, 1857–1861. IEEE
46. Hamilton-Wright A (2005) Transparent Decision Support Using Statistical Evidence. Ph.D. Thesis, Systems Design Engineering, University of Waterloo
47. Hamilton-Wright A, Stashuk DW (2005) Comparing 'Pattern Discovery' and Back-Propagation Classifiers, In: Proc. of the Int. J. Conf. Neural Networks (IJCNN), vol. 2, 1286–1291. IJCNN '05, Montréal, Québec
48. Hamilton-Wright A, Stashuk DW (2006) Transparent decision support using statistical reasoning and fuzzy inference, IEEE Trans Knowledge Data Eng 18(8):1125–1137
49. Hamilton-Wright A, Stashuk DW, Tizhoosh HR (2007) Fuzzy classification using pattern discovery, IEEE Trans Fuzzy Syst 15(5):772–783
50. Hamilton-Wright A, Stashuk DW (2006) Clinical Characterization of Electromyographic Data Using Computational Tools, In: Proc. Symp. Comp. Intel. in Bioinf. & Comp. Biol. CIBCB, Toronto
51. Hamilton-Wright A, Stashuk DW (2006) Clinical Decision Support By Fuzzy Logic Analysis of Quantitative Electromyographic Data, In: Proc. XVIth Int. Soc. of Electromyog. and Kinesiol. ISEK '06, Torino, Italy
52. Bishop CM (1995) Neural Networks for Pattern Recognition. Oxford
53. Rumelhart DE, Hinton GE, Williams RJ (1986) Learning representations by back-propagating errors, Nature 323:533–536
54. Minsky ML, Papert SA (1988) Perceptrons : An Introduction to Computational Geometry. MIT Press, 2nd edn.
55. Simpson PK (1991) Artificial Neural Systems. Windcrest/McGraw-Hill
56. Hertz J, Krogh A, Palmer RG (1991) Introduction to the Theory of Neural Computation. Santa Fe Institute Studies in the Sciences of Complexity
57. Cheeseman P (1985) In defense of probability, In: Proc. Ninth Int. Conf. A.I. (IJCAI-85), 1002–1009. Morgan Kaufmann, Santiago, Chile
58. Cheeseman P, Self M, Kelly J, Stutz J (1988) Bayesian classification, In: Proc. Seventh Nat. Conf. A.I. (AAAI-88), vol. 2, 607–611. Morgan Kaufmann, St. Paul, Minnesota
59. Cheeseman P, Kelly J, Self M, Stutz J, Taylor W, Freeman D (1993) Readings in Knowledge Acquisition and Learning: Automating the Construction and Improvement of Expert Systems, chap. AutoClass: a Bayesian classification system, 431–441. Morgan Kaufmann, San Mateo, California
60. Duda RO, Hart PE, Stork DG (2001) Pattern Classification. John Wiley & Sons, 2nd edn.

61. Haberman SJ (1973) The analysis of residuals in cross-classified tables, Biometrics 29(1):205–220
62. Haberman SJ (1979) Analysis of Qualitative Data, vol. 1 of *Springer Series in Statistics*, 78–79,82–83. Academic Press, Toronto
63. Antonie ML, Zaïane OR (2004) Knowledge Discovery In Databases, vol. 3202 of *Lecture Notes in Computer Science*, chap. Mining Positive and Negative Association Rules: An Approach for Confined Rules, 27–38. Springer
64. Wu X, Zhang C, Zhang S (2002) Mining Both Positive and Negative Association Rules, In: Proc. 19th Int. Conf. on Mach. Learn., 658–665. ICML '02, Morgan Kaufmann Publishers Inc., San Francisco
65. Hamilton-Wright A, Stashuk DW (2006) Fuzzy Rule Based Decision Making For Electromyographic Characterization, In: IPMU '06 [79]
66. Hamilton-Wright A, Stashuk DW, Pino L (2006) On Weight Of Evidence Based Reliability In 'Pattern Discovery', In: IPMU '06 [79]
67. Hamilton-Wright A, Stashuk DW, Pino L (2006) Internal Measures of Reliability in 'Pattern Discovery' Based Fuzzy Inference, In: IPMU '06 [79]
68. Gokhale DV (1999) On joint and conditional entropies, Entropy 1(2):21–24
69. Chau T (2001) Marginal maximum entropy partitioning yields asymptotically consistent probability density functions, IEEE Trans Pattern Anal Machine Intell 23(4):414–417
70. Bryson N, Joseph A (2001) Optimal techniques for class-dependent attribute discretization, J Op Res Soc 52(10):1130–1143
71. Ching JY, Wong AKC, Chan KCC (1995) Class-dependent discretization for inductive learning from continuous and mixed-mode data, IEEE Trans Pattern Anal Machine Intell 17(7):641–651
72. Liu L, Wong AKC, Wang Y (2004) A global optimal algorithm for class-dependent discretization of continuous data, J Int Data Anal 8(2):151–170
73. Bellman R (1961) Adaptive Control Processes: A Guided Tour. Princeton University Press, New Jersey
74. Guyon I, Elisseef A (2003) An introduction to variable and feature selection, J Mach Learn Research 3(7/8):1157–1182
75. Ye J, Janardan R, Li Q, Park H (2006) Feature reduction via generalized uncorrelated linear discriminant analysis, IEEE Trans Knowledge Data Eng 18(10):1312–1322
76. Newman DJ, Hettich S, Blake CL, Merz CJ (1998). UCI repository of machine learning databases
URL http://www.ics.uci.edu/~mlearn/MLRepository.html
77. FUZZ-IEEE '02 (2002) Proc. 11th IEEE Int. Conf. Fuzzy Sys., FUZZ-IEEE'02. Published in [78]
78. WCCI '02 (2002) Proc. 2002 World Congr. Comp. Int. WCCI 2002. IEEE Press, Honolulu
79. IPMU '06 (2006) Proc. 11th Int. Conf. on Info. Proc. and Mgt. of Uncertainty

Rough Sets In Data Analysis: Foundations and Applications

Lech Polkowski[1,2] and Piotr Artiemjew[2]

[1] Polish-Japanese Institute of Information Technology, Koszykowa 86, 02008
Warszawa, Poland
`polkow@pjwstk.edu.pl`
[2] Department of Mathematics and Computer Science, University of Warmia and
Mazury, Żołnierska 14, Olsztyn, Poland
`artem@matman.uwm.edu.pl`

Summary. Rough sets is a paradigm introduced in order to deal with uncertainty
due to ambiguity of classification caused by incompleteness of knowledge. The idea
proposed by Z. Pawlak in 1982 goes back to classical idea of representing uncertain
and/or inexact notions due to the founder of modern logic, Gottlob Frege: uncertain
notions should possess around them a region of uncertainty consisting of objects
that can be qualified with certainty neither into the notion nor to its complement.
The central tool in realizing this idea in rough sets is the relation of uncertainty
based on the classical notion of indiscernibility due to Gottfried W. Leibniz: objects
are indiscernible when no operator applied to each of them yields distinct values.

In applications, knowledge comes in the form of data; those data in rough sets
are organized into an information system: a pair of the form (U, A) where U is a
set of objects and A is a set of attributes, each of them a mapping $a : U \to V_a$, the
value set of a.

Each attribute a does produce the *a-indiscernibility relation* $IND(a) = \{(u, v) :
a(u) = a(v)\}$. Each set of attributes B does induce the *B-indiscernibility relation*
$IND(B) = \bigcap IND(a) : a \in B$. Objects u, v that are in the relation $IND(B)$ are
B-indiscernible. Classes $[u]_B$ of the relation $IND(B)$ form *B–elementary granules
of knowledge.*

Rough sets allow for establishing dependencies among groups of attributes: a
group B depends functionally on group C when $IND(C) \subseteq IND(B)$: in that case
values of attributes in B are functions of values of attributes in C.

An important case is when data are organized into a decision system: a triple
(U, A, d) where d is a new attribute called the *decision*. The decision gives a classification of object due to an expert, an external oracle; establishing dependencies
between groups B of attributes in A and the decision is one of tasks of rough set
theory.

The language for expressing dependencies is the descriptor logic. A *descriptor* is
a formula $(a = v)$ where $v \in V_a$, interpreted in the set U as $[a = v] = \{u : a(u) = v\}$.
Descriptor formulas are obtained from descriptors by means of connectives $\vee, \wedge, \neg, \Rightarrow$

L. Polkowski and P. Artiemjew: *Rough Sets In Data Analysis: Foundations and Applications*,
Studies in Computational Intelligence (SCI) **122**, 33–54 (2008)
`www.springerlink.com`

of propositional calculus; their semantics is: $[\alpha \vee \beta] = [\alpha] \cup [\beta]$, $[\alpha \wedge \beta] = [\alpha] \cap [\beta]$, $[\neg\alpha] = U \setminus [\alpha]$, $[\alpha \Rightarrow \beta] = [\neg\alpha] \cup [\beta]$.

In the language of descriptors, dependency between a group B of attributes and the decision is expressed as a decision rule: $\bigwedge_{a \in B}(a = v_a) \Rightarrow (d = v)$; a set of decision rules is a decision algorithm. There exist a number of algorithms for inducing decision rules.

Indiscernibility relations proved to be too rigid for classification, and the search has been in rough sets for more flexible similarity relations. Among them one class that is formally rooted in logic is the class of rough inclusions. They allow for forming granules of knowledge more robust than traditional ones. Algorithms based on them allow for a substantial knowledge reduction yet with good classification quality.

The problem that is often met in real data is the problem of missing values. Algorithms based on granulation of knowledge allow for solving this problem with a good quality of classification.

In this Chapter, we discuss:

- Basics of rough sets;
- Language of descriptors and decision rules;
- Algorithms for rule induction;
- Examples of classification on real data;
- Granulation of knowledge;
- Algorithms for rule induction based on granulation of knowledge;
- Examples of classification of real data;
- The problem of missing values in data.

2.1 Basics of rough sets

Introduced by Pawlak in [20], rough set theory is based on ideas that – although independently fused into a theory of knowledge – borrow some thoughts from Gottlob Frege, Gottfried Wilhelm Leibniz, Jan Łukasiewicz, Stanislaw Leśniewski, to mention a few names of importance.

Rough set approach rests on the assumption that knowledge is classification of entities into concepts (notions). To perform the classification task, entities should be described in a formalized symbolic language.

In case of the rough set theory, this language is the language of attributes and values. The formal framework for allowing this description is an information system, see Pawlak [21].

2.1.1 Information systems: formal rendering of knowledge

An information system is a pair (U, A), in which U is a set of *objects* and A is a set of *attributes*. Each attribute $a \in A$ is a mapping $a : U \to V_a$ from the universe U into the *value set* V_a of a. A variant of this notion is a basic in data mining notion of a *decision system*: it is a pair $(U, A \cup \{d\})$, where $d \notin A$ is the *decision*. In applications, decision d is the attribute whose value is set by an expert whereas attributes in A, called in this case *conditional attributes*, are selected and valued by the system user. Description of entities is done in the attribute–value language.

2.1.2 Attribute–value language. Indiscernibility

Attribute–value language is built from elementary formulas called *descriptors*; a descriptor is a formula of the form $(a = v)$, where $v \in V_a$. From descriptors, complex formulas are formed by means of connectives $\vee, \wedge, \neg, \Rightarrow$ of propositional calculus: if α, β are formulas then $\alpha \vee \beta$, $\alpha \wedge \beta$, $\neg \alpha$, $\alpha \Rightarrow \beta$ are formulas. These formulas and no other constitute the syntax of the *descriptor logic*.

Semantics of descriptor logic formulas is defined recursively: for a descriptor $(a = v)$, its meaning $[a = v]$ is defined as the set $\{u \in U : a(u) = v\}$. For complex formulas, one adopts the recursive procedure, given by the following identities:

- $[\alpha \vee \beta] = [\alpha] \cup [\beta]$.
- $[\alpha \wedge \beta] = [\alpha] \cap [\beta]$.
- $[\neg \alpha] = U \setminus [\alpha]$.
- $[\alpha \Rightarrow \beta] = [\neg \alpha] \cup [\beta]$.

Descriptor logic allows for coding of objects in the set U as sets of descriptors: for an object $u \in U$, the *information set* $Inf_A(u)$ is defined as the set $\{(a = a(u)) : a \in A\}$. It may happen that two objects, u and v, have the same information set: $Inf_A(u) = Inf_A(v)$; in this case, one says that u and v are *A–indiscernible*. This notion maybe relativized to any set $B \subseteq A$ of attributes: the *B–indiscernibility* relation is defined as $IND(B) = \{(u, v) : Inf_B(u) = Inf_B(v)\}$, where $Inf_B(u) = \{(a = a(u)) : a \in B\}$ is the information set of u restricted to the set B of attributes.

A more general notion of a *template* was proposed and studied in [18]: a template is a formula of the form $(a \in W_a)$, where $W_a \subseteq V_a$ is a set of values of the attribute a; the meaning $[a \in W_a]$ of the template $(a \in W_a)$ is the set $\{u \in U : a(u) \in W_a\}$. Templates can also (like descriptors) be combined by means of propositional connectives with semantics defined as with descriptors.

The indiscernibility relations are very important in rough sets: one easily may observe that for $u \in U$, and the formula in descriptor logic: $\phi_u^B : \bigwedge_{a \in B}(a = a(u))$, the meaning $[\phi_u^B]$ is equal to the equivalence class $[u]_B = \{v \in U : (u, v) \in IND(B)$ of the equivalence relation $IND(B)$.

The moral is: classes $[u]_B$ are *definable*, i.e., they have descriptions in the descriptor logic; also unions of those classes are definable: for a union $X = \bigcup_{j \in J}[u_j]_{B_j}$ of such classes, the formula $\bigvee_{j \in J} \phi_{u_j}^{B_j}$ has the meaning equal to X.

Concepts $X \subseteq U$ that are definable are also called *exact*; other concepts are called *rough*. The fundamental difference between the two kinds of concepts is that only exact concepts are "seen" in data; rough concepts are "blurred" and they can be described by means of exact concepts only; to this aim, rough sets offer the notion of an approximation.

2.1.3 Approximations

Due to Fregean idea [6], an inexact concept should possess a boundary into which objects that can be classified with certainty neither to the concept nor to its complement fall. This boundary to a concept is constructed from indiscernibility relations induced by attributes (features) of objects.

To express the B–boundary of a concept X induced by the set B of attributes, approximations over B are introduced, i.e.,
$\underline{B}X = \bigcup\{[u]_B : [u]_B \subseteq X\}$ (the B–lower approximation)

$\overline{B}X = \bigcup\{[u]_B : [u]_B \cap X \neq \emptyset\}$ (the B–upper approximation).

The difference $Bd_B X = \overline{B}X \setminus \underline{B}X$ is the B–boundary of X; when non–empty it does witness that X is rough.
For a rough concept X, one has the double strict inclusion: $\underline{B}X \subset X \subset \overline{B}X$ as the description of X in terms of two nearest to it exact concepts.

2.1.4 Knowledge reduction. Reducts

Knowledge represented in an information system (U, A) can be reduced: a reduct B of the set A of attributes is a minimal subset of A with the property that $IND(B) = INDd(A)$. Thus, reducts are minimal with respect to inclusion sets of attributes which preserve classification, i.e., knowledge.

Finding all reducts is computationally hard: the problem of finding a minimal length reduct is NP–hard, see [35].

An algorithm for finding reducts based on Boolean Reasoning technique was proposed in [35]; the method of Boolean Reasoning consists in solving a problem by constructing a Boolean function whose prime implicants would give solutions to the problem [3].

The Skowron–Rauszer algorithm for reduct induction: a case of Boolean Reasoning

In the context of an information system (U, A), the method of Boolean Reasoning for reduct finding proposed by Skowron and Rauszer [35], given input (U, A) with $U = \{u_1, ..., u_n\}$, starts with the *discernibility matrix*,
$M_{U,A} = [c_{i,j} = \{a \in A : a(u_i) \neq a(u_j)\}]_{1 \geq i,j \leq n}$,
and builds the Boolean function in the CNF form,
$f_{U,A} = \bigwedge_{c_{i,j} \neq \emptyset, i<j} \bigvee_{a \in c_{i,j}} \overline{a}$, where $\overline{a}$ is the Boolean variable assigned to the attribute $a \in A$.

The function $f_{U,A}$ is converted to its DNF form: $f_{U,A}^* : \bigvee_{j \in J} \bigwedge_{k \in K_j} \overline{a_{j,k}}$.

Then: sets of the form $R_j = \{a_{j,k} : k \in K_j\}$ for $j \in J$, corresponding to prime implicants $\bigwedge_{k \in K_j} \overline{a_{j,k}}$ are all reducts of A.

On the soundness of the algorithm We give here a proof of the soundness of the algorithm in order to acquaint the reader with this method which is also exploited in a few variants described below; the reader will be able to supply own proofs in those cases on the lines shown here.

We consider a set B of attributes and the valuation val_B on the Boolean variable set $\{\bar{a} : a \in A\}$: $val_B(\bar{a}) = 1$ in case $a \in B$ and 0, otherwise.

Assume that the Boolean function $f_{U,A}$ is satisfied under this valuation: $val_B(f_{U,A}) = 1$. This means that $val_B(\bigvee_{a \in c_{i,j}} \bar{a}) = 1$ for each $c_{i,j} \neq \emptyset$. An equivalent formula to this statement is: $\forall i, j.c_{i,j} \neq \emptyset \Rightarrow \exists a \in c_{i,j}.a \in B$. Applying tautology $p \Rightarrow q \Leftrightarrow \neg q \Rightarrow \neg p$ to the last implication, we obtain: $\forall a \in B.a \notin c_{i,j} \Rightarrow \forall a \in A.a \notin c_{i,j}$ for each pair i, j. By definition of the set $c_{i,j}$, the last implication reads: $IND(B) \subseteq IND(A)$. This means $IND(B) = IND(A)$ as $IND(A) \subseteq IND(B)$ always because $B \subseteq A$.

Now, we have $val_B(f_{U,A}^*) = 1$ as well; this means that $val_B(\bigwedge_{k \in K_j} \overline{a_{j,k}}) = 1$ for some $j_o \in J$. In turn, by definition of val_B, this implies that $B \subseteq \{a_{j_o,k} : k \in K_{j_o}\}$.

A conclusion from the comparison of values of val_B on $f_{U,A}$ and $f_{U,A}^*$ is that : $IND(B) = IND(A)$ if and only if $B \subseteq \{a_{j,k} : k \in K_j\}$ for the $j - th$ prime implicant of $f_{U,A}$. Thus, any minimal with respect to inclusion set B of attributes such that $IND(B) = IND(A)$ coincides with a set of attributes $\{a_{j,k} : k \in K_j\}$ corresponding to a prime implicant of the function $f_{U,A}$.

Choosing a reduct R, and forming the reduced information system (U, R) one is assured that no information encoded in (U, A) has been lost.

2.1.5 Decision systems. Decision rules: an introduction

A decision system $(U, A \cup \{d\})$ encodes information about the external classification d (by an oracle, expert etc.). Methods based on rough sets aim at finding a description of the concept d in terms of conditional attributes in A in the language of descriptors. This description is fundamental for expert systems, knowledge based systems and applications in Data Mining and Knowledge Discovery.

Formal expressions for relating knowledge in conditional part (U, A) to knowledge of an expert in (U, d) are *decision rules*; in descriptor logic they are of the form $\phi_U^B \Rightarrow (d = w)$, where $w \in V_d$, the value set of the decision.

Semantics of decision rules is given by general rules set in sect. 2.1.2: the rule $\phi_U^B \Rightarrow (d = w)$ is *certain* or *true* in case $[\phi_u^B] \subseteq [d = w]$, i.e., in case when each object v that satisfies ϕ_u^B, i.e., $(u, v) \in IND(B)$, satisfies also $d(v) = w$; otherwise the rule is said to be *partial*.

The simpler case is when the decision system is *deterministic*, i.e., $IND(A) \subseteq IND(d)$. In this case the relation between A and d is *functional*, given by the unique assignment $f_{A,d} : Inf_A(u) \to Inf_d(u)$, or, in the decision rule form as the set of rules: $\bigwedge_{a \in A}(a = a(u)) \Rightarrow (d = d(u))$. Each of these rules is clearly certain.

In place of A any reduct R of A can be substituted leading to shorter certain rules.

In the contrary case, some classes $[u]_A$ are split into more than one decision class $[v]_d$ leading to ambiguity in classification. In order to resolve the ambiguity, the notion of a δ–reduct was proposed in [35]; it is called a relative reduct in [2].

To define δ– reducts, first the generalized decision δ_B is defined for any $B \subseteq A$: for $u \in U$, $\delta_B(u) = \{v \in V_d : d(u') = v \land (u, u') \in IND(B)$ for some $u' \in U\}$. A subset B of A is a δ–reduct to d when it is a minimal subset od A with respect to the property that $\delta_B = \delta_A$.

δ–reducts can be obtained from the modified Skowron and Rauszer algorithm [35]: it suffices to modify the entries $c_{i,j}$ to the discernibility matrix, by letting $c_{i,j}^d = \{a \in A \cup \{d\} : a(u_i) \neq a(u_j)\}$ and then setting $c_{i,j}' = c_{i,j}^d \setminus \{d\}$ in case $d(u_i) \neq d(u_j)$ and $c_{i,j}' = \emptyset$ in case $d(u_i) = d(u_j)$. The algorithm described above input with entries $c_{i,j}'$ forming the matrix $M_{U,A}^\delta$ outputs all δ–reducts to d encoded as prime implicants of the associated Boolean function $f_{U,A}^\delta$.

For any δ–reduct R, rules of the form $\phi_u^R \Rightarrow \delta = \delta_R(u)$ are certain.

An example of reduct finding and decision rule induction

We conclude the first step into rough sets with a simple example of a decision system, its reducts and decision rules.

Table 2.1 shows a simple decision system.

Table 2.1. Decision system Simple

obj.	a_1	a_2	a_3	a_4	d
u_1	1	0	0	1	0
u_2	0	1	0	0	1
u_3	1	1	0	0	1
u_4	1	0	0	1	1
u_5	0	0	0	1	1
u_6	1	1	1	1	0

Reducts of the information system $(U, A = \{a_1, a_2, a_3, a_4\})$ can be found from the discernibility matrix $M_{U,A}$ in Table 2.2; by symmetry, cells $c_{i,j} = c_{j,i}$ with $i > j$ are not filled. Each attribute a_i is encoded by the Boolean variable i.

After reduction by means of absorption rules of sentential calculus: $(p \lor q) \land p \Leftrightarrow p$, $(p \land q) \lor p \Leftrightarrow p$, the DNF form $f_{U,A}^*$ is $1 \land 2 \land 3 \lor 1 \land 2 \land 4 \lor 1 \land 3 \land 4$. Reducts of A in the information system (U, A) are : $\{a_1, a_2, a_3\}$, $\{a_1, a_2, a_4\}$, $\{a_1, a_3, a_4\}$.

δ–reducts of the decision d in the decision system Simple, can be found from the modified discernibility matrix $M_{U,A}^\delta$ in Table 2.3.

Table 2.2. Discernibility matrix $M_{U,A}$ for reducts in (U, A)

obj.	u_1	u_2	u_3	u_4	u_5	u_6
u_1	$\emptyset$	$\{1,2,4\}$	$\{2,4\}$	$\emptyset$	$\{1\}$	$\{2,3\}$
u_2	–	$\emptyset$	$\{1\}$	$\{1,2,3\}$	$\{2,4\}$	$\{1,3,4\}$
u_3	–	–	$\emptyset$	$\{2,4\}$	$\{2,4\}$	$\{3,4\}$
u_4	–	–	–	$\emptyset$	$\{1\}$	$\{2,3\}$
u_5	–	–	–	–	$\emptyset$	$\{1,2,3\}$
u_6	–	–	–	–	–	$\emptyset$

Table 2.3. Discernibility matrix $M_{U,A}^{\delta}$ for δ–reducts in (U, A, d)

obj.	u_1	u_2	u_3	u_4	u_5	u_6
u_1	$\emptyset$	$\{1,2,4\}$	$\{2,4\}$	$\emptyset$	$\{1\}$	$\emptyset$
u_2	–	$\emptyset$	$\emptyset$	$\emptyset$	$\emptyset$	$\{1,3,4\}$
u_3	–	–	$\emptyset$	$\emptyset$	$\emptyset$	$\{3,4\}$
u_4	–	–	–	$\emptyset$	$\emptyset$	$\{2,3\}$
u_5	–	–	–	–	$\emptyset$	$\{1,2,3\}$
u_6	–	–	–	–	–	$\emptyset$

From the Boolean function $f_{U,A}^{\delta}$ we read off δ–reducts $R_1 = \{a_1, a_2, a_3\}$, $R_2 = \{a_1, a_2, a_4\}$, $R_3 = \{a_1, a_3, a_4\}$.

Taking R_1 as the reduct for inducing decision rules, we read the following certain rules:

$r_1 : (a_1 = 0) \wedge (a_2 = 1) \wedge (a_3 = 0) \Rightarrow (d = 1)$;
$r_2 : (a_1 = 1) \wedge (a_2 = 1) \wedge (a_3 = 0) \Rightarrow (d = 1)$;
$r_3 : (a_1 = 0) \wedge (a_2 = 0) \wedge (a_3 = 0) \Rightarrow (d = 1)$;
$r_4 : (a_1 = 1) \wedge (a_2 = 1) \wedge (a_3 = 1) \Rightarrow (d = 0)$;

and two possible rules

$r_5 : (a_1 = 1) \wedge (a_2 = 0) \wedge (a_3 = 0) \Rightarrow (d = 0)$;
$r_6 : (a_1 = 1) \wedge (a_2 = 0) \wedge (a_3 = 0) \Rightarrow (d = 1)$,

each with certainty factor $=.5$ as there are two objects with d=0.

2.1.6 Decision rules: advanced topics

In order to precisely discriminate between certain and possible rules, the notion of a *positive region* along with the notion of a *relative reduct* was proposed and studied in [35].

Positive region $pos_B(d)$ is the set $\{u \in U : [u]_B \subseteq [u]_d\} = \bigcup_{v \in V_d} \underline{B}[(d = v)]$; $pos_B(d)$ is the greatest subset X of U such that $(X, B \cup \{d\})$ is deterministic; it generates certain rules. Objects in $U \setminus pos_B(d)$ are subjected to ambiguity: given such u, and the collection $v_1, .., v_k$ of decision d values on the class $[u]_B$, the decision rule describing u can be formulated as, $\bigwedge_{a \in B}(a = a(u)) \Rightarrow \bigvee_{i=1,...,k}(d = v_i)$; each of the rules $\bigwedge_{a \in B}(a = a(u)) \Rightarrow (d = v_i)$ is possible but not certain as only for a fraction of objects in the class $[u]_B$ the decision takes the value v_i on.

Relative reducts are minimal sets B of attributes with the property that $pos_B(d) = pos_A(d)$; they can also be found by means of discernibility matrix $M^*_{U,A}$ [90]: $c^*_{i,j} = c^d_{i,j} \setminus \{d\}$ in case either $d(u_i) \neq d(u_j)$ and $u_i, u_j \in pos_A(d)$ or $pos(u_i) \neq pos(u_j)$ where pos is the characteristic function of $pos_A(d)$; otherwise, $c^*_{i,j} = \emptyset$.

For a relative reduct B, certain rules are induced from the deterministic system $(pos_B(d), A \cup \{d\})$, possible rules are induced from the non–deterministic system $(U \setminus pos_B(d), A \cup \{d\})$. In the last case, one can find δ–reducts to d in this system and turn the system into a deterministic one $(U \setminus pos_B(d), A, \delta)$ inducing certain rules of the form $\bigwedge_{a \in B}(a = a(u)) \Rightarrow \bigvee_{v \in \delta(u)}(d = v)$.

A method for obtaining decision rules with minimal number of descriptors [22], [34], consists in reducing a given rule $r : \phi/B, u \Rightarrow (d = v)$ by finding a set $R_r \subseteq B$ consisting of irreducible attributes in B only, in the sense that removing any $a \in R_r$ causes the inequality $[\phi/R_r, u \Rightarrow (d = v)] \neq [\phi/R_r \setminus \{a\}, u \Rightarrow (d = v)]$ to hold. In case $B = A$, reduced rules $\phi/R_r, u \Rightarrow (d = v)$ are called optimal basic rules (with minimal number of descriptors). The method for finding of all irreducible subsets of the set A [34], consists in considering another modification of discernibility matrix: for each object $u_k \in U$, the entry $c'_{i,j}$ into the matrix $M^\delta_{U,A}$ for δ–reducts is modified into $c^k_{i,j} = c'_{i,j}$ in case $d(u_i) \neq d(u_j)$ and $i = k \vee j = k$, otherwise $c^k_{i,j} = \emptyset$. Matrices $M^k_{U,A}$ and associated Boolean functions $f^k_{U,A}$ for all $u_k \in U$ allow for finding all irreducible subsets of the set A and in consequence all basic optimal rules (with minimal number of descriptors).

Decision rules are judged by their quality on the basis of the training set and by quality in classifying new unseen as yet objects, i.e., by their performance on the test set. Quality evaluation is done on the basis of some measures: for a rule $r : \phi \Rightarrow (d = v)$, and an object $u \in U$, one says that u matches r in case $u \in [\phi]$. $match(r)$ is the number of objects matching r. Support $supp(r)$ of r is the number of objects in $[\phi] \cap [(d = v)]$; the fraction $cons(r) = \frac{supp(r)}{match(r)}$ is the consistency degree of r: $cons(r) = 1$ means that the rule is certain.

Strength, $strength(r)$, of the rule r is defined, as the number of objects correctly classified by the rule in the training phase [15], [1], [8]; relative strength is defined as the fraction $rel - strength(r) = \frac{supp(r)}{|[(d=v)]|}$. Specificity of the rule r, $spec(r)$, is the number of descriptors in the premise ϕ of the rule r.

In the testing phase, rules vie among themselves for object classification when they point to distinct decision classes; in such case, negotiations among rules or their sets are necessary. In these negotiations, rules with better characteristics are privileged.

For a given decision class $c : d = v$, and an object u in the test set, the set $Rule(c, u)$ of all rules matched by u and pointing to the decision v, is characterized globally by $Support(Rule(c, u)) = \sum_{r \in Rule(c,u)} strength(r) \cdot$

$spec(r)$. The class c for which $Support(Rule(c,u))$ is the largest wins the competition and the object u is classified into the class $c : d = v$.

It may happen that no rule in the available set of rules is matched by the test object u and partial matching is necessary, i.e., for a rule r, the matching factor $match-fact(r,u)$ is defined as the fraction of descriptors in the premise ϕ of r matched by u to the number $spec(r)$ of descriptors in ϕ. The rule for which the partial support $Part - Support(Rule(c,u)) = \sum_{r \in Rule(c,u)} match - fact(r,u) \cdot strength(r) \cdot spec(r)$ is the largest wins the competition and it does assign the value of decision to u.

2.2 Discretization of continuous valued attributes

The important problem of treating continuous values of attributes has been resolved in rough sets with the help of discretization of attributes technique, common to many paradigms like decision trees, etc.; for a decision system (U, A, d), a cut is a pair (a, c), where $a \in A, c$ in reals. The cut (a, c) induces the binary attribute $b_{a,c}(u) = 1$ if $a(u) \geq c$ and it is 0, otherwise. Given a finite sequence $p_a = c_0^a < c_1^a < .... < c_m^a$ of reals, the set V_a of values of a is split into disjoint intervals: $(\leftarrow, c_0^a), [c_0^a, c_1^a), ..., [c_m^a, \rightarrow)$; the new attribute $D_a(u) = i$ when $b_{c_{i+1}^a} = 0, b_{c_i^a} = 1$, is a discrete counterpart to the continuous attribute a. Given a collection $P = \{p_a : a \in A\}$ (a cut system), the set $D = \{D_a : a \in A\}$ of attributes transforms the system (U, A, d) into the discrete system (U, D_P, d) called the P–segmentation of the original system. The set P is consistent in case generalized decision in both systems is identical, i.e., $\delta_A = \delta_{D_P}$; a consistent P is irreducible if P' is not consistent for any proper subset $P' \subset P$; P is optimal if its cardinality is minimal among all consistent cut systems, see [16], [17].

2.3 Classification

Classification methods can be divided according to the adopted methodology, into classifiers based on reducts and decision rules, classifiers based on templates and similarity, classifiers based on descriptor search, classifiers based on granular descriptors, hybrid classifiers.

For a decision system (U, A, d), classifiers are sets of decision rules. Induction of rules was a subject of research in rough set theory since its beginning. In most general terms, building a classifier consists in searching in the pool of descriptors for their conjuncts that describe decision classes sufficiently well. As distinguished in [37], there are three main kinds of classifiers searched for: *minimal*, i.e., consisting of the minimum possible number of rules describing decision classes in the universe, *exhaustive*, i.e., consisting of all possible rules, *satisfactory*, i.e., containing rules tailored to a specific use. Classifiers are evaluated globally with respect to their ability to properly classify objects, usually by *error* which is the ratio of the number of correctly classified objects to the number of test objects, *total accuracy* being the ratio of the number of correctly classified cases to the number of recognized cases, and *total coverage*, i.e, the ratio of the number of recognized test cases to the number of test cases.

Minimum size algorithms include LEM2 algorithm due to Grzymala–Busse [9] and covering algorithm in the RSES package [33]; exhaustive algorithms include, e.g., LERS system due to Grzymala–Busse [7], systems based on discernibility matrices and Boolean reasoning [34], see also [1], [2], implemented in the RSES package [33].

Minimal consistent sets of rules were introduced in Skowron and Rauszer [35]. Further developments include dynamic rules, approximate rules, and relevant rules as described in [1], [2], as well as local rules (op. cit.) effective in implementations of algorithms based on minimal consistent sets of rules. Rough set based classification algorithms, especially those implemented in the RSES system [33], were discussed extensively in [2].

In [1], a number of techniques were verified in experiments with real data, based on various strategies:

`discretization` of attributes (codes: N-no discretization, S-standard discretization, D-cut selection by dynamic reducts, G-cut selection by generalized dynamic reducts);

`dynamic selection of attributes` (codes: N-no selection, D-selection by dynamic reducts, G-selection based on generalized dynamic reducts);

`decision rule choice` (codes: A-optimal decision rules, G-decision rules on basis of approximate reducts computed by Johnson's algorithm, simulated annealing and Boltzmann machines etc., N-without computing of decision rules);

`approximation of decision rules` (codes: N-consistent decision rules, P-approximate rules obtained by descriptor dropping);

`negotiations among rules` (codes: S-based on strength, M-based on maximal strength, R-based on global strength, D-based on stability).

Any choice of a strategy in particular areas yields a compound strategy denoted with the alias being concatenation of symbols of strategies chosen in consecutive areas, e.g., NNAND etc.

We record here in Table 2.4 an excerpt from the comparison (Table 8, 9, 10 in [1]) of best of these strategies with results based on other paradigms in classification for two sets of data: Diabetes and Australian credit from UCI Repository [40].

An adaptive method of classifier construction was proposed in [43]; reducts are determined by means of a genetic algorithm, see [2], and in turn reducts induce subtables of data regarded as classifying agents; choice of optimal ensembles of agents is done by a genetic algorithm.

Table 2.4. A comparison of errors in classification by rough set and other paradigms

paradigm	*system/method*	*Diabetes*	*Austr.credit*
Stat.Methods	*Logdisc*	0.223	0.141
Stat.Methods	*SMART*	0.232	0.158
Neural Nets	*Backpropagation2*	0.248	0.154
Neural Networks	*RBF*	0.243	0.145
Decision Trees	*CART*	0.255	0.145
Decision Trees	*C4.5*	0.270	0.155
Decision Trees	*ITrule*	0.245	0.137
Decision Rules	*CN2*	0.289	0.204
Rough Sets	*NNANR*	0.335	0.140
Rough Sets	*DNANR*	0.280	0.165
Rough Sets	*best result*	$0.255(DNAPM)$	$0.130(SNAPM)$

2.4 Approaches to classification in data based on similarity

Algorithms mentioned in sect. 2.3 were based on indiscernibility relations which are equivalence relations. A softer approach is based on similarity relations, i.e., relations that are reflexive and possibly symmetric but need not be transitive. Classes of these relations provide coverings of the universe U instead of its partitions.

2.4.1 Template approach

Classifiers of this type were constructed by means of templates matching a given object or closest to it with respect to a certain distance function, or on coverings of the universe of objects by tolerance classes and assigning the decision value on basis of some of them [18]; we include in Table 2.5 excerpts from classification results in [18].

Table 2.5. Accuracy of classification by template and similarity methods

paradigm	*system/method*	*Diabetes*	*Austr.credit*
Rough Sets	*Simple.templ./Hamming*	0.6156	0.8217
Rough Sets	*Gen.templ./Hamming*	0.742	0.855
Rough Sets	*Simple.templ./Euclidean*	0.6312	0.8753
Rough Sets	*Gen.templ./Euclidean*	0.7006	0.8753
Rough Sets	*Match.tolerance*	0.757	0.8747
Rough Sets	*Clos.tolerance*	0.743	0.8246

A combination of rough set methods with the k–nearest neighbor idea is a further refinement of the classification based on similarity or analogy in [42]. In this approach, training set objects are endowed with a metric, and the test objects are classified by voting by k nearest training objects for some k that is subject to optimization.

2.4.2 Similarity measures based on rough inclusions

Rough inclusions offer a systematic way for introducing similarity into object sets. A rough inclusion $\mu(u, v, r)$ (read: u is a part of v to the degree of at least r) introduces a similarity that is not symmetric.

Rough inclusions in an information system (U, A) can be induced in some distinct ways as in [25], [27]. We describe here just one method based on using Archimedean t–norms, i.e., t–norms $t(x, y)$ that are continuous and have no idempotents, i.e., values x with $t(x, x) = x$ except $0, 1$ offer one way; it is well–known, see, e.g., [23], that up to isomorphism, there are two Archimedean t–norms: the Lukasiewicz t–norm $L(x, y) = max\{0, x + y - 1\}$ and the product (Menger) t–norm $P(x, y) = x \cdot y$. Archimedean t–norms admit a functional characterization, see, e.g, [23]: $t(x, y) = g(f(x) + f(y))$, where the function $f : [0, 1] \to R$ is continuous decreasing with $f(1) = 0$, and $g : R \to [0, 1]$ is the pseudo–inverse to f, i.e., $f \circ g = id$. The t–induced rough inclusion μ_t is defined [24] as $\mu_t(u, v, r) \Leftrightarrow g(\frac{|DIS(u,v)|}{|A|}) \geq r$ where $DIS(u, v) = \{a \in A : a(u) \neq a(v)\}$. With the Lukasiewicz t–norm, $f(x) = 1 - x = g(x)$ and $IND(u, v) = U \times U \setminus DIS(u, v)$, the formula becomes: $\mu_L(u, v, r) \Leftrightarrow \frac{|IND(u,v)|}{|A|} \geq r$; thus in case of Lukasiewicz logic, μ_L becomes the similarity measure based on the Hamming distance between information vectors of objects reduced modulo $|A|$; from probabilistic point of view, it is based on the relative frequency of descriptors in information sets of u, v. This formula permeates data mining algorithms and methods, see [10].

2.5 Granulation of knowledge

The issue of granulation of knowledge as a problem on its own, has been posed by L.A. Zadeh [44]. Granulation can be regarded as a form of clustering, i.e., grouping objects into aggregates characterized by closeness of certain parameter values among objects in the aggregate and greater differences in those values from aggregate to aggregate. The issue of granulation has been a subject of intensive studies within rough set community in, e.g., [14], [29], [31].

Rough set context offers a natural venue for granulation, and indiscernibility classes were recognized as *elementary granules* whereas their unions serve as *granules of knowledge*.

For an information system (U, A), and a rough inclusion μ on U, granulation with respect to similarity induced by μ is formally performed by exploiting the class operator Cls of mereology [13]. The class operator is applied to any non–vacuous property F of objects (i.e. a distributive entity) in the universe U and produces the object $ClsF$ (i.e., the collective entity) representing wholeness of F. The formal definition of Cls is: assuming a part relation in U and the associated ingredient relation ing, $ClsF$ does satisfy conditions,

1. if $u \in F$ then u is ingredient of $ClsF$.

2. if v is an ingredient of $ClsF$ then some ingredient w of v is an ingredient as well of a T that is in F;

in plain words, each ingredient of $ClsF$ has an ingredient in common with an object in F. An example of part relation is the proper subset $\subset$ relation on a family of sets; then the subset relation $\subseteq$ is the ingredient relation, and the class of a family F of sets is its union $\bigcup F$. The merit of class operator is in the fact that it always projects hierarchies onto the collective entity plane containing objects.

For an object u and a real number $r \in [0, 1]$, we define the granule $g_\mu(u, r)$ about u of the radius r, relative to μ, as the class $ClsF(u, r)$, where the property $F(u, r)$ is satisfied with an object v if and only if $\mu(v, u, r)$ holds.

It was shown [24] that in case of a transitive μ, v is an ingredient of the granule $g_\mu(u, r)$ if and only if $\mu(v, u, r)$. This fact allows for writing down the granule $g_\mu(u, r)$ as a distributive entity (a set, a list) of objects v satisfying $\mu(v, u, r)$.

Granules of the form $g_\mu(u, r)$ have regular properties of a neighborhood system [25]. Granules generated from a rough inclusion μ can be used in defining a compressed form of the decision system: a granular decision system [25]; for a granulation radius r, and a rough inclusion μ, we form the collection $U^G_{r,\mu} = \{g_\mu(u, r)\}$. We apply a strategy $\mathcal{G}$ to choose a covering $Cov^G_{r,\mu}$ of the universe U by granules from $U^G_{r,\mu}$. We apply a strategy $\mathcal{S}$ in order to assign the value $a^*(g)$ of each attribute $a \in A$ to each granule $g \in Cov^G_{r,\mu}$: $a^*(g) = \mathcal{S}(\{a(u) : u \in g\})$. The granular counterpart to the decision system (U, A, d) is a tuple $(U^G_{r,\mu}, \mathcal{G}, \mathcal{S}, \{a* : a \in A\}, d^*)$. The heuristic principle that H: *objects, similar with respect to conditional attributes in the set A, should also reveal similar (i.e., close) decision values, and therefore, granular counterparts to decision systems should lead to classifiers satisfactorily close in quality to those induced from original decision systems* that is at the heart of all classification paradigms, can be also formulated in this context [25]. Experimental results bear out the hypothesis [28].

The granulated data set offers a compression of the size of the training set and a fortiori, a compression in size of the rule set. Table 2.6 shows this on the example of Pima Indians Diabetes data set [40]. Exhaustive algorithm of RSES [33] has been applied as the rule inducting algorithm. Granular covering has been chosen randomly, majority voting has been chosen as the strategy $\mathcal{S}$. Results have been validated by means of 10–fold cross validation, see, e.g., [5]. The radii of granulation have been determined by the chosen rough inclusion μ_L: according to its definition in sect.2.4.2, an object v is in the granule $g_r(u)$ in case at least r fraction of attributes agree on u and v; thus, values of r are multiplicities of the fraction $\frac{1}{|A|}$ less or equal to 1. The radius "nil" denotes the results of non–granulated data analysis.

Table 2.6. 10-fold CV; Pima; exhaustive algorithm. r=radius, macc=mean accuracy, mcov=mean coverage, mrules=mean rule number, mtrn=mean size of training set

r	$macc$	$mcov$	$mrules$	$mtrn$
nil	0.6864	0.9987	7629	692
0.125	0.0618	0.0895	5.9	22.5
0.250	0.6627	0.9948	450.1	120.6
0.375	0.6536	0.9987	3593.6	358.7
0.500	0.6645	1.0	6517.6	579.4
0.625	0.6877	0.9987	7583.6	683.1
0.750	0.6864	0.9987	7629.2	692
0.875	0.6864	0.9987	7629.2	692

For the exhaustive algorithm, the accuracy in granular case exceeds or equals that in non–granular case from the radius of .625 with slightly smaller sizes of training as well as rule sets and it reaches 95.2 percent of accuracy in non–granular case, from the radius of .25 with reductions in size of the training set of 82.6 percent and in the rule set size of 94 percent. The difference in coverage is less than .4 percent from $r = .25$ on, where reduction in training set size is 82.6 percent, and coverage in both cases is the same from the radius of .375 on with reductions in size of both training and rule set of 48, resp., 53 percent.

The fact of substantial reduction in size of the training set as well in size of the rule set coupled with the fact of a slight only decrease in classification accuracy testifies to validity of the idea of granulated data sets; this can be of importance in case of large biological or medical data sets which after granulation would become much smaller and easier to analyze.

2.5.1 Concept–dependent granulation

A variant of granulation idea is the concept-dependent granulation [28] in which granules are computed relative to decision classes, i.e., the restricted granule $g\mu^d(u, r)$ is equal to the intersection $g_\mu(u, r) \cap [d = d(u)]$ of the granule $g_\mu(u, r)$ with the decision class $[d = d(u)]$ of u. At the cost of an increased number of granules, the accuracy of classification is increased. In Table 2.7, we show the best results of classification obtained by means of various rough set methods on Australian credit data set [40]. The best result is obtained with concept–dependent granulation.

Table 2.7. Best results for Australian credit by some rough set based algorithms; in case ∗, reduction in object size is 49.9 percent, reduction in rule number is 54.6 percent; in case ∗∗, resp., 19.7, 18.2; in case ∗ ∗ ∗, resp., 3.6, 1.9

source	*method*	*accuracy*	*coverage*
[1]	$SNAPM(0.9)$	$error = 0.130$	–
[18]	*simple.templates*	0.929	0.623
[18]	*general.templates*	0.886	0.905
[18]	*closest.simple.templates*	0.821	1.0
[18]	*closest.gen.templates*	0.855	1.0
[18]	*tolerance.simple.templ.*	0.842	1.0
[18]	*tolerance.gen.templ.*	0.875	1.0
[43]	*adaptive.classifier*	0.863	–
[28]	$granular^*.r = 0.642$	0.8990	1.0
[28]	$granular^{**}.r = 0.714$	0.964	1.0
[28]	$granular^{***}.concept.r = 0.785$	0.9970	0.9995

2.6 Missing values

Incompleteness of data sets is an important problem in data especially biological and medical in which case often some attribute values have not been recorded due to difficulty or impossibility of obtaining them. An information/decision system is *incomplete* in case some values of conditional attributes from A are not known; some authors, e.g., Grzymala–Busse [8], [9], make distinction between values that are *lost* (denoted ?), i.e., they were not recorded or were destroyed in spite of their importance for classification, and values that are *missing* (denoted ∗) as those values that are not essential for classification. Here, we regard all lacking values as missing without making any distinction among them denoting all of them with ∗. Analysis of systems with missing values requires a decision on how to treat such values; Grzymala–Busse in his work [8], analyzes nine such methods known in the literature, among them, *1. most common attribute value, 2. concept–restricted most common attribute value, (...), 4. assigning all possible values to the missing location, (...), 9. treating the unknown value as a new valid value.* Results of tests presented in [8] indicate that methods *4,9* perform very well among all nine methods. For this reason we adopt these methods in this work for the treatment of missing values and they are combined in our work with a modified method *1:* the missing value is defined as the most frequent value in the granule closest to the object with the missing value with respect to a chosen rough inclusion.

Analysis of decision systems with missing data in existing rough set literature relies on an appropriate treatment of indiscernibility: one has to reflect in this relation the fact that some values acquire a distinct character and must be treated separately; in case of missing or lost values, the relation of indiscernibility is usually replaced with a new relation called a *characteristic relation*. Examples of such characteristic functions are given in, e.g., Grzymala–Busse [9]: the function ρ is introduced, with $\rho(u, a) = v$ meaning that the attribute a takes on u the value v. Semantics of descriptors is changed, viz., the meaning $[(a = v)]$ has as elements all u such that $\rho(u, a) = v$, in case $\rho(u, a) = ?$ the entity u is not included into $[(a = v)]$, and in case $\rho(u, a) = *$, the entity u is included into $[(a = v)]$ for all values $v \neq *, ?$. Then the characteristic relation is $R(B) = \{(u, v) : \forall.a \in B.\rho(u, a) = ? \Rightarrow (\rho(u, a) = \rho(v, a) \vee \rho(u, a) = * \vee \rho(v, a) = *)\}$, where $B \subseteq A$. Classes of the relation $R(B)$ are then used in defining approximations to decision classes from which certain and possible rules are induced, see [9]. Specializations of the characteristic relation $R(B)$ were defined in [38] (in case of only lost values) and in [11] (in case of only "*don't care*" missing values). An analysis of the problem of missing values along with algorithms *IApriori Certain* and *IAprioriPossible* for certain and possible rule generation was given in [12].

We will use the symbol $*$ commonly used for denoting the missing value; we will use two methods *4, 9* for treating $*$, i.e, either $*$ is a "*don't care*" symbol meaning that any value of the respective attribute can be substituted for $*$,thus $* = v$ for each value v of the attribute, or $*$ is a new value on its own, i.e., if $* = v$ then v can be only $*$.

Our procedure for treating missing values is based on the granular structure $(U^G_{r,\mu}, \mathcal{G}, \mathcal{S}, \{a^* : a \in A\})$; the strategy $\mathcal{S}$ is the majority voting, i.e., for each attribute a, the value $a^*(g)$ is the most frequent of values in $\{a(u) : u \in g\}$, with ties broken randomly. The strategy $\mathcal{G}$ consists in random selection of granules for a covering.

For an object u with the value of $*$ at an attribute a, and a granule $g = g(v, r) \in U^G_{r,\mu}$, the question whether u is included in g is resolved according to the adopted strategy of treating $*$: in case $* = don't\ care$, the value of $*$ is regarded as identical with any value of a hence $|IND(u, v)|$ is automatically increased by 1, which increases the granule; in case $* = *$, the granule size is decreased. Assuming that $*$ is sparse in data, majority voting on g would produce values of a^* distinct from $*$ in most cases; nevertheless the value of $*$ may appear in new objects g^*, and then in the process of classification, such value is repaired by means of the granule closest to g^* with respect to the rough inclusion μ_L, in accordance with the chosen method for treating $*$.

In plain words, objects with missing values are in a sense absorbed by close to them granules and missing values are replaced with most frequent values in objects collected in the granule; in this way the method *4* or *9* in [8] is combined with the idea of the most frequent value *1*, in a novel way.

We have thus four possible strategies:

- Strategy A: in building granules $*=don't\ care$, in repairing values of $*$, $*=don't\ care$;
- Strategy B: in building granules $*=don't\ care$, in repairing values of $*$, $* = *$;
- Strategy C: in building granules $* = *$, in repairing values of $*$, $*=don't\ care$;
- Strategy D: in building granules $* = *$, in repairing values of $*$, $* = *$.

2.7 Case of real data with missing values

We include results of tests with Breast cancer data set [40] that contains missing values. We show in Tables 2.8, 2.9, 2.10, 2.11, results for intermediate values of radii of granulation for strategies A,B,C,D and exhaustive algorithm of RSES [33]. For comparison, results on error in classification by the endowed system LERS from [8] for approaches similar to our strategies A and D (methods 4 and 9, resp., in Tables 2 and 3 in [8]) in which $*$ is either always $*$ (method 9) or $*$ is always *don't care* (method 4) are recalled in Tables 2.8 and 2.11. We have applied here the 1-train–and–9 test, i.e., the data set is split randomly into 10 equal parts and training set is one part whereas the rules are tested on each of remaining 9 parts separately and results are averaged.

Table 2.8. Breast cancer data set with missing values. Strategy A: r=granule radius, mtrn=mean granular training sample size, macc=mean accuracy, mcov=mean covering, gb=LERS method 4, [8]

r	$mtrn$	$macc$	$mcov$	gb
0.555556	9	0.7640	1.0	0.7148
0.666667	14	0.7637	1.0	
0.777778	17	0.7129	1.0	
0.888889	25	0.7484	1.0	

Table 2.9. Breast cancer data set with missing values. Strategy B: r=granule radius, mtrn=mean granular training sample size, macc=mean accuracy, mcov=mean covering

r	$mtrn$	$macc$	$mcov$
0.555556	7	0.0	0.0
0.666667	13	0.7290	1.0
0.777778	16	0.7366	1.0
0.888889	25	0.7520	1.0

Table 2.10. Breast cancer data set with missing values. Strategy C: r=granule radius, mtrn=mean granular training sample size, macc=mean accuracy, mcov=mean covering

r	$mtrn$	$macc$	$mcov$
0.555556	8	0.7132	1.0
0.666667	14	0.6247	1.0
0.777778	17	0.7328	1.0
0.888889	25	0.7484	1.0

Table 2.11. Breast cancer data set with missing values. Strategy D: r=granule radius, mtrn=mean granular training sample size, macc=mean accuracy, mcov=mean covering, gb=LERS method 9 [8]

r	$mtrn$	$macc$	$mcov$	gb
0.555556	9	0.7057	1.0	0.6748
0.666667	16	0.7640	1.0	
0.777778	17	0.6824	1.0	
0.888889	25	0.7520	1.0	

A look at Tables 2.8–2.11 shows that granulated approach gives with Breast cancer data better results than obtained earlier with the LERS method. This strategy deserves therefore attention.

2.8 Applications of rough sets

A number of software systems for inducing classifiers were proposed based on rough set methodology, among them LERS by Grzymala–Busse ; TRANCE due to Kowalczyk; RoughFamily by Słowiński and Stefanowski; TAS by Suraj; PRIMEROSE due to Tsumoto; KDD-R by Ziarko; RSES by Skowron et al; ROSETTA due to Komorowski, Skowron et al; RSDM by Fernandez–Baizan et al; GROBIAN due to Duentsch and Gediga RoughFuzzyLab by Swiniarski. All these systems are presented in [30].

Rough set techniques were applied in many areas of data exploration, among them in exemplary areas:

`Processing of audio signals:` [4].

`Pattern recognition:` [36].

`Signal classification:` [41].

`Image processing:` [39].

`Rough neural computation modeling:` [26].

`Self organizing maps:` [19].

`Learning cognitive concepts:` [32].

2.9 Concluding remarks

Basic ideas, methods and results obtained within the paradigm of rough sets by efforts of many researchers, both in theoretical and application oriented aspects, have been recorded in this Chapter. Further reading, in addition to works listed in References, may be directed to the following monographs or collections of papers:

A.   Polkowski L, Skowron, A (eds.) (1998) Rough Sets in Knowledge Discovery, Vols. 1 and 2, Physica Verlag, Heidelberg

B.   Inuiguchi M, Hirano S, Tsumoto S (eds.) (2003) Rough Set Theory and Granular Computing, Springer, Berlin

C.   Transactions on Rough Sets I. Lecture Notes in Computer Science (2004) 3100, Springer, Berlin

D.   Transactions on Rough Sets II. Lecture Notes in Computer Science (2004) 3135, Springer Verlag, Berlin

E.   Transactions on Rough Sets III. Lecture Notes in Computer Science (2005) 3400, Springer, Berlin

F.   Transactions on Rough Sets IV. Lecture Notes in Computer Science (2005) 3700, Springer Verlag, Berlin

G.   Transactions on Rough Sets V. Lecture Notes in Computer Science (2006) 4100, Springer, Berlin

H.   Transactions on Rough Sets VI. Lecture Notes in Computer Science (2006) 4374, Springer, Berlin

References

1. Bazan JG (1998) A comparison of dynamic and non–dynamic rough set methods for extracting laws from decision tables. In: Polkowski L, Skowron A (eds.), Rough Sets in Knowledge Discovery 1. Physica, Heidelberg 321–365
2. Bazan JG, Synak P, Wróblewski J, Nguyen SH, Nguyen HS (2000) Rough set algorithms in classification problems. In: Polkowski L, Tsumoto S, Lin TY (eds.) Rough Set Methods and Applications. New Developments in Knowledge Discovery in Information Systems, Physica , Heidelberg 49–88
3. Brown MF (2003) Boolean Reasoning: The Logic of Boolean Equations, 2nd ed., Dover, New York
4. Czyżewski A, et al. (2004) Musical phrase representation and recognition by means of neural networks and rough sets, Transactions on Rough Sets I. Lecture Notes in Computer Science 3100, Springer, Berlin 254–278

5. Duda RO, Hart PE, Stork DG (2001) Pattern Classification, John Wiley and Sons, New York
6. Frege G (1903) Grundlagen der Arithmetik II, Jena
7. Grzymala–Busse JW (1992) LERS – a system for learning from examples based on rough sets. In: Słowiński R (ed.) Intelligent Decision Support: Handbook of Advances and Applications of the Rough Sets Theory. Kluwer, Dordrecht 3–18
8. Grzymala–Busse JW, Ming H (2000) A comparison of several approaches to missing attribute values in data mining, Lecture Notes in AI 2005, Springer, Berlin, 378–385
9. Grzymala–Busse JW (2004) Data with missing attribute values: Generalization of indiscernibility relation and rule induction, Transactions on Rough Sets I. Lecture Notes in Computer Science 3100, Springer, Berlin 78–95
10. Klösgen W, Żytkow J (eds.) (2002) Handbook of Data Mining and Knowledge Discovery, Oxford University Press, Oxford
11. Kryszkiewicz M (1999) Rules in incomplete information systems, Information Sciences 113:271–292
12. Kryszkiewicz M, Rybiński H (2000) Data mining in incomplete information systems from rough set perspective. In: Polkowski L, Tsumoto S, Lin TY (eds.) Rough Set Methods and Applications, Physica Verlag, Heidelberg 568–580
13. Leśniewski S (1916) Podstawy Ogólnej Teoryi Mnogosci (On the Foundations of Set Theory), in Polish. See English translation (1982) Topoi 2:7–52
14. Lin TY (2005) Granular computing: Examples, intuitions, and modeling. In: Proceedings of IEEE 2005 Conference on Granular Computing GrC05, Beijing, China. IEEE Press 40–44
15. Michalski RS, et al (1986) The multi–purpose incremental learning system AQ15 and its testing to three medical domains. In: Proceedings of AAAI-86, Morgan Kaufmann, San Mateo CA 1041–1045
16. Nguyen HS (1997) Discretization of Real Valued Attributes: Boolean Reasoning Approach, PhD Dissertation, Warsaw University, Department of Mathematics, Computer Science and Mechanics
17. Nguyen HS, Skowron A (1995) Quantization of real valued attributes: Rough set and Boolean reasoning approach, In: Proceedings 2nd Annual Joint Conference on Information Sciences, Wrightsville Beach NC 34–37
18. Nguyen SH (2000) Regularity analysis and its applications in Data Mining, In: Polkowski L, Tsumoto S, Lin TY (eds.), Physica Verlag, Heidelberg 289–378
19. Pal S K, Dasgupta B, Mitra P (2004) Rough–SOM with fuzzy discretization. In: Pal SK, Polkowski L, Skowron A (eds.), Rough – Neural Computing. Techniques for Computing with Words. Springer, Berlin 351–372
20. Pawlak Z (1982) Rough sets, Int. J. Computer and Information Sci. 11:341–356
21. Pawlak Z (1991) Rough sets: Theoretical Aspects of Reasoning About Data. Kluwer, Dordrecht
22. Pawlak Z, Skowron A (1993) A rough set approach for decision rules generation. In: Proceedings of IJCAI'93 Workshop W12. The Management of Uncertainty in AI; also ICS Research Report 23/93, Warsaw University of Technology, Institute of Computer Science
23. Polkowski L (2002) Rough Sets. Mathematical Foundations, Physica Verlag, Heidelberg
24. Polkowski L (2004) Toward rough set foundations. Mereological approach. In: Proceedings RSCTC04, Uppsala, Sweden, Lecture Notes in AI 3066, Springer, Berlin 8–25

25. Polkowski L (2005) Formal granular calculi based on rough inclusions. In: Proceedings of IEEE 2005 Conference on Granular Computing GrC05, Beijing, China, IEEE Press 57–62
26. Polkowski L (2005) Rough–fuzzy–neurocomputing based on rough mereological calculus of granules, International Journal of Hybrid Intelligent Systems 2:91–108
27. Polkowski L (2006) A model of granular computing with applications. In: Proceedings of IEEE 2006 Conference on Granular Computing GrC06, Atlanta, USA. IEEE Press 9–16
28. Polkowski L, Artiemjew P (2007) On granular rough computing: Factoring classifiers through granular structures. In: Proceedings RSEISP'07, Warsaw, Lecture Notes in AI 4585, Springer, Berlin, 280–289
29. Polkowski L, Skowron A (1997) Rough mereology: a new paradigm for approximate reasoning, International Journal of Approximate Reasoning 15:333–365
30. Polkowski L, Skowron A (eds.) (1998) Rough Sets in Knowledge Discovery 2. Physica Verlag, Heidelberg
31. Polkowski L, Skowron A (1999) Towards an adaptive calculus of granules. In: Zadeh L A, Kacprzyk J (eds.) Computing with Words in Information/Intelligent Systems 1. Physica Verlag, Heidelberg 201–228
32. Semeniuk–Polkowska M (2007) On conjugate information systems: A proposition on how to learn concepts in humane sciences by means of rough set theory, Transactions on Rough Sets VI. Lecture Notes in Computer Science 4374:298–307, Springer, Berlin
33. Skowron A et al (1994) RSES: A system for data analysis. Available: `http:\\logic.mimuw.edu.pl/~rses/`
34. Skowron A (1993) Boolean reasoning for decision rules generation. In: Komorowski J, Ras Z (eds.), Proceedings of ISMIS'93. Lecture Notes in AI 689:295–305. Springer, Berlin
35. Skowron A, Rauszer C (1992) The discernibility matrices and functions in decision systems. In: Słowiński R (ed) Intelligent Decision Support. Handbook of Applications and Advances of the Rough Sets Theory. Kluwer, Dordrecht 311–362
36. Skowron A, Swiniarski RW (2004) Information granulation and pattern recognition. In: Pal S K, Polkowski L, Skowron A (eds.), Rough – Neural Computing. Techniques for Computing with Words. Springer, Berlin 599–636
37. Stefanowski J (2006) On combined classifiers, rule induction and rough sets, Transactions on Rough Sets VI. Lecture Notes in Computer Science 4374:329–350. Springer, Berlin
38. Stefanowski J, Tsoukias A (2001) Incomplete information tables and rough classification, Computational Intelligence 17:545–566
39. Swiniarski RW, Skowron A (2004) Independent component analysis, principal component analysis and rough sets in face recognition, Transactions on Rough Sets I. Lecture Notes in Computer Science 3100:392–404. Springer, Berlin
40. UCI Repository: `http://www.ics.uci.edu./~mlearn/databases/`
41. Wojdyłło P (2004) WaRS: A method for signal classification. In: Pal S K, Polkowski L, Skowron A (eds.), Rough – Neural Computing. Techniques for Computing with Words. Springer, Berlin 649–688
42. Wojna A (2005) Analogy–based reasoning in classifier construction, Transactions on Rough Sets IV. Lecture Notes in Computer Science 3700:277–374. Springer, Berlin

43. Wróblewski J (2004) Adaptive aspects of combining approximation spaces. In: Pal S K, Polkowski L, Skowron A (eds.), Rough – Neural Computing. Techniques for Computing with Words. Springer, Berlin 139–156
44. Zadeh LA (1979) Fuzzy sets and information granularity. In: Gupta M, Ragade R, Yaeger RR (eds.) Advances in Fuzzy Set Theory and Applications. North–Holland, Amsterdam 3–18

3

Evolving Solutions: The Genetic Algorithm and Evolution Strategies for Finding Optimal Parameters

Thomas McTavish and Diego Restrepo

University of Colorado at Denver and Health Sciences Center
Aurora, CO 80045, USA
{Thomas.McTavish,Diego.Restrepo}@uchsc.edu

Summary. This chapter provides an introduction to evolutionary algorithms (EAs) and their applicability to various biological problems. There is a focus on EAs' use as an optimization technique for fitting parameters to a model. A number of design issues are discussed including the data structure being operated on (chromosome), the construction of robust fitness functions, and intuitive breeding strategies. Two detailed biological examples are given. The first example demonstrates the EA's ability to optimize parameters of various ion channel conductances in a model neuron by using a fitness function that incorporates the dynamic range of the data. The second example shows how the EA can be used in a hybrid technique with classification algorithms for more accuracy in the classifier, feature pruning, and for obtaining relevant combinations of features. This hybrid technique allows researchers to glean an understanding of important features and relationships embedded in their data that might otherwise remain hidden.

3.1 Introduction

Evolutionary algorithms (EAs) provide a means of finding an optimal solution to models of data or systems. This chapter provides an overview and some general design principles of two classes of EAs: The *genetic algorithm* (GA), and *evolution strategy* (ES). We will also briefly discuss *genetic programming*.

Unlike many areas of biology and computer science, as a form of biologically inspired computing, EAs give computer scientists and biologists an opportunity to speak the same language. EAs demonstrate how the concepts and mechanics of genetics and biological evolution, which has spawned an extreme diversity of life forms, can also be used to find solutions to fit the niche of an abstract computational problem. EAs and evolutionary computational strategies therefore go far beyond enabling models of artificial life and biological evolution and have proved useful in several disparate areas of finance, engineering, and science.

T. McTavish and D. Restrepo: *Evolving Solutions: The Genetic Algorithm and Evolution Strategies for Finding Optimal Parameters*, Studies in Computational Intelligence (SCI) **122**, 55–78 (2008)
www.springerlink.com   © Springer-Verlag Berlin Heidelberg 2008

While some researchers developed computational models of evolution in the 1950s and 60s, it was primarily the book by John Holland, *Adaptation in Natural and Artificial Systems* [1], that described the GA and how evolutionary principles could solve a large number of abstract problems. Concurrently, students at the Technical University of Berlin were demonstrating the applicability of ES to optimize aircraft wing parameters in wind tunnels [2].

EAs are therefore somewhat old in terms of computational techniques. They are fairly straightforward to understand and implement; and they are also applicable to a broad number of disparate problems in science and engineering. However, most biologists have not heard of the genetic algorithm or an evolutionary algorithm, and if they have, from their names they assume the algorithms are used to assemble genomes or phylogenetic trees. We will discuss the algorithms in the context of how they can be used to find optimal sets of parameters to describe all sorts of data—even biological problems unrelated to evolution or genetics.

3.2 Searching Parameter Space

To provide a deeper understanding of data and enable predictions for future data, a frequent goal of scientific and engineering efforts is to describe data in terms of a parameterized model. For example, if two-dimensional data could be modeled as a line of the form $y = mx + b$, the data is given as the points (X, Y) and the parameters of the model are m and b.

The difficulty of describing data with a model is that if the model$\leftrightarrow$data mapping is nonlinear, then one cannot simply solve the problem. One has to *search* through the space of possible parameter values and choose the best set that reasonably describes the data. Sampling all possible parameter sets is often an impossible task, even when the range of parameters is small or finitely discretized. For this reason, approximating methods have to be employed, allowing the search to sample *parameter space* in such a way to hopefully find a globally optimal solution in the process.

When a model is tested against the data, it can report its error. Parameter space can therefore map to an *error landscape* which provides an intuition of how regions in parameter space are likely to perform. A search method therefore employs *heuristics* using the error from any number of points in parameter space to help direct the search toward a global minimum, the solution with least error. The difficulty for most methods is avoiding being tricked into pursuing a local minimum, a solution that is not the global minimum, but where the neighbors in parameter space deliver worse solutions. Avoiding local minima requires adequate sampling of parameter space and the ability to pop out of a local minimum if a proposed solution becomes trapped. The genetic algorithm provides a good strategy for searching parameter space effectively, combining the elements of following a gradient but also with the ability to jump out of local minima and sample disparate areas in parameter space.

3.3 The Evolutionary Algorithm

The EA is a search technique that employs the ideas of organisms evolving and being naturally selected for to fill a niche, and applies these principles to a number of proposed solutions that are allowed to modify and borrow from each other in such a way that they evolve to fit data.

In pseudocode, an EA can be simply described as

EVOLUTIONARY-ALGORITHM()
1   Start with an initial population of possible solutions.
2   Repeat for some number of generations or until a solution is found,
3   Select those individual solutions which will reproduce
4   Breed a new generation of candidate solutions from those individual
            solutions using mutation and recombination

Let us look deeper at this simple, but effective search method. This requires the introduction and definition of several terms.

3.3.1 The individual

An *individual* is simply a proposed solution. Each individual has a *genotype* and a *phenotype*. An individual's genotype is the set of values for the parameters being optimized. An individual's phenotype are those parameter values plugged into the model.

The chromosome

GAs and ESs follow the algorithm outlined above. A primary difference between GAs and ESs begins with the way each algorithm encodes the parameters being optimized which subsequently determines the type of operations on those parameters that can be performed. In both cases, the data structure of the genotype is called the *chromosome.*

In the classical GA, the chromosome is a binary string of 0s and 1s, the presence of each *gene* represented as a bit. If the model can be set up to look for the best combinations of objects, the presence of the object would be determined by a single gene. However, as in biology where several genes define a feature, a given parameter may be described by several genes. For example, if a parameter can take on one of 16 possible values, we might encode this parameter in 4 bits ($2^4 = 16$). Therefore, for use in GAs, real values (floating point numbers and integers) have to be converted to binary representations. For example, if a parameter was allowed to vary between 0.0 and 1.0 and this parameter was represented in 4 bits, the binary value 0110 might correlate with 0.267, one of the 16 discretized values in the set $[0.0, 0.067, ...0.933, 1.0]$.

Another common use for the GA is to obtain an optimal sorting for some list. In this case, a chromosome is the permutation. For example, in the list ABCDEF, the permutation BEDAFC would be a valid chromosome. Here the string is obviously not binary, but contains indexes to items in a list.

Encodings of real values, each parameter as a single gene on the chromosome, is the domain of ES. (There may be a bit of confusion that arises with the term "real-valued" or "real-coded" GA. From the classical definitions of the GA and ES, real-valued GAs are actually ESs. Much of the literature, however, combines GA and ES under one umbrella, simply calling them both GA). Chromosomes of ESs/real-valued GAs is straightforward: They are typically represented as a vector of floating point or integer values, for example, [0.4 6].

Many real-valued parameters can be set up to be represented as binary strings. There is broad debate about which is better: Traditional GAs that convert parameters to binary or ESs which maintain the real-value representation. GA↔ES comparisons of real-valued parameters often conclude that ES implementations are faster and more accurate [3, 4]. If you do not have the ability to try both approaches, it is probably easiest to avoid issues with binary translation and stick to an ES when evolving real-valued parameters.

3.3.2 Population genetics

If we have a notion of how well a phenotype describes the data, then through the agglomeration of several phenotypes, we should be able to assemble some heuristics to discover and navigate the error landscape. Indeed, that is what EAs do. Their heuristic is natural selection. An EA works to progressively select better solutions through the combination and modification of those solutions that were better than other solutions previously evaluated. That is, from the *population* of potential solutions in one *generation*, the EA selects those individual solutions that were better than others and which also sample disparate areas in parameter space (line 3 in the algorithm), and from that subset, combines aspects from those individuals or makes modifications to those solutions so that the solutions in the next generation might be even closer to an optimum solution (line 4). In so doing, the EA implicitly maintains a history to help direct its search. Let us evaluate each line of the algorithm in more detail.

3.3.3 Line 1: Start with an initial population of possible solutions

It is typical to begin the search with a random population of individuals, each individual with a unique genotype. That is, each parameter of the genotype is randomly assigned, drawn from that parameter's range and probability distribution. We therefore begin with a shotgun approach of some number of sample solutions in parameter space.

Most implementations of EAs maintain the same population size throughout their operation. Studies that have measured the optimal size of the population in GAs indicate that this number should be about the size of the length of the binary string encoding, but depending on the mutation and crossover operators, complexity of the error landscape, complexity of the model, and if the number of parameters is large, this guideline becomes variable [5]. While an EA can often jump out of local minima, it may still get trapped. Therefore, it may take a large population to help ensure better coverage of the error landscape and increase the likelihood that the global minimum is found.

3.3.4 Line 2: Repeat for some number of generations or until a solution is found,

In most cases, due to noise or simplicity of the model, models will never exactly describe all of the data. For this reason, the error will nearly always be nonzero, even for the global minimum. A terminating condition is therefore necessary to prevent the infinite search for zero error. Such a condition might be to report the best solution after N generations and/or to run until the error is within an acceptable range.

3.3.5 Line 3: Select those individual solutions which will reproduce

Natural selection states that those individuals of a species that are more fit to reproduce will have a reproductive advantage over their contemporaries. An EA also employs the notion of fitness. Fitness is the reciprocal of error, so an EA seeks to maximize fitness or, equivalently, to minimize error. An EA minimizes error by repeatedly biasing the selection of the solutions it will operate on to form new candidate solutions toward the fitter solutions in the current generation.

3.3.6 The fitness function

An individual's fitness is quantified through the *fitness function* which is a measure of how close a phenotype is to the optimum solution. The fitness function is also known as the *evaluation function* or *objective function*. The phenotype is compared to the data, and in the simplest case, a scalar value quantifies how well the model matches the data.

Errors can be more complicated than simple distance measures of the phenotype's model output to the data. For example, say our model contains the variables A and B. If we know A and B can never both be high because of some physical or biological law, then even if high values of A and B plugged into our model describes our data well, then it is wrong. In short, the fitness function can be improved with an understanding of the model parameters.

A clear understanding of the data is also important. How much noise does it contain? Is there bias in the data? Should some features be weighted more heavily than others? Are some features correlated? Are there enough data points for a thorough analysis? Can the data be transformed into other spaces (for example, time-domain data into the spectral domain) for other measures? Obviously, the better we understand the data and the parameters of the model, the better we can construct a more robust fitness function.

Care should be taken so that the error function can create broad gradients and minimize local minima. Consider a binary measure—that either a solution works or does not. The error landscape created in this case is like a golf putting green, nothing but flat terrain and then a small hole. In this case, the error function provides no information to guide the search process. Any optimization algorithm would be as effective as an exhaustive search, or worse! Broader gradients can be obtained, then, through more complex descriptors and larger measurements of error. If the EA is not converging toward a solution, the error landscape is probably too flat and/or choppy with local minima. If this is the case, it is common to experiment with different population sizes, breeding strategies, and recombination operators (discussed below), but it may be more effective to experiment with a different error measure.

3.3.7 Beyond the fitness function: Other selection considerations

The role of the fitness function is to provide a measure of how each individual compares with others in the generation with respect to the global objective. However, consider the goal of the algorithm: To find a global minimum (which implicitly means avoiding local minima). The best candidates are therefore those solutions which minimize error *while sampling disparate and broad areas of parameter space.* In biological natural selection, the genes or regulators of genes that are more fit can rather quickly infiltrate a population. This is something we want to partially avoid in the EA to prevent the search from falling into local minima.

As an example, say we have a population of 100 individuals and that we choose the 20 phenotypes with the least error to act on. Selection of the top n individuals like this is known as *truncation selection.* (The new generation will contain 100 individuals (offspring) which are modifications of the 20). If those 20 genotypes are clustered in parameter space, their offspring will also be clustered in this same area. This may persist for several generations or even indefinitely. Steps to avoid such *population convergence* or *crowding* while performing truncation selection can include having a high mutation rate and broad mutation range (Section 3.3.11) or the introduction of new random individuals into the breeding pool. Such aggressive steps may be unwanted while still being ineffective at popping out of local minima. For these reasons, other selection techniques are more commonly used.

Many EA implementations stochastically select individuals using *roulette-wheel* or *tournament selection.* Those solutions with higher fitness are biased to be selected over those that have low fitness, but low-fitness individuals still have the possibility of being selected. The reason one may want to include some low-fitness individuals in the breeding population is that it can be assumed that these low-fitness individuals are far from the other individuals in parameter space. Therefore, they maintain variation in the new generation and help ensure that disparate areas in parameter space are being sampled.

Diversity maintenance in the population can also be obtained through *sharing* [6, 7] which decreases the fitness of an individual if it is close to other individuals in parameter space. There are also other methods that deal with crowding by replacing individuals only if the children are indeed better solutions than their parents [8, 9]. This technique keeps the original shotgun sampling diffuse, delaying convergence, but does not explicitly instill diversity. Additionally, multiple populations could be allowed. The idea is that each population is likely to converge to a different local minimum, or *niche*. If the populations are allowed to interact sparsely, exchanging very few individuals and then only every n^{th} generation, then the aspects from each population might collectively combine to form a new niche closer to the global optimum.

In summary, selection needs to balance the fitness of the solution with other population characteristics to sample multiple areas in parameter space.

3.3.8 Line 4: Breed a new generation of candidate solutions from those individual solutions using mutation and recombination

Individuals chosen to be operated on go through a process of *reproduction* to form a new generation of candidate solutions. While the *breeding strategy* includes the subset selected through the fitness function and the previously mentioned selection criteria, it also directs reproduction. Reproduction can include the cloning of a previous solution, the blending of previous solutions, or the modification of previous solutions in making a new candidate. The breeding strategy therefore includes methods for choosing which previous solutions will be cloned, which previous solutions merge and how, and which previous solutions are modified and how.

3.3.9 Elitism

The method of choosing the fittest n individuals from the breeding pool as the ones to be cloned is known as *elitism*. These elite solutions from the previous generation proceed to the next generation without any modification. Since their phenotype has already been evaluated, their fitness function does not need to be calculated again, so few computational resources need to be devoted to them.

The reason you may want to have a few individuals in each generation form an elite, is that they guarantee that the progeny in the new generation will have error at least as small as the previous generation. That is, there is no guarantee that operations on the previous solutions that form new candidate solutions will result in decreased error for any of the new solutions. In fact, like biological natural selection itself, it is the rare modification that results in less error, so without forming an elite, it may be possible to create a new generation that has individuals who do not have less error than their parents. Therefore, by maintaining a small set of the "best so far" in each generation, we guarantee that those individuals will remain in the gene pool.

3.3.10 Recombination

Recombination involves the merging of parameters of two solutions from the previous generation into new solutions known as *offspring* or *children* in the new generation. In the simplest case, the selection of pairs is entirely random within the population of those selected to breed, including elites. Pairing can be more directed, however. For example, incest prevention prohibits the mating of similar individuals in parameter space to maintain diversity [10]. At the same time, if individuals are too different from each other, their children may likely be poor solutions. In short, it is often best to employ several breeding strategies.

Recombination often employs the typical meiotic form of genetic *crossover* where offspring will receive some number of genes from one parent and the remainder from the other parent (Figures 3.1 and 3.2 middle). It is normal to generate two offspring and for the offspring to be complements of each other. This keeps the population size constant and also provides balance of sampling in parameter space.

Generally speaking, crossover is more important than mutation in the GA whereas the converse is true for ES. This is primarily because crossover and mutation are performed differently and have different effects in each.

In the GA, crossover operates to quickly weed out which bits should be 1 and which bits should be 0. Here's how. Firstly, we can say, obviously, that if a gene is the same in each parent, then crossover of that bit has no effect. Therefore, children will only *possibly differ* from their parents at those genes that are different between the parents. After obtaining the fitness of the children, we have the potential of gaining a lot of information. We know the fitness of the parents and the bits that are different between them. We also know the fitness of the children who only vary by a different combination of those same bits. Crossover therefore functions to compare the roles of those bits that are different between the parents and begins to characterize the appropriate settings of those bits. We potentially glean the most information when crossover occurs such that half of the different genes go to one child and half go to the other. This is the motivation behind the *uniform crossover* [11] and the *half-uniform crossover* (HUX) [12].

Another common GA crossover operator is the *one-point crossover* where one child will consist of the left portion of one parent and the right portion of the other parent (Figure 3.1 right). The crossover point is usually randomly selected and can therefore bisect a parameter that may span some number

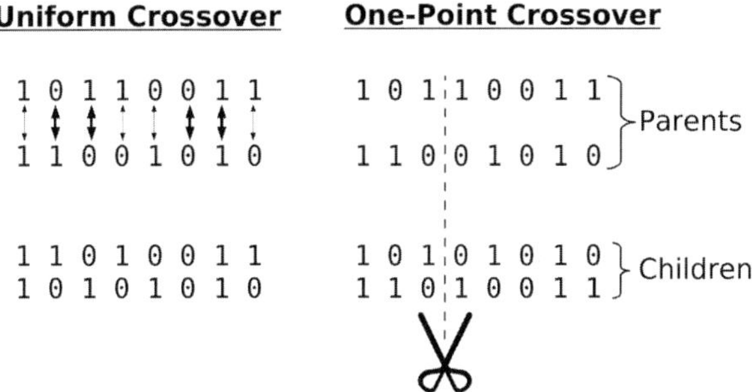

Fig. 3.1. Two types of crossover in the GA. Parents are in the top rows and children are in the bottom. In uniform crossover, each bit has the same probability of flipping. Bold arrows show where a bit flip occurred. Bold arrows to the right have no bearing in the children since the parents have the same bits set at those positions. In one-point crossover, the top child receives the left portion of the top parent and the right portion of the bottom parent. The converse is true for the bottom child.

of genes. The idea behind one-point crossover can be extended to allow n-point crossovers. When utilizing point crossovers, one has to be careful of the placement of the parameters on the chromosome. Genes that are close to each other will tend to crossover together. Therefore, if genes are known to vary together, they should be placed close to each other on the chromosome.

In ES/real-valued GA crossover, children also receive some set of parameters from one parent and the remaining set from the other (Figure 3.2 middle). The role of crossover in real-valued strategies, however, is to find the strongest parameter values that are contributing to each parent's fitness and to incorporate those into the children. It assumes that two parents probably contribute differently to the error—that some genes are good at reducing error in one parent and other genes are probably the ones largely responsible for reducing error in the other parent. With the right swapping of genes, then, a child could potentially reduce the error to even a greater degree.

The reason crossover is not such a prevalent director of evolution in real-valued scenarios is that it is really difficult to determine the combination of parameters to swap that might actually yield fitter offspring. The most common crossover operators randomly select which genes to swap, but real-valued encodings offer a number of customized scenarios. For example, one could easily employ a crossover operator to swap the n most different genes between two parents. The motivation for such an approach is that the most disparate genes between two parents might best describe their different ways of reducing error. Another crossover operator applied on another set of parents might swap their most similar genes to hone those parameters. (Also see Section 3.4.1). It is common to employ a few crossover techniques on a generation, but only one crossover operation is typically applied on a set of parents.

A more complicated form of real-valued crossover is the *blended crossover* [13]. With the blended crossover (also known as BLX-α), a new candidate parameter could be 25% of parent P_A and 75% of parent P_B. Indeed, this is the case for Gene A in the bottom child in the bottom portion of Figure 3.2. Conversely, to maintain balance in parameter space, the top child is 75% of P_A and 25% of P_B.

One needs to be careful when naïvely applying blended crossover, however. It is easy to think that blending within the range of $0 - 100\%$ is the best way to merge two values of a parameter. Blending a parameter in this fashion, though, will work to constrain the range of the parameter. At the population level, this blending will push a given parameter toward the centroid of the gene's value for that generation. Over generations, the centroid might shift, but the range of the variable will continually shrink. To avoid introducing such bias, it is necessary to expand the range of the blend to something like -50% to 150%. This is the case for Gene D. The top child of the bottom row is 150% of the top parent and -50% of the bottom parent. Conversely, the bottom child is -50% of the top parent and 150% of the bottom parent. While this may seem inappropriate, it maintains the dynamic range of the parameter, the gene constricting between some parents and expanding between others.

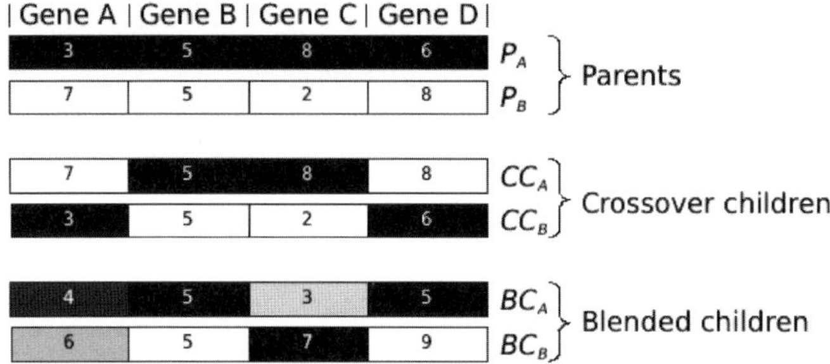

Fig. 3.2. Recombination of real values via crossover (middle) and blended crossover (bottom). The genes from the top parent, P_A, are colored black and the genes from the bottom parent, P_B, are white. In normal crossover of children (CC), genes are randomly swapped between the parents. In blended crossover (BC), a gene in child BC_A will contain $(-50 + n)\%$ of P_A and $(150 - n)\%$ of P_B where n is a random variable. Conversely, this same gene in child BC_B will contain $(-50 + n)\%$ of P_B and $(150 - n)\%$, n being the same value as applied to child BC_A. Grayscale values of the gene indicate the relative contribution of each parent, black or white, to the child's gene.

It is important to note that not all parents have to breed. The crossover rate is typically subject to experimentation and may, in fact, be quite low. In this case, parents can clone themselves and then mutate to introduce variability into the children.

3.3.11 Mutation

A *mutation* is a modification to a gene. The reason for introducing mutation is to add variation to the population, giving the algorithm other sample points to measure, and to knock individuals out of local minima.

In the GA, a common simple mutation operator is the uniform mutator. Each bit of the chromosome has a low (usually about 1%) probability of flipping at each generation.

One problem with binary chromosomes is the Hamming cliff problem [14]. Consider the strings 01111 and 10000. If these bits represented one numeric integer, the value of these strings varies by 1. (01111 in binary equals 15 and 10000 in binary equals 16). Also consider the strings 01111 and 11111. These strings vary by one bit, but their values are different by 16. (01111 in binary equals 15 and 11111 in binary equals 31). To help ensure that a single point mutation has minimal, and in a sense, incremental response, it is rare to give parameters their normal binary representations. Instead, it is common to encode binary representations using Gray encoding [15] which makes it so that all adjacent numbers, like 15 and 16, only require one bit change.

There are a number of mutation strategies one can employ with ES. A common, simple mutation strategy is to let each gene have some probability of mutating and when it does, modify that real parameter by some amount— usually the addition of a draw from a normal (Gaussian) distribution centered at zero. Of course, uniform and other probability distributions can also be used. The smaller the mutation, the better it can pursue a gradient and the larger, the better it can pop out of local minima. It may therefore make sense to employ a number of mutators, some with tight *mutation range* and some with a broad range, each mutator with a different *mutation rate*.

In ES, mutation strategy is often tightly coupled with *replacement strategy*, choosing when to remove an individual from the breeder pool. Because many mutations may be deleterious, resulting in poorer fitness in the children, a common replacement strategy in ES is to discard parents from the breeder pool only when they produce fitter offspring.

There is nothing to say that binary representations cannot also employ mutation strategies used in ES. Given that several bits may encode a parameter, it would be easy to decode the bits to capture the value of the parameter, mutate the value as in ES, and then re-encode the binary representation.

3.3.12 Finding optimal parameters *and* the model

In many cases, it may not only be parameters of the model that we wish to optimize, but it may also include the operators of the model itself! In this case, EAs can also be used. This is the domain of *genetic programming* (GP). In this case, chromosomes are computer programs, usually assemblies of tree-like structures. Each terminal node (leaf) in the tree is a parameter and each branching node is an operator (like add, subtract, multiply, divide, log, exponent, etc.). Operators are subject to mutation, changing from one type to another, but also branches (edges) between nodes are subject to appearing and being removed.

As a simple example of a genetic program, suppose a model to describe the data is $y = 3x + 2$. Figure 3.3 illustrates an optimal genetic program's chromosome to describe this data. Given (X, Y) input data pairs, the EA would figure out the "2" and "3" leaves as well as the "+" and "*" internal nodes to best-fit the data.

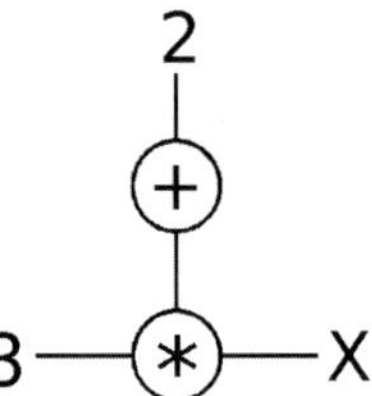

Fig. 3.3. Genetic programming example. The model and parameters, $y = 3x + 2$.

It is, however, difficult to know how to evolve a program toward an end goal. In natural systems, the goal is and always has been survival and reproduction, but the factors influencing survival and reproduction are constantly changing as a given species evolves to exploit one niche after another. If we wanted to evolve a very large sea creature from an ancestral fish, would we first have it develop lungs and walk on land? Nature demonstrates through the blue whale that evolution does not take a linear course!

GPs, like natural systems, are subject to phylogenetic constraints. The developmental program of horses is such that they can never evolve wings. Analogously, GP may be able to compose programs that bifurcate between "birds" and "horses" but are never able to devise a pegasus because to do so would require specific fundamental changes to the program which would render the individual so unfit compared to the horses and birds already in the population.

3.4 Advanced topics

We now concentrate on further strategies to improve the running time of an EA and/or its results.

3.4.1 Errors as vectors

Error is typically reported as one number—the sum of all of the discrete measures of error. Instead, the error of a phenotype could be reported as a vector in *error-parameter space* where each of these discrete measures represents a separate dimension. This can perhaps be best understood with an example. Say that my data is in the frequency domain, and I am searching for parameters that can deliver a specific frequency and a specific amplitude. As a single number, error would be reported as the sum of the distance from the target frequency plus the distance to the target amplitude. If error is instead represented as a vector in error parameter space, one dimension could be error in frequency and one dimension could be error in amplitude. Doing this can improve the search significantly because it can direct the phenotypes we are interested in pairing for reproduction.

Consider the three vectors in Figure 3.4. These all have the same magnitude so in the traditional error landscape, they would have equal weight. Vector A, however, has no error in amplitude and vector C has no error in frequency. When error is expressed as vectors, we can better target breeding strategies. In this case, vectors A and C are good candidates for breeding because we would like to take whatever it is in A that makes it hit the target amplitude and whatever it is in C that makes it find the target frequency and hopefully discover offspring that can hit both the target frequency and amplitude. We could also preferentially select parameters that were more dissimilar to be selected for crossover. Additionally, EAs are not bound by natural laws of sexual reproduction. If we know one individual reduces error along one dimension, another reduces error in another, and a third reduces error in yet another dimension, then we may seek to breed all three parents.

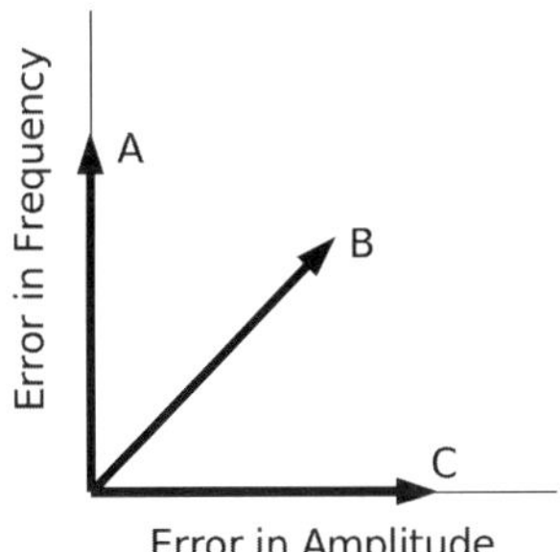

Fig. 3.4. Error represented as vectors in error parameter space. Error is measured in distance from target frequency along the Y axis and distance from the amplitude target along the X axis. A, B, C are individual phenotypes with the same magnitude of error.

Expressing errors as vectors can also make for more robust selection criteria. I may have a phenotype that reduces error in one dimension better than any other phenotype, but gives horrible error along other dimensions. Even though this phenotype has an overall high error, I may choose to keep this phenotype in my reproductive pool simply because it may describe one dimension better than any other phenotype.

Treating errors as vectors is the subject of Multi-Objective Evolutionary Algorithms (MOEAs) which are covered in more detail in Chapter 4.

3.4.2 Evolution of evolution

While EAs can drastically improve search time by pruning parameter space, several factors such as population size, crossover operators, crossover rates, mutators, and mutation rates will impact the speed and accuracy of the results. We are unlikely, however, to know what the best operators and settings for the EA might be. In a sense, it would be nice to throw the kitchen sink at the problem and somehow let the EA determine appropriate settings. This can be accomplished by maintaining with a child, the operations that gave rise to it. Therefore, when comparing the fitness of individuals in a generation, we can also compare the successful and unsuccessful operators. For example, if we note that small mutations gave rise to fitter individuals than those that received large mutations, then the mutation range may be deemed too aggressive and dynamically change. Likewise, if a particular crossover operator yields fitter children than another crossover operator, we can bias toward that stronger operator. When the EA modifies itself from such acquired history, this is known as *self-adaptation*. A popular mechanism in ES for self-adaptation is the Covariance Matrix Adaptation Evolution Strategy (CMA-ES) [16, 17].

3.4.3 Been there, done that

It may become apparent that an individual or population of individuals is stuck in a local minimum. Rather than letting them continue to exhaust computational resources and corrupt the gene pool, it may be time for them to die. One can annihilate these candidate solutions by discarding parents that do not produce fitter offspring after n generations. This is known as *aging*. It may also be important to annihilate all of the individuals in the region so that they do not continue to find the same local minimum. Furthermore, it may be useful to dynamically tag this region in parameter space as off-limits so that new individuals do not migrate down the same valley.

3.4.4 Lamarckism

Recombination in the natural world is not goal-oriented. Operations randomly happen and the fitter characteristics are simply more likely to proceed to the next generation. There is nothing to prevent us, however, from employing Lamarckism and employing operators which are goal directed in EAs. This is the domain of *memetic algorithms* which allow children to perform local searches before undergoing mutation and recombination by the EA.

As an example, the EA could work with individuals of local minima. That is, with each generation, individual solutions could follow gradient descent or some greedy, localized search until they became trapped in a local minimum. At that point, the EA could inject mutations and recombination to these local minima. The presumption with this technique is that through the ensuing crossover and mutation of several local minima, better local minima will likely be found in each subsequent generation.

3.4.5 Where EAs fail

A number of factors can make an EA unable to converge on a solution. We have talked about the importance of robust fitness measures and various breeding strategies. The complexity of the model and the interdependence of the variables may also cause a very rough fitness landscape. It may be, however, that the formulation of the problem is *deceptive*. That is, there is a hill surrounding a tiny steep hole that is the global optimum. Samples near the top of the hill will migrate away from the global optimum. Resolving deceptive problems can include the application of different fitness measures, data transformations, model modifications, or dropping the EA for another optimization technique.

3.5 Examples

3.5.1 Optimizing parameters in a computer simulation model

Mitral cells are neurons in the main olfactory bulb responsible for propagating signals from olfactory sensory neurons to pyramidal cell neurons in the olfactory cortex. Like many neurons, active mitral cells fire action potentials with a certain periodicity and often synchronize with other mitral cells. Additionally, it has been reported in the literature that the resting membrane potential of isolated mitral cells which are not receiving any other synaptic input often exhibit subthreshold oscillations [18]. That is, when they are not firing, the difference of electrical charge inside the cell compared to outside the cell fluctuates in a rhythmic fashion. When this charge is centered at -65 mV, the frequency of oscillations averages 13 Hz. At a more positive charge centered at -59 mV, just below firing threshold, the frequency of oscillations is 39 Hz. The amplitude of the oscillations averages 1.85 mV. These findings are summarized in Table 3.1. It is presumed that transmembrane ion channel proteins are responsible for such electrical fluctuations, but the presence and density of such channels have not been empirically determined.

Two neuroscientists, Rubin and Cleland, took an existing mitral cell computational model [19] and predicted that the addition of six particular ion channels could give rise to subthreshold oscillations [20]. They could not say, however, what the conductance (i.e. density) of each channel should be to mimic this oscillatory behavior. Constraining the variability of each channel's

Table 3.1. Subthreshold Oscillation Data. (Amplitude = 1.85 mV for all samples).

Membrane Potential (mV)	Subthreshold Frequency
-65	13
-64	20
-63	26
-62	31
-61	35
-60	38
-59	40

density to a small range, they ran over 60,000 simulations with different values of the channels' densities. They then chose the set which elicited the closest response. Because each ion channel is mathematically modeled with differential equations, each simulation takes a number of seconds to compute. The problem is also therefore highly nonlinear. We were curious to see if an EA which allowed for a broader density range would find the same or even more accurate channel densities while reducing the number of simulations.

This mitral cell model was built with the NEURON simulation package [21] so we modified NEURON to incorporate the Open BEAGLE EA library [22]. The goal was to determine the six channel densities which could match the data as summarized in Table 3.1. Therefore, the genotype was the density of each of these six channels. The phenotype was the behavior of the computational neuron model with each particular set of channel densities plugged into it. A fit individual should behave such that when the cell's membrane has a base potential of -65 mV it exhibits a subthreshold oscillation frequency of 13 Hz, and when its membrane potential has a base of -64 mV it should oscillate at 20 Hz, etc. We therefore wanted 7 samples from each individual where each sample would uniquely correspond with a row in Table 3.1. Each of the 7 samples reported an error value of how close it matched its corresponding row and the overall error for the individual was the sum of these values.

Because each set of ion channel densities impacts membrane dynamics differently, it is not known how much current is necessary to depolarize the individual so that its base membrane potential can be -65 mV, then -64 mV, etc. Yet, we wanted 7 samples from each individual which matched the data. We therefore created 11 possible samples by injecting 11 incremental steps of depolarizing current. We then set this up as a *weighted bipartite graph / optimal assignment* problem and employed the *Kuhn-Munkres (Hungarian) algorithm* [23] to choose the 7 best matches of these 11 steps. Figure 3.5 provides an illustration of a matching and our error function (described below).

Now we could have modeled the data in Table 3.1 with a simple equation and for each individual asked how close it mapped to this target function. However, this could have made it such that an individual fit very well, but only to a small portion of the data. Setting the problem up as an optimal assignment problem ensured adequate coverage of the target data.

For our error measure, we took the Euclidean distance of each of the 11 samples to each of the 7 target rows along three dimensions: 1) The base membrane potential 2) the oscillatory frequency, and 3) the amplitude of the oscillations to 1.85 mV. Because frequency had the most variability and dynamic range, we scaled the other measures by a factor of 3 to provide a better balance between the individually reported error vectors. The sum of the error vectors from a sample to a target row first provided the weight of the edge in the bipartite graph for the optimal assignment. After the optimal assignment, the sum of the matched edges was the error reported for an individual.

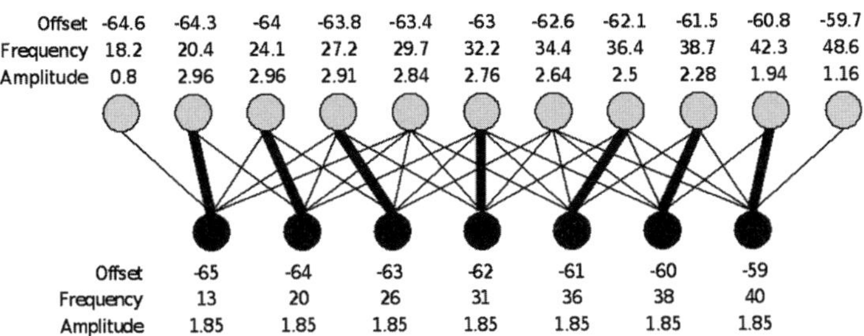

Fig. 3.5. Example of an optimal assignment of 11 sorted samples (top row) to 7 sorted target values (bottom). The weight of the edge is the error of a sample to a target row in Table 3.1 where error is defined as the Euclidean distance $\sqrt{(3 * offset_s - 3 * offset_t)^2 + (freq_s - freq_t)^2 + (3 * amp_s - 3 * 1.85)^2}$ and where s subscripts denote the sample and t subscripts denote the target. The heavy edges denote an optimal assignment or matching. The error reported for an individual to the EA is the sum of these matched edges.

We employed four types of EAs: 1) A binary encoding and single composite error measure, 2) a binary encoding with multiple error measures (i.e. a multi-objective EA) where the sum of the 7 offsets' distances were reported as a value, the sum of the errors in frequency as another value, the sum of the amplitude errors as a third value, and the composite of all three as another value[1] 3) real-valued with a single error measure, and 4) real-valued with multiple error measures. Three trials were performed with each type of EA. The results are highlighted in Figure 3.6 and summarized in Table 3.2. The results show that most EAs were able to converge on solutions that provided slightly better results than Rubin and Cleland's within a couple thousand simulations. The real-valued single error measure was least liable to get stuck in a local minimum, even though one run with a bitstring GA did better than any of the real-valued approaches. Those runs that got stuck could have

---

[1] Multiobjective EAs deliver the "Pareto front", describing trade-offs between errors. That is, for some error value along one dimension, the MOEA will provide the smallest values for the other errors. MOEAs do not directly reduce the composite score. By including the composite score, the EA could seek to reduce it.

perhaps been assisted with population migration, but that was not employed for these simulations to keep the runs segregated. It is also important to highlight that our results largely resemble Rubin and Cleland's, providing further support for their findings as well as the use of EAs to also solve the problem.

Table 3.2. Summary of model simulations. Optimal coefficients for the six channels from the best individual from each type of simulation. Abbreviations: S is for ion channels at the soma. D is for ion channels in the lateral dendrite. kA is an inactivating Potassium channel, kCa is a Calcium-dependent Potassium channel. NaP is a persistent Sodium channel. Ih is a hyperpolarization-activated cation current. Bottom two rows describe the allowed dynamic range of each variable.

| | Evolved channel coefficients | | | | | | Total | |
	S_{kA}	S_{kCa}	D_{NaP}	D_{kA}	D_{kCa}	D_{Ih}	runs	Error
Rubin/Cleland	0.012	0.12	0.00042	0.0074	0.10	0.0024	60000	14.6
Bit/Single	0.010	0.10	0.00042	0.0090	0.10	0.0022	1910	9.87
Bit/Multi	0.016	0.08	0.00039	0.0023	0.13	0.0027	2300	13.9
Real/Single	0.007	0.13	0.00038	0.0040	0.11	0.0025	2910	10.7
Real/Multi	0.020	0.04	0.00041	0.0062	0.11	0.0022	3060	11.4
Min value	0.005	0.01	0.00010	0.0010	0.01	0.0010		
Max value	0.100	0.50	0.00100	0.0500	0.50	0.0100		

3.5.2 Feature reduction and selection

Biological data is frequently laden with several signals and noise. It is often difficult or impossible to extract the important features or combinations of features to explain the data. *Feature reduction* algorithms such as Principal Components Analysis (PCA) transform several features into a few meta-features. Employing such a technique has proven useful in pre-filtering data and as a "wrapper" to clustering and classification algorithms such as K-Means, Artificial Neural Networks (ANNs), and Support Vector Machines (SVMs) to make them more robust [24–26].

By and large, machine learning techniques such as ANNs and SVMs operate as black boxes, filtering and transforming the input data into a particular output classification. They do not, however, make it easy or perhaps even possible to describe the particular input features which give rise to a particular output. While feature reducers can make a stronger classifier, by collapsing several features into a meta-feature, feature reducers often further cloud the ability to reverse-engineer the black box. This is unfortunate because it is especially meaningful when it can also be known what features and combinations of features really explain the output.

EAs have been employed as feature reducers to great effect [24, 25], but they can also be used as a *feature selector*, permitting a reverse-engineering

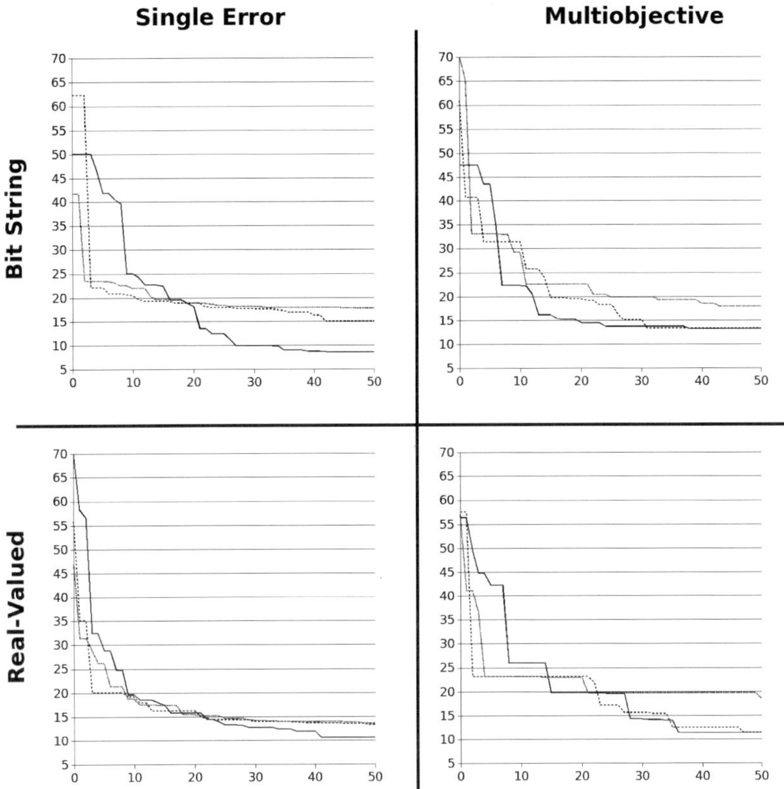

Fig. 3.6. Evolution of 4 different types of EAs to determine the best density values for six ion channels to exhibit subthreshold oscillations in a computational mitral cell neuron model. 3 runs with 60 individuals for 50 generations was performed with each EA type. Generations are plotted on the X-axis. Y-axis is on a logarithmic scale and indicates the error value of the individual from each generation with the lowest error. Solid line indicates the run finishing with the lowest error. The single-error scenarios employed an elite of 3 individuals per generation. Multiobjective scenarios used NSGA-II [8] for their replacement strategy. Tournament selection was used on all scenarios. The bitstring scenarios encoded each parameter with 12 bits and used a uniform crossover operator with a rate of 50% for each bit and with a 30% chance of a given individual being selected for crossover. They also used a uniform mutation operator set at 2%. The bitstring single-error strategy processed about 1925 individuals and the multiobjective strategy processed 2285 individuals in each run. The real-valued scenarios employed a blended crossover (BLX-α) [13] with $\alpha = 0.5$ such that a variable would blend between -50% and 150% of its parents. Individuals had a 30% chance of being selected for crossover with each generation. A Gaussian mutation operator was used with σ values for each parameter set to 20% of the range of the parameter. Each gene of each individual had a 10% chance of being mutated in each generation. The real-valued single-error scenario processed 2910 individuals and the multiobjective strategy processed 3060 in each run.

of the black box clusterer or classifier [26, 27]. When EAs are coupled with clusterers or classifiers, features are the genes being optimized. Therefore, the feature landscape is equivalent to the fitness landscape. A niche in parameter space, then, will describe those combinations of features which give rise to a particular classification. EAs can therefore provide information about input features in ways other feature reducers cannot.

As an example of an EA used for feature selection, Lavine and Vora used the GA to discriminate European honeybees from Africanized honeybees through gas chromatograms of the bees' secretions [27]. Figure 3.7 shows an example of a gas chromatogram. Within each trace are 65 peaks labeled with a letter and a number.

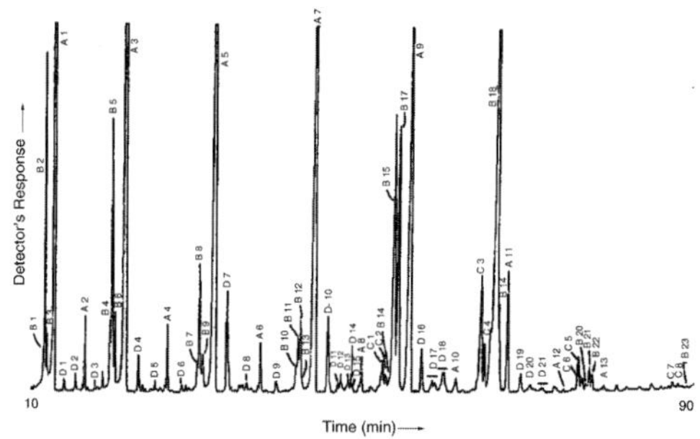

Fig. 3.7. Gas chromatographic trace of the hydrocarbon extracts obtained from the wax gland, cuticle, and exocrine gland of a heavily Africanized forager. A: normal alkanes; B: alkenes; C: dienes; and D: branched chain alkanes. Reprinted with kind permission from Lavine et al. [27].

Figure 3.8 shows a plot of the two principal components of these 65 peaks. European honeybees are labeled with a 1 and Africanized bees are labeled with a 2. What is apparent in the PCA plot is that there is not a clear distinction between the two sets. At least two reasons exist for the lack of separation. Either the data does not contain enough information to segregate the bees into distinct categories, or the existing data contains too many features clouding the most pertinent ones.

Assuming that relevant features were being masked by unnecessary features (noise), Lavine and Vora employed the GA to search for the relevant peaks in the chromatogram that could best discriminate the bees. They let the presence of a peak be determined by a single bit on a 65-bit chromosome. In this capacity, the chromosome effectively served as a filter on the chromatograms to either keep a peak at its original value (gene present) or set the peak to zero (gene not present). The fitness of an individual was assessed by perfoming PCA on 238 bee chromatograms filtered by the chromosome and then scoring the plot by how well it segregated the two classes. Figure 3.8 therefore shows the PCA plot with the chromosome of all ones: [111...111].

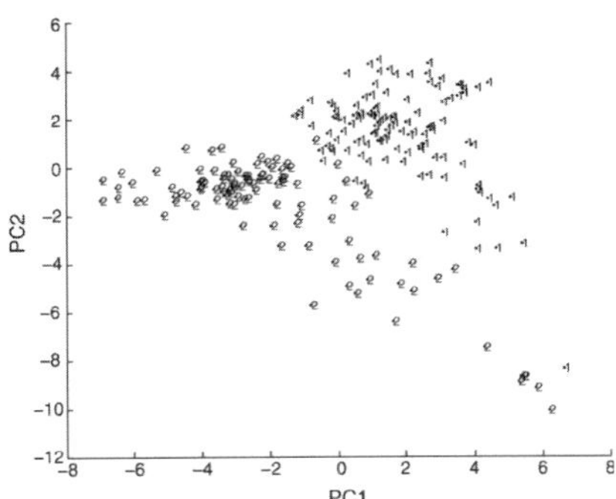

Fig. 3.8. Plot of the two largest principal components of the 65 gas chromatography peaks and 238 European and Africanized honeybee gas chromatograms that comprise the training set. Each bee is represented as a point in the principal component plot: 1 represents European honeybees, and 2 represents moderately and heavily Africanized honeybees. Reprinted with kind permission from Lavine et al. [27].

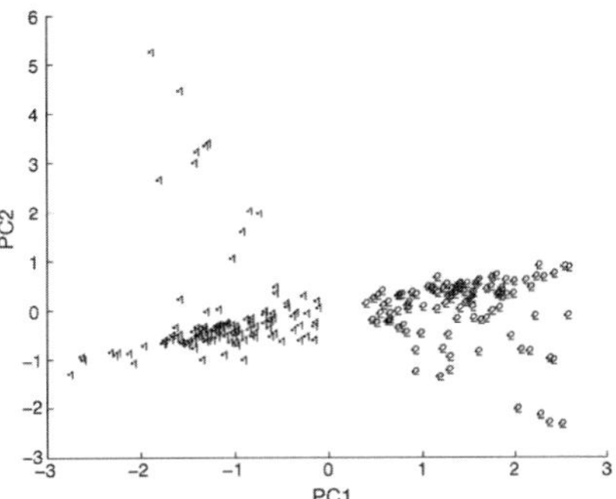

Fig. 3.9. Plot of the two largest principal components developed from the four gas chromatography peaks identified by the pattern recognition GA. Each bee is represented as a point in the principal component plot: 1 represents European honeybees, and 2 represents moderately and heavily Africanized honeybees. Clustering of the honeybees by genotype is evident. Reprinted with kind permission from Lavine et al. [27].

Figure 3.9 shows the result of the chromosome that displayed the best separation of the classes in the PCA plot. In this chromosome, only 4 genes were used (corresponding to peaks B11, B14, B15, and B22) and all other peaks were set to zero. The GA had therefore been able to select the 4 peaks which could discriminate the two classes.

The gas chromatogram looks quite similar to the output of mass spectrometers (MS) where, indeed, EAs have been used frequently in proteomics [24, 28]. For more EA examples in bioinformatics, see the review by Pal, Bandyopadhyay, and Ray [29] which highlight how EAs have been applied to fragment assembly, gene mapping, sequence alignment, gene finding and promoter identification, microarray analysis, molecular structure prediction, and protein-ligand docking problems.

3.6 Summary

In the words of the genetic algorithm's author, John Holland, "Genetic algorithms [are] computer programs that "evolve" in ways that resemble natural selection [that] can solve complex problems even their creators do not fully understand" [30]. There is a rigorous theory behind real-valued and bit-string strategies which describes how sampling individuals and combinations of individuals implicitly samples, to some degree, large regions in parameter space, but that is beyond the scope of this chapter. That is, even though mutations are performed on single individuals and recombination only combines two or a few individuals, the effect is felt at the population level. Natural selection and sexual reproduction operate to modify the population as a whole, and by operating on the population, the EA employs a broad, collective heuristic to direct its search.

The field of evolutionary computation, while still somewhat young, has a solid foundation of theory, datasets for comparative testing, a plethora of techniques, and a number of software packages available. Several books have been written on EAs. There are also peer-reviewed journals, professional societies, and international conferences on the subject. Therefore, there are a number of resources for exploring the subject further and for employing and modifying an EA in almost any application.

For further information, the book *Genetic Algorithms + Data Structures = Evolution Programs* by Michalewicz [3] provides a rich introduction to the subject. Additionally, the search term "evolutionary algorithm" in Wikipedia [31] or other search engines can point you to several examples, tutorials, and other internet resources. As mentioned before, the review by Pal, Bandyopadhyay, and Ray [29] contains several references to bioinformatics examples. Finally, the search term, "genetic algorithm" in PubMed [32] will present several examples of the GA's use in biomedicine.

EAs have proven useful in several engineering and scientific disciplines including Biology. People have exploited EAs in a number of creative ways and have hybridized them with other techniques. However, there remain many outstanding problems in Biology that may exploit EAs to great effect. When the fitness landscape is complex and unknown, the time to search needs to be reduced, as is often the case in computer simulations and complex models with several variables, or when features need to be reduced or selected in classification and clustering problems, then an EA might provide a solution.

3.7 Acknowledgments

We would like to thank Christian Gagne for his assistance with Open BEAGLE [22] and Michael Hines for his assistance with integrating Open BEAGLE into the NEURON simulation environment [21] for our computational simulations. We would also like to thank Nathan Schoppa for sharing his intimate knowledge of the mitral cell and Larry Hunter for his technical critique and advice. Our simulation work was funded by NIH grants DC004657, DC006070, 5R01-LM008111-03, and DC006640.

References

1. Holland JH (1975) Adaptation in Natural and Artificial Systems. University of Michigan Press, Ann Arbor, MI

2. Beyer HG, Schwefel HP (2002) Evolution strategies - a comprehensive introduction, Natural Computing 1:3–52
3. Michalewicz Z (1996) Genetic Algorithms + Data Structures = Evolution Programs. Springer
4. Okabe T, Jin Y, Sendhoff B (2005) Theoretical comparisons of search dynamics of genetic algorithms and evolution strategies, In: Evolutionary Computation, 2005. The 2005 IEEE Congress on, vol. 1, 382– 389 Vol.1
5. Alander JT (1992) On optimal population size of genetic algorithms, CompEuro'92'Computer Systems and Software Engineering', Proceedings 65–70
6. Goldberg DE, Richardson J (1987) Genetic algorithms with sharing for multimodal function optimization, Proceedings of the Second International Conference on Genetic Algorithms on Genetic algorithms and their application table of contents 41–49
7. Holland JH (1992) Adaptation in Natural and Artificial Systems: An Introductory Analysis with Applications to Biology, Control, and Artificial Intelligence. MIT Press
8. Deb K, Agrawal S, Pratap A, Meyarivan T (2000) A fast elitist non-dominated sorting genetic algorithm for multi-objective optimization: Nsga-ii, Proceedings of the Parallel Problem Solving from Nature VI Conference 849–858
9. De Jong KA (1975). An analysis of the behavior of a class of genetic adaptive systems
10. Eshelman LJ, Schaffer JD (1991) Preventing premature convergence in genetic algorithms by preventing incest, Proceedings of the Fourth International Conference on Genetic Algorithms 115–122
11. Sywerda G (1989) Uniform crossover in genetic algorithms, Proceedings of the third international conference on Genetic algorithms table of contents 2–9
12. Eshelman LJ (1991) The chc adaptive search algorithm: How to have safe search when engaging in nontraditional genetic recombination, Foundations of Genetic Algorithms 1:265–283
13. Eshelman LJ, Schaffer JD (1993) Real-coded genetic algorithms and interval-schemata, Foundations of Genetic Algorithms 2:187–202
14. Schaffer JD, Caruana RA, Eshelman LJ, Das R (1989) A study of control parameters affecting online performance of genetic algorithms for function optimization, Proceedings of the third international conference on Genetic algorithms table of contents 51–60
15. Gray F (1953). Pulse code communication
16. Hansen N, Ostermeier A (1996) Adapting arbitrary normal mutation distributions in evolution strategies: the covariance matrix adaptation, In: Evolutionary Computation, 1996., Proceedings of IEEE International Conference on, 312–317
17. Hansen N, Ostermeier A (2001) Completely derandomized self-adaptation in evolution strategies, Evolutionary Computation 9:159–195
18. Desmaisons D, Vincent JD, Lledo PM (1999) Control of action potential timing by intrinsic subthreshold oscillations in olfactory bulb output neurons, Journal of Neuroscience 19:10727–10737
19. Davison AP, Feng J, Brown D (2000) A reduced compartmental model of the mitral cell for use in network models of the olfactory bulb, Brain Res Bull 51:393–9
20. Rubin DB, Cleland TA (2006) Dynamical mechanisms of odor processing in olfactory bulb mitral cells, Journal of Neurophysiology 96:555

21. Hines ML, Carnevale NT (1997) The neuron simulation environment, Neural Comp 9:1179–1209
22. Gagne C, Parizeau M (2002) Open beagle: A new versatile c++ framework for evolutionary computation, Late-Breaking Papers of the Genetic and Evolutionary Computation Conference (GECCO); New York City 161–168
23. Kuhn HW (2005) The hungarian method for the assignment problem, Naval Research Logistics 52:7–21
24. Li L, Tang H, Wu Z, Gong J, Gruidl M, Zou J, Tockman M, Clark RA (2004) Data mining techniques for cancer detection using serum proteomic profiling, Artificial Intelligence In Medicine 32:71–83
25. Wang M, Zhou X, King RW, Wong STC (2007) Context based mixture model for cell phase identification in automated fluorescence microscopy, BMC Bioinformatics 8:32
26. Yang J, Honavar V (1998) Feature subset selection using a genetic algorithm, Intelligent Systems and Their Applications, IEEE [see also IEEE Intelligent Systems] 13:44–49
27. Lavine BK, Vora MN (2005) Identification of africanized honeybees, J Chromatogr A 1096:69–75
28. Jeffries NO (2005) Performance of a genetic algorithm for mass spectrometry proteomics, feedback
29. Pal SK, Bandyopadhyay S, Ray SS (2005) Evolutionary computation in bioinformatics: A review, IEEE Transactions on Systems, Man, and Cybernetics, Part-C
30. Holland JH (1992) Genetic algorithms: Computer programs that "evolve" in ways that resemble natural selection can solve complex problems even their creators do not fully understand, Scientific American 267:66–72
31. Wikipedia. Evolutionary algorithm — Wikipedia, The Free Encyclopedia URL `http://en.wikipedia.org/w/index.php?title=Evolutionary_algorithm&oldid=138840243`
32. NCBI. Pubmed home URL `http://www.ncbi.nlm.nih.gov/sites/entrez?db=PubMed`

4

An Introduction to Multi-Objective Evolutionary Algorithms and Some of Their Potential Uses in Biology

Antonio López Jaimes and Carlos A. Coello Coello

CINVESTAV-IPN
Evolutionary Computation Group (EVOCINV)
Departamento de Computación
Av. IPN No. 2508
Col. San Pedro Zacatenco
México, D.F. 07360
MEXICO
tonio.jaimes@gmail.com, ccoello@cs.cinvestav.mx

Summary. This chapter provides a brief introduction to the use of evolutionary algorithms in the solution of problems with two or more (normally conflicting) objectives (called "multi-objective optimization problems"). The chapter provides some basic concepts related to multi-objective optimization as well as a short description of the main features of the multi-objective evolutionary algorithms most commonly used nowadays. In the last part of the chapter, some applications of multi-objective evolutionary algorithms in Biology (mainly within Bioinformatics) will be reviewed. The chapter will conclude with some promising paths for future research, aiming to identify areas of opportunity for those interested in the intersection of these two disciplines: multi-objective evolutionary algorithms and Biology.

4.1 Introduction

Many real-world problems in most disciplines have two or more objectives that we aim to optimize at the same time. Such problems are called "multi-objective", and their solution implies finding good trade-offs among the objectives. Traditionally, multi-objective optimization problems have been dealt with using a variety of mathematical programming techniques that have been developed over the years [34, 63]. However, in recent years, the use of

A.L. Jaimes and C.A. Coello Coello: *An Introduction to Multi-Objective Evolutionary Algorithms and Some of Their Potential Uses in Biology*, Studies in Computational Intelligence (SCI) **122**, 79–102 (2008)
www.springerlink.com
© Springer-Verlag Berlin Heidelberg 2008

metaheuristics[1] to solve such problems has become increasingly popular (see for example [22, 28]).

Evolutionary algorithms (EAs) are a metaheuristic inspired on the "survival of the fittest" principle from Darwin's evolutionary theory [39]. EAs have become very popular as multi-objective optimizers because of their ease of use (and implementation) and flexibility (e.g., EAs are less sensitive than mathematical programming techniques to the initial search points and to the specific features of a problem). Additionally, the fact that EAs are population-based techniques makes it possible to simultaneously manage a set of solutions, instead of one at a time, as normally happens with mathematical programming techniques.

Multiobjective evolutionary algorithms (MOEAs) date back to the mid-1980s [47, 83], although they became popular in the mid-1990s. Today, it is possible to find applications of MOEAs in practically every discipline, including biology [18].

The rest of this chapter is organized as follows. In Section 4.2, we provide some basic multi-objective optimization concepts required to make the chapter self-contained. Section 4.3 contains a brief description of the main MOEAs in current use. Section 4.4 contains a survey of some of the most representative applications of MOEAs in biology. Section 4.5 indicates some potential paths for future research in this area. Finally, our conclusions are provided in Section 4.6.

4.2 Basic Concepts

This chapter deals with the solution of the **Multiobjective Optimization Problem** (MOP) (also called multicriteria optimization, multiperformance or vector optimization problem), which can then be defined (in words) as the problem of finding [69]:

> "a vector of decision variables which satisfies constraints and optimizes a vector function whose elements represent the objective functions. These functions form a mathematical description of performance criteria which are usually in conflict with each other. Hence, the term "optimize" means finding such a solution which would give the values of all the objective functions acceptable to the decision maker."

The **decision variables** are the numerical quantities for which values are to be chosen in an optimization problem. In most optimization problems there are always restrictions imposed by the particular characteristics of the

---

[1] A **metaheuristic** is a high level strategy for exploring search spaces by using different methods [7]. Metaheuristics have both a diversification (i.e., exploration of the search space) and an intensification (i.e., exploitation of the accumulated search experience) procedure.

environment or available resources (e.g., physical limitations, time restrictions, etc.). These restrictions must be satisfied in order to consider a certain solution acceptable. All these restrictions in general are called **constraints**, and they describe dependences among decision variables and constants (or parameters) involved in the problem.

In multiobjective optimization, the goal is to optimize a set of objective functions (i.e., more than two) simultaneously. Thus, in this context, the notion of "optimum" changes, because in MOPs, the aim is to find good compromises (or "trade-offs") rather than a single solution as in global optimization (in which we aim to optimize a single objective function). The notion of "optimum" most commonly adopted is that originally proposed by Francis Ysidro Edgeworth [33] and later generalized by Vilfredo Pareto [72]. Although some authors call this notion the *Edgeworth-Pareto optimum*, the most commonly accepted term is *Pareto optimum*.

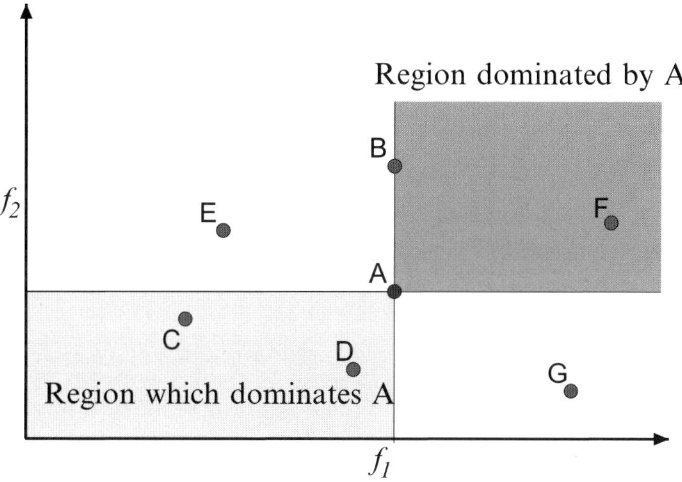

Fig. 4.1. Graphical representation of some solutions that are dominated and others that dominate a reference point (**A** in this case). Note that both **E** and **G** are **nondominated** with respect to **A**. So, **A** is better than **B** or **F**, but it is equally good than **E** or **G**.

A solution is Pareto optimal if there exists no other feasible solution (i.e., one which satisfies all the constraints of the problem) which would decrease some criterion without causing a simultaneous increase in at least one other criterion (assuming minimization). The vectors corresponding to these Pareto optimal solutions are called **nondominated**. Figure 4.1 provides a graphical representation of solutions that dominate and solutions that are dominated by a reference point for a problem with two objective functions. When plotted in objective function space, these nondominated vectors are collectively known as the **Pareto front** (Figure 4.2 shows the graphical representation of a Pareto front).

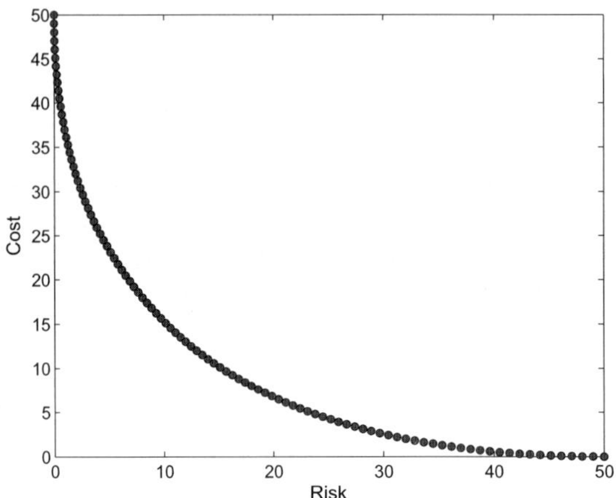

Fig. 4.2. Pareto front of a hypothetical problem with two objectives: risk and cost.

Although it is normally assumed that a MOEA will generate the entire Pareto front (or as many elements of it, as possible), in practice the entire front is rarely needed. This can be easily understood with an example. In Figure 4.2 the solutions lying at the extreme right of the Pareto front represent the lowest possible cost, but with the highest risk. Conversely, solutions lying at the top left of the Pareto front, represent the lowest possible risk, but with the highest cost. Normally, solutions that represent the best possible trade-offs among the objectives are the aim of the search (in the case of Figure 4.2, solutions lying on the "knee" of the Pareto curve).

4.3 MOEAs in Current Use

Although the first reference on the use of EAs for solving multi-objective problems dates back to the late 1960s [80], the first actual implementations was introduced in the mid-1980s [47, 82, 83]. For several years (up to the first half of the 1990s), most of the MOEAs developed had a relatively simple design and were mostly based on linear aggregating functions [88], lexicographic ordering [38], and target-vector approaches [15, 98].

However, four MOEAs are considered the most representative of this early period: the Vector Evaluated Genetic Algorithm (VEGA) [83], the Multi-Objective Genetic Algorithm (MOGA) [37], the Niched-Pareto Genetic Algorithm (NPGA) [42], and the Nondominated Sorting Genetic Algorithm

(NSGA) [86]. Details of these algorithms can be found in their original publications and in other sources (see for example [17]).

During the mid-1990s, elitism was formally introduced in MOEAs [104], and became a standard mechanism for the algorithms developed since then. In single-objective EAs, elitism simply consists of retaining the best individual from the current generation, and passing it without any changes to the next generation. In contrast, in multi-objective optimization, elitism involves retaining the solutions that are nondominated with respect to all the individuals that have been evaluated so far. Thus, instead of retaining only one individual, several must be kept. This introduces additional issues that need to be taken into account (e.g., should we bound the number of individuals to be retained? If so, how do we decide which individuals must be removed?). Elitism is an important mechanism, not only because it allows to keep the globally nondominated individuals (as opposed to handling only the locally nondominated individuals, as done with early MOEAs), but also because it is a requirement to prove convergence [96].

Despite the high number of elitist MOEAs developed from the mid-1990s to date (see for example [5, 21, 23, 24, 94, 100, 102]), three of them are normally considered to be the most representative in the current literature[2]:

1. The **Strength Pareto Evolutionary Algorithm** (SPEA): Developed by Zitzler and Thiele [104], this approach integrates ideas from the different MOEAs previously mentioned (i.e., MOGA [37], NPGA [42] and NSGA [86]). SPEA incorporates elitism through the usage of an external archive containing nondominated solutions previously found (the so-called external nondominated set). At each generation, nondominated individuals are copied to this external nondominated set, and are retained only if they are nondominated with respect to the contents of the set. If they dominated any individuals previously stored in the external set, such dominated individuals are deleted. For each individual in this external set, a *strength* value is computed. This strength is similar to the ranking value of MOGA [37], since it is proportional to the number of solutions to which a certain individual dominates. The fitness of each member of the current population is computed according to the strengths of all external nondominated solutions that dominate it (i.e., the external set plays a role in the selection process). The fitness assignment process of SPEA considers both closeness to the true Pareto front and even distribution of solutions at the same time. However, instead of using niches based on distance (as done in earlier MOEAs such as MOGA [37]), Pareto dominance is used to ensure that the solutions are properly distributed along the Pareto front. SPEA does not require a niche radius, but its effectiveness relies on the size of the external nondominated set. In fact, since the external nondominated set

---

[2] For more information on MOEAs, interested readers can refer to the EMOO repository, which is located at: `http://delta.cs.cinvestav.mx/~ccoello/EMOO/`

participates in the selection process of SPEA, if its size grows too large, it might reduce the selection pressure, thus slowing down the search. Because of this, the authors decided to adopt a technique that prunes the contents of the external nondominated set so that its size remains below a certain threshold. In 2001, a revised version of SPEA (called **SPEA2**) was introduced. SPEA2 has three main differences with respect to its predecessor [103]: (1) it incorporates a fine-grained fitness assignment strategy which takes into account for each individual the number of individuals that dominate it and the number of individuals by which it is dominated; (2) it uses a nearest neighbor density estimation technique which guides the search more efficiently, and (3) it has an enhanced archive truncation method that guarantees the preservation of boundary solutions.

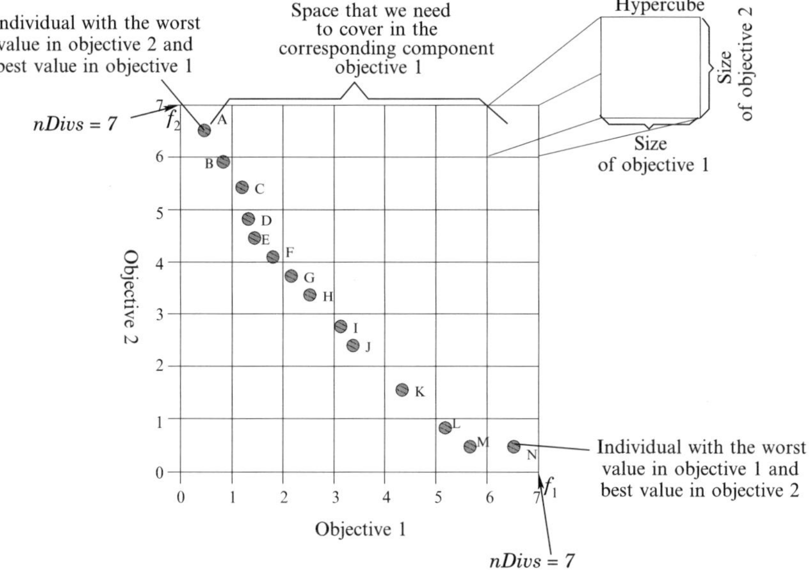

Fig. 4.3. Graphical illustration of the adaptive grid used by PAES.

2. The **Pareto Archived Evolution Strategy** (PAES): Developed by Knowles and Corne [52, 53], this is probably the most simple MOEA that can be conceived. It consists of a $(1+1)$ evolution strategy (i.e., a single parent that generates a single offspring), combined with an external archive that records the nondominated solutions found along the search. As in SPEA, this external archive is used to compare each new individual produced. An interesting aspect of PAES is its procedure to maintain diversity which consists of a crowding mechanism that divides objective space in a recursive manner. Each solution is placed in a certain

grid location based on the values of its objectives (which are used as its "coordinates" or "geographical location") as indicated in Figure 4.3. A map of such grid is maintained, indicating the number of solutions that reside in each grid location.

3. The **Nondominated Sorting Genetic Algorithm II** (NSGA-II): This MOEA is described in Deb et al. [29, 31], and it consists of a considerably improved version of the NSGA [86]. The NSGA-II estimates the density of solutions surrounding a particular solution in the population by computing the average distance of two points on either side of this point along each of the objectives of the problem. This value is called *crowding distance* and its computation is not only efficient, but requires no extra parameters. During selection, the NSGA-II uses a crowded-comparison operator which takes into consideration both the nondomination rank of an individual in the population and its crowding distance (i.e., nondominated solutions are preferred over dominated solutions, but between two solutions with the same nondomination rank, the one that resides in the less crowded region is preferred). This introduces a total ordering (instead of the partial ordering that traditional Pareto ranking generates), and facilitates the selection process. That is the reason why the NSGA-II combines the population of parents with the population of offspring and selects the best half of them. This sort of selection scheme is implicitly elitist and, therefore, no external archive is required in this case. Due to its ease of use, efficacy, and efficiency, the NSGA-II has become a landmark against which other MOEAs are often compared.

4. **Coevolutionary MOEAs**: In evolutionary computation, the term **coevolution** is used to refer to a change in the genetic composition of a species (or group of species) as a response to a genetic change of another one. In a more general sense, coevolution refers to a reciprocal evolutionary change between species that interact with each other. The term "coevolution" is usually attributed to Ehrlich and Raven who published a paper on their studies performed with butterflies and plants in the mid-1960s [35]. The relationships between the populations of two different species can be described considering all their possible types of interactions. Such interaction can be positive or negative depending on the consequences that such interaction produces on the population. Evolutionary computation researchers have developed several coevolutionary approaches in which normally two or more species relate to each other in different forms [70]. The key issue in these coevolutionary algorithms is that the fitness of an individual in a population depends on the individuals of a different population. In fact, we can say that an algorithm is coevolutionary if it has such property.

There are two main classes of coevolutionary algorithms in the evolutionary computation literature:

a) Those based on competition relationships (called **competitive co-evolution**): In this case, the fitness of an individual is the result of a series of "encounters" with other individuals [71, 81]. This sort of coevolutionary scheme has been normally adopted for games.

b) Those based on cooperation relationships (called **cooperative coevolution**): In this case, the fitness of an individual is the result of a collaboration with individuals of other species (or populations) [74, 77]. This sort of coevolutionary scheme has been normally adopted for solving optimization problems.

A variety of coevolutionary MOEAs have been proposed in the specialized literature (see for example [3, 20, 46, 50, 51, 58, 61, 73, 89, 90]), but a detailed description of them is beyond the scope of this chapter. Interested readers may refer to Chapter 3 in [19] for more information on this topic.

4.4 Applications of MOEAs in Biology

The use of MOEAs in Biology has raised an increasing interest in the last few years, mainly within Bioinformatics [40, 65]. An analysis of the literature shows five main types of applications of MOEAs in Biology:

1. **System optimization:** This refers to some applications in which it is of interest to determine the degree of optimality of a certain biological system.

2. **Classification:** A wide variety of problems in bioinformatics rely on performing classification tasks (either supervised, unsupervised or combinations of both).

3. **Sequence and structure alignment:** Here, the aim is to assess the structural similarities between a certain macromolecule and a sequence available from a database. The search is done through a series of alignments.

4. **Structure prediction and design:** In this case, the goal is to predict the structure of a macromolecule, given that the function properties of macromolecules derive from their three-dimensional shape. This three-dimensional shape is, in turn, mainly determined by the sequence of bases or amino acids.

5. **Inverse problems:** These are problems in which we have certain information that was generated by a biological process and our goal is to infer the original system using such available information.

Next, we will briefly review some of the most representative work within each of these types of applications.

4.4.1 System optimization

A single nucleotide polymorphism (SNP) is a variation that occurs at only one single nucleotide of two deoxyribonucleic acid (DNA) sequences (e.g., GA**AC**CT and GA**GC**CT). Geneticists carry out projects using a set of SNPs in order to, for example, search for genes responsible for a disease. Thus, prior to project initiation, geneticists need to select a subset of SNPs from large databases. Hubley et al. [44] formulated this task as a bi-objective optimization problem and proposed an algorithm called Multiobjective Analyzer for Genetic Marker Acquisition (MAGMA). The desired goals of a mapping project are to maximize the probability of locating a disease gene while minimizing the total project cost. However, as the authors point out, these goals may be subjective, difficult to describe, or may require excessive computation. To overcome this problem, the authors make use of the so called proxy objectives. That is to say, objectives that only capture certain aspects of the actual objectives [40]. In this case, the first proxy objective is to search for evenly spaced high-quality SNPs with an average spacing. Thus, the objectives are: minimize the average deviation from the ideal gap length between two SNPs and maximize the average quality of the SNPs. A solution in this problem is represented by a binary string where a bit is set to 1 only if the corresponding SNP is in the solution. The proposed algorithm was tested using two real SNP selection problems with a relatively small library of SNPs, and a constructed problem with a large library containing a vast number of SNPs. The Pareto front in all cases had a concave shape, and MAGMA was able to discover the true Pareto front in the three problems.

In a later study, Hubley et al. [43] proposed two new proxy objectives that reflect more precisely the actual goals of a project. Here the cost is modeled in a straightforward manner, as the sum of the cost associated with each SNP. The probability of project success is treated as the quality of a SNP, which is a heuristic combination of (i) allele frequency, (ii) database reliability, and (iii) biochemical suitability.

Lee et al. [55] formulate the probe design for DNA microarrays as a multi-objective problem, which is then solved by the NSGA-II [31]. The resulting problem has four objectives and one constraint. The authors use a thermodynamic criteria to assist the decision maker to choose a solution from the generated Pareto front. The melting temperature can be used to determine if a candidate probe hybridizes to the wrong target gene or not. This way, one can choose the set of probes which have the least mis-hybridizing probes from the obtained Pareto front. Based on the specificity of hybridization of each probe, the proposed method achieved more reliable probe sets than those pre-existing oligonucleotide microarray for HPV (human papilloma virus) detection.

4.4.2 Classification

Deb and Reddy [32] address the classification of two-class cancer data using the NSGA-II [31]. Here, the authors formulate a two-objective and a three-objective problem. The first problem consists of minimizing the size of the gene-subset and minimizing the sum of misclassifications in the training and test samples. In the second problem, the misclassifications in the training and in test samples are considered as two different objectives. As some solutions with desirable subset sizes do not belong to the Pareto front, when using the standard Pareto domination concept, the authors introduce a variant called biased dominance. This modified concept allows that multiple solutions lying in parallel to a f_i axis do not dominate each other. An interesting finding is that in this problem a vector in the objective space can be produced by more than one solution in the decision variable space. The authors modified the NSGA-II so that it could take into account these types of solutions.

Usually, microarray data contain a large number of features (genes) from which most of them are non essential to carry out data classification. Banerjee et al. [1] proposed a MOEA that employs rough sets to reduce the number of features in order to ease the classification of gene expression patterns in a microarray. The set of genes is modeled as a rough set in such a way that the essential features are represented by the reduct of the information system. Thus, the objectives of this feature selection problem are: (i) to obtain a reduct of small cardinality and simultaneously (ii) to still classify all the elements of the universe with the same accuracy as with the entire attribute set. The feature reduction is carried out in a two stage process. The first stage generates an initial crude redundancy reduction among features by normalizing the expression values (attributes) and eliminating constantly expressed genes and ambiguously expressed genes (i.e., those with average expression value). In the second stage, the crude reduced data set is optimized by the NSGA-II [31] to achieve a refined minimal feature set. The formulation of this multi-objective problem includes two objectives: minimization of the number of attributes in the reduct set and maximization of the capacity to distinguish objects in order to achieve an acceptable classification. The only variable considered is the reduct which is represented by a binary string of length m (where m is the number of attributes). In this string, 1 indicates that the corresponding attribute is present in the reduct while 0 indicates the contrary. The proposed method was validated using microarray data sets consisting of three different cancer samples, namely colon cancer, lymphoma and leukemia. The Pareto front obtained in the three data sets is a typical convex front where the reduct cardinality decreases as the number of misclassifications increases. The proposed MOEA was compared against other approaches which include a probabilistic neural network, a *t*-test-based feature selection with a fuzzy neural network, a saliency analysis to support vector machines and a linear aggregating function approach. Considering the available results, the MOEA achieved a better correct classification percentage than the other approaches using the three datasets.

Liu et al. [57] have proposed an entropy-based method to select genes related to the different cancer classes, simultaneously reducing the redundancy among the genes. This bi-objective problem is solved using an aggregation approach solved by a greedy algorithm.

Bleuler et al. [6] proposed an evolutionary framework for bi-clustering of gene expression data in a single-objective context. The main idea of the framework is to explore the search space by an EA and refine the solutions found by using a local search bi-clustering method. The framework was implemented using the bi-clustering method proposed by Cheng and Church [14]. The results showed that the EA coupled with a local search performs significantly better than the Cheng and Church's bi-clustering algorithm alone.

Recently, Mitra and co-workers [2, 64, 65] proposed a similar framework to that of Bleuler et al. [6] but in a multi-objective context. The two objectives considered were the maximization of the bi-cluster size and the maximization of the homogeneity. According to the results, this framework achieves better results than some other methods available in the literature [6, 14, 99, 101].

Prelic et al. [78] carried out a systematic comparison of five salient bi-clustering methods based on greedy search techniques, namely: the algorithm of Cheng and Church [14], Samba [92], the Order Preserving Submatrix Algorithm [4], the Iterative Signature Algorithm [45] and *xMotif*, [67]. Madeira and Oliveira [59] provides a survey on bi-clustering methods that, besides greedy search techniques, includes clustering methods based on strategies such as divide-and-conquer, and exhaustive enumeration to mention a few. The authors adopted only external indices to assess the performance of the algorithms. External indices are based on additional data in order to validate the achieved results. Moreover, the comparison study considered both synthetic and real datasets. The former has the advantage that the true optimal solutions are known *a priori*. As a reference algorithm, the authors proposed a fast and exact algorithm that uses a simple data model yet reflecting the fundamental idea of bi-clustering.

4.4.3 Sequence and Structure alignment

Malard et al. [60] formulate the *de novo* peptide identification as a constrained multi-objective optimization problem. The objectives considered in the study are the maximization of the similarity between portions of two peptides, and the maximization of the likelihood ratio between the null hypothesis and the alternative hypothesis. The former is that spectral peaks match ion fragments only by chance, whereas the latter is that spectral peaks match ion fragments because the candidate solution is in the sample. Constraints are treated as an objective function in a similar way as the Constrained Multi-objective Optimization by Genetic Algorithm (COMOGA) proposed by Surry and Radcliffe [87]. The algorithm was implemented using the island parallel model [22, 68, 95], in which some subpopulations evolve independently of each other, although individuals periodically migrate between neighboring islands.

Boisson et al. [8] also studied the protein sequencing using the *de novo* peptide sequencing approach, although using a single-objective genetic algorithm. As the evaluations of the objective function involved in the problem are time consuming, Boisson et al. [9] decided to use a parallel genetic algorithm to discover the sequence of an experimental protein. The algorithm was implemented on a grid of computers.

Calonder et al. [12] address the problem of identifying gene modules on the basis of different types of biological data such as gene expression (GE), protein-protein interactions (PPI) and metabolic pathways (MP). Module identification refers to the identification of groups of genes similar with respect to its function or regulation mechanism. The particular problem addressed in this work is to identify the best module containing some user defined query genes with respect to n biological data sets. Some single-objective approaches for the identification of modules have been proposed including a co-clustering approach where a combined distance function is used as the objective function. Another approach combines distances on the Gene Ontology graph with gene expression data and applies a memetic algorithm[3] for identifying high scoring clusters.

The proposed multi-objective approach has some advantages over a single-objective aggregation approach. First, it is not required to define an overall similarity measure, which is often difficult since we need to aggregate measures (i.e., objective function values) with different scales and interpretations. With a multi-objective approach each similarity measure can be treated as an independent objective. Also, it offers a way to study the interactions and conflicts between the data sets. That is, the visual inspection of the trade-offs in the Pareto front allow us to determine, for instance, if accepting a slightly worse similarity on one data type could increase the similarity on the other data types substantially. Finally, as the objectives are treated independently, it is possible to easily integrate arbitrary data types and similarity measures.

In this formulation, each data type is associated with a distinct objective which is defined as the mean distance from all genes to the query genes on the corresponding data set. For each objective, a suitable measure of distance is computed. Each solution (module) is represented by a binary string of length m (where m is the number of genes) where a value of 1 indicates that the corresponding gene is included in the module. The MOEA employed in this work is the indicator-based evolutionary algorithm (IBEA) [102].

---

[3] Pablo Moscato [66] introduced the concept of "memetic algorithm" to denote the use of local search heuristics with a population-based strategy.

In the experimental study, the authors considered three bi-objective problems using different data types: GE-GE data on Arabidopsis, GE-MP data on Arabidopsis, and a yeast GE-PPI data set. In the experimental study, a local search heuristic was added to the evolutionary algorithm. However, the results revealed that the local search imposed a noticeable bias toward one of the objectives. The performance of the algorithm was compared against that of a single-objective aggregation approach and that of a k-means algorithm. In the first case, to generate the Pareto front, the single-objective optimizer was run repeatedly with 21 different weight vectors. The comparison of the resulting Pareto fronts using the ε-indicator [105] revealed that the multi-objective approach achieved approximation sets better than those obtained by the single-objective approach. In order to compare the multi-objective approach with the k-means algorithm, the authors ran the k-means using only the GE data, and then they selected at random a query gene from one of the clusters. For this cluster, they calculated the value of the two objectives, GE and PPI, to get a Pareto front consisting of a single solution. The same query gene was used as input to the EA to get the approximation set. Again, the ε-indicator showed that the EA performs better than k-means.

Zwir et al. [106] presented a two-level methodology for the elicitation and summarization of qualitative features in DNA sequences of *Tripanosoma cruzi*. The first stage had the goal of recognizing instances of interesting features through a multi-objective genetic-based clustering method. Here, the clustering problem was formulated as a multi-objective problem that takes into account, independently, the multiple measures of cluster quality. At this stage, Pareto local dominance was adopted. That is, a solution is locally non-dominated if there does not exist a neighboring solution that dominates it. At the second stage, the Pareto front obtained in the first stage was summarized in order to obtain a compact description of the set of interesting features.

4.4.4 Structure Prediction and Design

Lee and co-workers [56, 84] used the controlled elitist NSGA-II [30] to generate a set of DNA sequences which can be used in microarray design or in DNA-based computing. The desirable properties of a DNA sequence are the quality measures achieved while satisfying certain constraints. The quality of a sequence can be achieved by minimizing four objectives: the similarity between two sequences in the set, the number of bases that can be hybridized between sequences in the set, the degree of successive occurrences of the same base and the probability to form a secondary structure. A good sequence should have similar physical and chemical properties. To guarantee these characteristics the authors use as constraints the number of bases 'G' and 'C' in the sequence and the melting temperature where more than half of the double strands start to break into single strands. These constraints are handled with a tournament selection that determine the winner using the following rules: a feasible solution is preferred over an infeasible solution; between two infeasible solutions the solution with the smaller constraint violation is preferred; between two feasible solutions the solution that dominates the other is preferred. The proposed approach was compared against three similar algorithms [27, 36, 91] using an instance problem of a set of 7 DNA sequences of length 20. The comparison was based on the average values of each objective over the generated set of DNA sequences. The results showed that the proposed method achieved smaller average values in all the objectives than the other approaches considered.

Day et al. [26] employed the multi-objective fast messy genetic algorithm (MO fmGA) [107] to solve the protein structure prediction problem. This study is based on an energy minimization technique which uses the CHARMm energy function. This function is composed of 10 major terms and in order to utilize a multi-objective framework, it was decomposed in two minimization objectives: (i) the sum of the connected and (ii) the sum of the non-connected atom energies. The decision variables for this problem are the dihedral angles for the protein being solved. The algorithm was applied to two proteins, [Met]-Enkephelin and Polyalanine. For both problems, a convex Pareto front was obtained. The results were compared against those obtained in a previous study using a single-objective fmGA (SO fmGA) [62]. To do so, for each vector of the obtained Pareto front the two objective values were added to obtain a single value. Then, the best objective value found was compared with the single value achieved by the SO fmGA. For [Met]-Enkephelin, the MO fmGA found the best solution, while for Polyalanine, the MO-fmGA compared favorably with respect to the SO fmGA.

Chen et al. [13] proposed a method to solve the structure alignment problem for homologous proteins. This problem can be formulated as a multi-objective optimization problem where the objectives are: maximize the number of aligned atoms and minimize their distance. The proposed method relies on a bipartite matching algorithm whose convergence is numerically stable and is also theoretically ensured.

4.4.5 Inverse problems

Phylogenetic inference is the construction of trees that represent the genealogical relationships between different species. Contrary to other kinds of taxonomy, phylogenetic classification is based on common ancestry and not mere similarity of appearance or function [75]. The reconstruction of phylogenetic trees relies on various types of data sets, for example nucleotide and amino acid sequences, protein shapes, anatomical characters, or behavioral traits to name a few.

Poladian and Jermiin [76] proposed using a multi-objective evolutionary approach to infer phylogenetic trees integrating many types of available data. As pointed out by the authors, MOEAs are especially suitable to obtain phylogenetic inferences for three reasons: (i) the large combinatorial space associated with all possible phylogenies, (ii) the conflicting results obtained by using different data sets and (iii) the fact that, a single best tree may not tell the whole story but a nearly-best trees may also reveal information about the relationship between two species.

One of the current problems in phylogenetic inference is how to assess, combine, modify or reject different types of data. *Total evidence* [75] is one of the two main lines of thought about how to integrate information from different data types, which advocates the use of all available data to infer a phylogenetic tree. Instead of combining all available information, a multi-objective approach allows us to manage each type of information as a different objective. Thus, the multi-objective approach of Poladian and Jermiin [76] yields a family of trees instead of the single tree obtained by a combined analysis. The authors employed a basic MOEA where each solution encodes the type of topology of a candidate tree and the length (inferred evolutionary distance between species) for each edge of the tree. In this formulation each objective of the problem corresponds to the maximization of the likelihood of the tree given a type of information. The method was applied to a simple four-species problem using two data sets. The authors concluded that the visual inspection of the resulting Pareto front will help the experienced biological practitioner to interpret the conflict between the data sets and decide a plan of action. Furthermore, with a multi-objective approach the practitioner does not need to determine *a priori* the relative importance of the data.

The inference of gene regulatory networks is other type of inverse problems. Some gene products determine where, when and how much another gene is expressed into proteins. Thus cellular processes like cell growth, differentiation and reproduction are a result of complex interactions between genes instead of an isolated reaction of few genes. Gene regulatory networks are used to represent these interactions between genes using a directed graph. The task of the bioinformatician is to model such networks from large amounts of microarray data.

Spieth et al. [85] address the problem of finding gene regulatory networks using an evolutionary algorithm combined with a local search method. The global optimizer is a genetic algorithm whereas an evolution strategy plays the role of the local optimizer. The performance assessment showed that the proposed memetic algorithm is superior to standard optimization approaches found in the literature.

Recently, Keedwell and Narayanan [49] combined a genetic algorithm with a neural network to elucidate gene regulatory networks. The genetic algorithm has the goal of evolving a population of genes, while the neural network is used to evaluate how well the expression of the set of genes affects the expression values of other genes.

4.5 Future Areas of Research

As we have seen, MOEAs have been applied to different problems in biology and bioinformatics. However, there are other possible paths for future research that may be worth exploring. For example:

- **Use of Hybrid Approaches:** The use of combinations of soft computing[4] techniques for solving multi-objective problems arising in biology may be an interest path of future research in the area. Currently, most applications of soft computing in areas such as bioinformatics, normally rely on the use of a single technique [65] (e.g., artificial neural networks for classification or evolutionary algorithms for optimization). However, the use of combinations of techniques may introduce greater benefits. For example, a MOEA can be used to evolve the topology of an artificial neural network which serves as a classifier, adopting accuracy and complexity as the optimization criteria.

- **Incorporation of User's Preferences:** Most MOEAs are commonly employed under the assumption that the entire Pareto optimal set is needed. However, in most practical applications, not all the solutions are required, since users normally identify regions of interest within the Pareto front [41]. There are several ways in which the user's preferences can be incorporated into a MOEA such that the search is narrowed to a certain portion of the Pareto front (see for example [19]). Although in recent years more MOEA researchers have become interested in this topic (see for example [10, 11, 16, 25, 79, 97]), it certainly requires much further work.

- **Use of Domain Knowledge:** The incorporation of domain knowledge may improve the performance of MOEAs adopted to solve complex problems. Such knowledge may be provided either *a priori* (when available) or can be extracted during the search [48, 54]. This knowledge may influence the operators of a MOEA or can be used to design heuristic procedures aimed to reduce the size of the search space.

4.6 Conclusions

In this chapter, we have explored the use of MOEAs in different biological and bioinformatics applications. First, the most popular MOEAs in current use were briefly described. Then, a simple taxonomy of applications was introduced and representative applications within each class were described. It is worth noting, however, that this review was presented from the perspective

---

[4] Soft computing refers to a collection of computational techniques in computer science which attempt to study, model, and analyze complex phenomena. Such techniques include evolutionary algorithms, neural networks and fuzzy logic [93].

of a computer scientist and not from a biologist's point of view. We hope, however, that biologists may find it useful in spite of its possible pitfalls.

Readers will also note that no attempt was made to be critical in the review, since the aim was to provide a wide view of the field rather than to introduce any potential bias in the current work being done in this area. Clearly, the interest from biologists for using MOEAs is increasing and we certainly hope that such trend is maintained in the years to come, since such has been the main goal of this chapter.

Acknowledgements

The authors thank the anonymous reviewers for their valuable comments, which greatly helped us to improve the contents of this chapter.

The first author acknowledges support from CONACyT through a scholarship to pursue graduate studies at the Computer Science Department from CINVESTAV-IPN. The second author acknowledges support from CONACyT project no. 45683-Y.

References

1. Banerjee M, Mitra S, Banka H (2007) Evolutionary Rough Feature Selection in Gene Expression Data. *IEEE Transactions on Systems, Man, and Cybernetics, Part C: Applications and Reviews*, 37:622–632
2. Banka H, Mitra S (2006) Evolutionary biclustering of gene expressions. *Ubiquity*, 7(42):1–12
3. Barbosa HJ, Barreto AM (2001) An interactive genetic algorithm with co-evolution of weights for multiobjective problems. In: Spector L, Goodman ED, Wu A, Langdon W, Voigt HM, Gen M, Sen S, Dorigo M, Pezeshk S, Garzon MH, Burke E (eds.) *Proceedings of the Genetic and Evolutionary Computation Conference (GECCO'2001)*, 203–210. Morgan Kaufmann Publishers, San Francisco, California
4. Ben-Dor A, Chor B, Karp R, Yakhini Z (2003) Discovering Local Structure in Gene Expression Data: The Order-Preserving Submatrix Problem. *Journal of Computational Biology*, 10(3-4):373–384
5. Beume N, Naujoks B, Emmerich M (2007) SMS-EMOA: Multiobjective selection based on dominated hypervolume. *European Journal of Operational Research*, 181(3):1653–1669
6. Bleuler S, Prelić A, Zitzler E (2004) An EA Framework for Biclustering of Gene Expression Data. In: *Congress on Evolutionary Computation (CEC 2004)*, 166–173. IEEE, Piscataway, NJ
7. Blum C, Roli A (2003) Metaheuristics in combinatorial optimization: Overview and conceptual comparison. *ACM Computing Surveys*, 35(3):268–308
8. Boisson J, Jourdan L, Talbi E, Rolando C (2006) A Preliminary Work on Evolutionary Identification of Protein Variants and New Proteins on Grids. *20th International Conference on Advanced Information Networking and Applications - Volume 2 (AINA'06)*, 02:583–587

9. Boisson J, Jourdan L, Talbi E, Rolando C (2006) Protein Sequencing with an Adaptive Genetic Algorithm from Tandem Mass Spectrometry. In: *IEEE Congress on Evolutionary Computation (CEC), Vancouver, BC, Canada*, 1412–1419

10. Branke J, Deb K (2005) Integrating User Preferences into Evolutionary Multi-Objective Optimization. In: Jin Y (ed.) *Knowledge Incorporation in Evolutionary Computation*, 461–477. Springer, Berlin Heidelberg. ISBN 3-540-22902-7

11. Branke J, Kaußler T, Schmeck H (2001) Guidance in Evolutionary Multi-Objective Optimization. *Advances in Engineering Software*, 32:499–507

12. Calonder M, Bleuler S, Zitzler E (2006) Module Identification from Heterogeneous Biological Data Using Multiobjective Evolutionary Algorithms. In: *Parallel Problem Solving from Nature (PPSN IX)*, no. 4193 in LNCS, 573–582. Springer

13. Chen L, Wu L, Wang R, Wang Y, Zhang S, Zhang X (2005) Comparison of Protein Structures by Multi-Objective Optimization. *Genome Informatics*, 16(2):114–124

14. Cheng Y, Church G (2000) Biclustering of expression data. *Proc Int Conf Intell Syst Mol Biol*, 8:93–103

15. Coello Coello CA (1996) *An Empirical Study of Evolutionary Techniques for Multiobjective Optimization in Engineering Design*. Ph.D. thesis, Department of Computer Science, Tulane University, New Orleans, LA

16. Coello Coello CA (2000) Handling Preferences in Evolutionary Multiobjective Optimization: A Survey. In: *2000 Congress on Evolutionary Computation*, vol. 1, 30–37. IEEE Service Center, Piscataway, New Jersey

17. Coello Coello CA (2000) An Updated Survey of GA-Based Multiobjective Optimization Techniques. *ACM Computing Surveys*, 32(2):109–143

18. Coello Coello CA, Lamont GB (eds.) (2004) *Applications of Multi-Objective Evolutionary Algorithms*. World Scientific, Singapore. ISBN 981-256-106-4

19. Coello Coello CA, Lamont GB, Van Veldhuizen DA (2007) *Evolutionary Algorithms for Solving Multi-Objective Problems*. Kluwer Academic Publishers, New York, second edn. ISBN 978-0-387-33254-3

20. Coello Coello CA, Reyes Sierra M (2003) A Coevolutionary Multi-Objective Evolutionary Algorithm. In: *Proceedings of the 2003 Congress on Evolutionary Computation (CEC'2003)*, vol. 1, 482–489. IEEE Press, Canberra, Australia

21. Coello Coello CA, Toscano Pulido G (2001) Multiobjective Optimization using a Micro-Genetic Algorithm. In: Spector L, Goodman ED, Wu A, Langdon W, Voigt HM, Gen M, Sen S, Dorigo M, Pezeshk S, Garzon MH, Burke E (eds.) *Proceedings of the Genetic and Evolutionary Computation Conference (GECCO'2001)*, 274–282. Morgan Kaufmann Publishers, San Francisco, California

22. Coello Coello CA, Van Veldhuizen DA, Lamont GB (2002) *Evolutionary Algorithms for Solving Multi-Objective Problems*. Kluwer Academic Publishers, New York. ISBN 0-3064-6762-3

23. Corne DW, Jerram NR, Knowles JD, Oates MJ (2001) PESA-II: Region-based Selection in Evolutionary Multiobjective Optimization. In: Spector L, Goodman ED, Wu A, Langdon W, Voigt HM, Gen M, Sen S, Dorigo M, Pezeshk S, Garzon MH, Burke E (eds.) *Proceedings of the Genetic and Evolutionary Computation Conference (GECCO'2001)*, 283–290. Morgan Kaufmann Publishers, San Francisco, California

24. Corne DW, Knowles JD, Oates MJ (2000) The Pareto Envelope-based Selection Algorithm for Multiobjective Optimization. In: Schoenauer M, Deb K, Rudolph G, Yao X, Lutton E, Merelo JJ, Schwefel HP (eds.) *Proceedings of the Parallel Problem Solving from Nature VI Conference*, 839–848. Springer. Lecture Notes in Computer Science No. 1917, Paris, France

25. Cvetković D, Parmee IC (2002) Preferences and their Application in Evolutionary Multiobjective Optimisation. *IEEE Transactions on Evolutionary Computation*, 6(1):42–57

26. Day RO, Zydallis JB, Lamont GB (2002) Solving the Protein structure Prediction Problem through a Multi-Objective Genetic Algorithm. In: *Proceedings of IEEE/DARPA International Conference on Computational Nanoscience (ICCN'02)*, 32–35

27. Deaton R, Chen J, Bi H, Rose J (2002) A software tool for generating non-crosshybridizing libraries of DNA oligonucleotides. *Preliminary Proceedings of the 8th International Meeting on DNA Based Computers, June*, 10–13

28. Deb K (2001) *Multi-Objective Optimization using Evolutionary Algorithms.* John Wiley & Sons, Chichester, UK. ISBN 0-471-87339-X

29. Deb K, Agrawal S, Pratab A, Meyarivan T (2000) A Fast Elitist Non-Dominated Sorting Genetic Algorithm for Multi-Objective Optimization: NSGA-II. In: Schoenauer M, Deb K, Rudolph G, Yao X, Lutton E, Merelo JJ, Schwefel HP (eds.) *Proceedings of the Parallel Problem Solving from Nature VI Conference*, 849–858. Springer. Lecture Notes in Computer Science No. 1917, Paris, France

30. Deb K, Goel T (2001) Controlled elitist non-dominated sorting genetic algorithms for better convergence. *Proceedings of the First International Conference on Evolutionary Multi-Criterion Optimization*, 67–81

31. Deb K, Pratap A, Agarwal S, Meyarivan T (2002) A Fast and Elitist Multi-objective Genetic Algorithm: NSGA–II. *IEEE Transactions on Evolutionary Computation*, 6(2):182–197

32. Deb K, Reddy A (2003) Reliable classification of two-class cancer data using evolutionary algorithms. *BioSystems*, 72(1):111–129

33. Edgeworth FY (1881) *Mathematical Psychics.* P. Keagan, London, England

34. Ehrgott M (2005) *Multicriteria Optimization.* Springer, Berlin, second edn. ISBN 3-540-21398-8

35. Ehrlich P, Raven P (1964) Butterflies and Plants: A Study in Coevolution. *Evolution*, 18:586–608

36. Faulhammer D, Cukras AR, Lipton RJ, Landweber LF (2000) Molecular computation: RNA solutions to chess problems. *Annals of the New York Academy of Sciences*, 97(4):1385–1389

37. Fonseca CM, Fleming PJ (1993) Genetic Algorithms for Multiobjective Optimization: Formulation, Discussion and Generalization. In: Forrest S (ed.) *Proceedings of the Fifth International Conference on Genetic Algorithms*, 416–423. University of Illinois at Urbana-Champaign, Morgan Kaufmann Publishers, San Mateo, California

38. Fourman MP (1985) Compaction of Symbolic Layout using Genetic Algorithms. In: Grefenstette JJ (ed.) *Genetic Algorithms and their Applications: Proceedings of the First International Conference on Genetic Algorithms*, 141–153. Lawrence Erlbaum, Hillsdale, New Jersey

39. Goldberg DE (1989) *Genetic Algorithms in Search, Optimization and Machine Learning.* Addison-Wesley Publishing Company, Reading, Massachusetts

40. Handl J, Kell DB, Knowles J (2007) Multiobjective optimization in bioinformatics and computational biology. *IEEE-ACM Transactions on Computational Biology and Bioinformatics*, 4(2):279–292

41. Handl J, Knowles J (2007) An Evolutionary Approach to Multiobjective Clustering. *IEEE Transactions on Evolutionary Computation*, 11(1):56–76

42. Horn J, Nafpliotis N, Goldberg DE (1994) A Niched Pareto Genetic Algorithm for Multiobjective Optimization. In: *Proceedings of the First IEEE Conference on Evolutionary Computation, IEEE World Congress on Computational Intelligence*, vol. 1, 82–87. IEEE Service Center, Piscataway, New Jersey

43. Hubley R, Zitzler E, Roach J (2003) Evolutionary algorithms for the selection of single nucleotide polymorphisms. *BMC Bioinformatics*, 4(30)

44. Hubley R, Zitzler E, Siegel A, Roach J (2002) Multiobjective Genetic Marker Selection. In: *Advances in Nature-Inspired Computation: The PPSN VII Workshops*, 32–33. University of Reading, UK

45. Ihmels J, Friedlander G, Bergmann S, Sarig O, Ziv Y, Barkai N (2002) Revealing modular organization in the yeast transcriptional network. *Nat Genet*, 31(4):370–7

46. Iorio AW, Li X (2004) A Cooperative Coevolutionary Multiobjective Algorithm Using Non-dominated Sorting. In: et al KD (ed.) *Genetic and Evolutionary Computation–GECCO 2004. Proceedings of the Genetic and Evolutionary Computation Conference. Part I*, 537–548. Springer-Verlag, Lecture Notes in Computer Science Vol. 3102, Seattle, Washington, USA

47. Ito K, Akagi S, Nishikawa M (1983) A Multiobjective Optimization Approach to a Design Problem of Heat Insulation for Thermal Distribution Piping Network Systems. *Journal of Mechanisms, Transmissions and Automation in Design (Transactions of the ASME)*, 105:206–213

48. Jin Y (ed.) (2005) *Knowledge Incorporation in Evolutionary Computation*. Springer, Berlin. ISBN 3-540-22902-7

49. Keedwell E, Narayanan A (2005) Discovering gene networks with a neural-genetic hybrid. *IEEE/ACM Transactions on Computational Biology and Bioinformatics*, 2(3):231–242

50. Keerativuttiumrong N, Chaiyaratana N, Varavithya V (2002) Multi-objective Co-operative Co-evolutionary Genetic Algorithm. In: Merelo Guervós JJ, Adamidis P, Beyer HG, nas JLFV, Schwefel HP (eds.) *Parallel Problem Solving from Nature—PPSN VII*, 288–297. Springer-Verlag. Lecture Notes in Computer Science No. 2439, Granada, Spain

51. Kleeman MP, Lamont GB (2006) Coevolutionary Multi-Objective EAs: The Next Frontier? In: *2006 IEEE Congress on Evolutionary Computation (CEC'2006)*, 6190–6199. IEEE, Vancouver, BC, Canada

52. Knowles JD, Corne DW (1999) The Pareto Archived Evolution Strategy: A New Baseline Algorithm for Multiobjective Optimisation. In: *1999 Congress on Evolutionary Computation*, 98–105. IEEE Service Center, Washington, D.C.

53. Knowles JD, Corne DW (2000) Approximating the Nondominated Front Using the Pareto Archived Evolution Strategy. *Evolutionary Computation*, 8(2):149–172

54. Landa Becerra R, Coello Coello CA (2006) Solving Hard Multiobjective Optimization Problems Using ε-Constraint with Cultured Differential Evolution. In: Runarsson TP, Beyer HG, Burke E, Merelo-Guervós JJ, Whitley LD, Yao X (eds.) *Parallel Problem Solving from Nature - PPSN IX, 9th International*

Conference, 543–552. Springer. Lecture Notes in Computer Science Vol. 4193, Reykjavik, Iceland

55. Lee I, Kim S, Zhang B (2004) Multi-objective Evolutionary Probe Design Based on Thermodynamic Criteria for HPV Detection. *Lecture Notes in Computer Science*, 3157:742–750

56. Lee I, Shin S, Zhang B (2003) DNA sequence optimization using constrained multi-objective evolutionary algorithm. *Evolutionary Computation, 2003. CEC'03. The 2003 Congress on*, 4

57. Liu X, Krishnan A, Mondry A (2005) An Entropy-based gene selection method for cancer classification using microarray data. *BMC Bioinformatics*, 6:76

58. Lohn JD, Kraus WF, Haith GL (2002) Comparing a Coevolutionary Genetic Algorithm for Multiobjective Optimization. In: *Congress on Evolutionary Computation (CEC'2002)*, vol. 2, 1157–1162. IEEE Service Center, Piscataway, New Jersey

59. Madeira S, Oliveira A (2004) Biclustering algorithms for biological data analysis: a survey. *IEEE/ACM Transactions on Computational Biology and Bioinformatics*, 1(1):24–45

60. Malard J, Heredia-Langner A, Baxter D, Jarman K, Cannon W (2004) Constrained de novo peptide identification via multi-objective optimization. *Parallel and Distributed Processing Symposium, 2004. Proceedings. 18th International*

61. Mao J, Hirasawa K, Hu J, Murata J (2001) Genetic Symbiosis Algorithm for Multiobjective Optimization Problems. In: *Proceedings of the 2001 Genetic and Evolutionary Computation Conference. Late-Breaking Papers*, 267–274. San Francisco, California

62. Michaud SR, Zydallis JB, Lamont GB, Pachter R (2001) Scaling a genetic algorithm to medium-sized peptides by detecting secondary structures with an analysis of building blocks. In: *Proceedings of the First International Conference on Computational Nanoscience*, 29–32

63. Miettinen KM (1999) *Nonlinear Multiobjective Optimization*. Kluwer Academic Publishers, Boston, Massachusetts

64. Mitra S, Banka H (2006) Multi-objective evolutionary biclustering of gene expression data. *Pattern Recognition*, 39(12):2464–2477

65. Mitra S, Banka H, Pal S (2006) A MOE framework for Biclustering of Microarray Data. *Proceedings of the 18th International Conference on Pattern Recognition (ICPR'06)-Volume 01*, 1154–1157

66. Moscato P (1989) On Evolution, Search, Optimization, Genetic Algorithms and Martial Arts. Towards Memetic Algorithms. Tech. Rep. 158–79, Caltech Concurrent Computation Program, California Institute of Technology, Pasadena, California

67. Murali T, Kasif S (2003) Extracting conserved gene expression motifs from gene expression data. *Proc. Pacific Symp. Biocomputing*, 8:77–88

68. Nedjah N, Alba E, de Macedo Mourelle L (2006) *Parallel Evolutionary Computations*. Springer-Verlag. ISBN 3-540-32837-8

69. Osyczka A (1985) Multicriteria optimization for engineering design. In: Gero JS (ed.) *Design Optimization*, 193–227. Academic Press

70. Paredis J (1995) Coevolutionary computation. *Artificial Life*, 2(4):355–375

71. Paredis J (1998) Coevolutionary algorithms. In: Bäck T, Fogel DB, Michalewicz Z (eds.) *The Handbook of Evolutionary Computation, 1st Supplement*, 225–238. Institute of Physics Publishing and Oxford University Press

72. Pareto V (1896) *Cours D'Economie Politique*, vol. I and II. F. Rouge, Lausanne
73. Parmee IC, Watson AH (1999) Preliminary Airframe Design Using Co-Evolutionary Multiobjective Genetic Algorithms. In: Banzhaf W, Daida J, Eiben AE, Garzon MH, Honavar V, Jakiela M, Smith RE (eds.) *Proceedings of the Genetic and Evolutionary Computation Conference (GECCO'99)*, vol. 2, 1657–1665. Morgan Kaufmann, San Francisco, California
74. Pimpawat C, Chaiyaratana N (2001) Using a co-operative co-evolutionary genetic algorithm to solve a three-dimensional container loading problem. In: *Proceedings of the Congress on Evolutionary Computation 2001 (CEC'2001)*, vol. 2, 1197–1204. IEEE Service Center, Piscataway, New Jersey
75. Poladian L, Jermiin L (2004) What might evolutionary algorithms (EA) and multi-objective optimization (MOO) contribute to phylogenetics and the total evidence debate? In: et al RP (ed.) *GECCO 2004 Workshop Proceedings*. Seattle, Washington, USA
76. Poladian L, Jermiin L (2006) Multi-objective evolutionary algorithms and phylogenetic inference with multiple data sets. *Soft Computing-A Fusion of Foundations, Methodologies and Applications*, 10(4):359–368
77. Potter MA, de Jong K (1994) A Cooperative Coevolutionary Approach to Function Optimization. In: Davidor Y, Schwefel HP, Männer R (eds.) *Parallel Problem Solving from Nature—PPSN III*, 249–257. Springer-Verlag. Lecture Notes in Computer Science Vol. 866, Jerusalem, Israel
78. Prelić A, Bleuler S, Zimmermann P, Wille A, Bühlmann P, Gruissem W, Hennig L, Thiele L, Zitzler E (2006) A Systematic Comparison and Evaluation of Biclustering Methods for Gene Expression Data. *Bioinformatics*, 22(9):1122–1129
79. Rachmawati L, Srinivasan D (2006) Preference Incorporation in Multi-objective Evolutionary Algorithms: A Survey. In: *2006 IEEE Congress on Evolutionary Computation (CEC'2006)*, 3385–3391. IEEE, Vancouver, BC, Canada
80. Rosenberg RS (1967) *Simulation of genetic populations with biochemical properties.* Ph.D. thesis, University of Michigan, Ann Arbor, Michigan, USA
81. Rosin C, Belew R (1996) New methods for competitive coevolution. *Evolutionary Computation*, 5(1):1–29
82. Schaffer JD (1984) *Multiple Objective Optimization with Vector Evaluated Genetic Algorithms.* Ph.D. thesis, Vanderbilt University, Nashville, Tennessee
83. Schaffer JD (1985) Multiple Objective Optimization with Vector Evaluated Genetic Algorithms. In: *Genetic Algorithms and their Applications: Proceedings of the First International Conference on Genetic Algorithms*, 93–100. Lawrence Erlbaum, Hillsdale, New Jersey
84. Shin S, Lee I, Kim D, Zhang B (2005) Multiobjective Evolutionary Optimization of DNA Sequences for Reliable DNA Computing. *Evolutionary Computation, IEEE Transactions on*, 9(2):143–158
85. Spieth C, Streichert F, Supper J, Speer N, Zell A (2005) Feedback Memetic Algorithms for Modeling Gene Regulatory Networks. *Computational Intelligence in Bioinformatics and Computational Biology, 2005. CIBCB'05. Proceedings of the 2005 IEEE Symposium on*, 1–7
86. Srinivas N, Deb K (1994) Multiobjective Optimization Using Nondominated Sorting in Genetic Algorithms. *Evolutionary Computation*, 2(3):221–248
87. Surry P, Radcliffe N (1997) The COMOGA Method: Constrained Optimisation by Multiobjective Genetic Algorithms. *Control and Cybernetics*, 26(3):391–412

88. Syswerda G, Palmucci J (1991) The Application of Genetic Algorithms to Resource Scheduling. In: Belew RK, Booker LB (eds.) *Proceedings of the Fourth International Conference on Genetic Algorithms*, 502–508. Morgan Kaufmann Publishers, San Mateo, California

89. Tan KC, Yang YJ, Goh CK (2006) A Distributed Cooperative Coevolutionary Algorithm for Multiobjective Optimization. *IEEE Transactions on Evolutionary Computation*, 10(5):527–549

90. Tan KC, Yang YJ, Lee TH (2003) A Distributed Cooperative Coevolutionary Algorithm for Multiobjective Optimization. In: *Proceedings of the 2003 Congress on Evolutionary Computation (CEC'2003)*, vol. 4, 2513–2520. IEEE Press, Canberra, Australia

91. Tanaka F, Nakatsugawa M, Yamamoto M, Shiba T, Ohuchi A (2002) Towards a general-purpose sequence design system in DNA computing. *Evolutionary Computation, 2002. CEC'02. Proceedings of the 2002 Congress on*, 73–78

92. Tanay A, Sharan R, Shamir R (2002) Discovering statistically significant biclusters in gene expression data. *Bioinformatics*, 18(Suppl 1):S136–S144

93. Tettamanzi A, Tomassini M (2001) *Soft Computing: Integrating Evolutionary, Neural and Fuzzy Systems*. Springer-Verlag. ISBN 978-3540422044

94. Toscano Pulido G, Coello Coello CA (2003) The Micro Genetic Algorithm 2: Towards Online Adaptation in Evolutionary Multiobjective Optimization. In: Fonseca CM, Fleming PJ, Zitzler E, Deb K, Thiele L (eds.) *Evolutionary Multi-Criterion Optimization. Second International Conference, EMO 2003*, 252–266. Springer. Lecture Notes in Computer Science. Volume 2632, Faro, Portugal

95. Veldhuizen DAV, Zydallis JB, Lamont GB (2003) Considerations in engineering parallel multiobjective evolutionary algorithms. *IEEE Transactions on Evolutionary Computation*, 7(3):144–173

96. Villalobos-Arias M, Coello Coello CA, Hernández-Lerma O (2006) Asymptotic Convergence of Metaheuristics for Multiobjective Optimization Problems. *Soft Computing*, 10(11):1001–1005

97. Wang J, Terpenny JP (2005) Interactive Preference Incorporation in Evolutionary Engineering Design. In: Jin Y (ed.) *Knowledge Incorporation in Evolutionary Computation*, 525–543. Springer, Berlin Heidelberg. ISBN 3-540-22902-7

98. Wienke PB, Lucasius C, Kateman G (1992) Multicriteria target optimization of analytical procedures using a genetic algorithm. *Analytical Chimica Acta*, 265(2):211–225

99. Yang J, Wang H, Wang W, Yu P (2003) Enhanced biclustering on expression data. *Bioinformatics and Bioengineering, 2003. Proceedings. Third IEEE Symposium on*, 00:321–327

100. Zeng SY, Kang LS, Ding LX (2004) An Orthogonal Multi-objective Evolutionary Algorithm for Multi-objective Optimization Problems with Constraints. *Evolutionary Computation*, 12(1):77–98

101. Zhang Z, Teo A, Ooi B, Tan K (2004) Mining deterministic biclusters in gene expression data. *Bioinformatics and Bioengineering, 2004. BIBE 2004. Proceedings. Fourth IEEE Symposium on*, 283–290

102. Zitzler E, Künzli S (2004) Indicator-based Selection in Multiobjective Search. In: et al XY (ed.) *Parallel Problem Solving from Nature - PPSN VIII*, 832–842. Springer-Verlag. Lecture Notes in Computer Science Vol. 3242, Birmingham, UK

103. Zitzler E, Laumanns M, Thiele L (2001) SPEA2: Improving the Strength Pareto Evolutionary Algorithm. In: Giannakoglou K, Tsahalis D, Periaux J, Papailou P, Fogarty T (eds.) *EUROGEN 2001. Evolutionary Methods for Design, Optimization and Control with Applications to Industrial Problems*, 95–100. Athens, Greece
104. Zitzler E, Thiele L (1999) Multiobjective Evolutionary Algorithms: A Comparative Case Study and the Strength Pareto Approach. *IEEE Transactions on Evolutionary Computation*, 3(4):257–271
105. Zitzler E, Thiele L, Laumanns M, Fonseca CM, da Fonseca VG (2003) Performance Assessment of Multiobjective Optimizers: An Analysis and Review. *IEEE Transactions on Evolutionary Computation*, 7(2):117–132
106. Zwir I, Zaliz R, Ruspini E (2002) Automated Biological Sequence Description by Genetic Multiobjective Generalized Clustering. *Annals of the New York Academy of Sciences*, 980(1):65–82
107. Zydallis JB, Veldhuizen DAV, Lamont GB (2001) A Statistical Comparison of Multiobjective Evolutionary Algorithms Including the MOMGA–II. In: Zitzler E, Deb K, Thiele L, Coello CAC, Corne D (eds.) *First International Conference on Evolutionary Multi-Criterion Optimization*, 226–240. Springer-Verlag. Lecture Notes in Computer Science No. 1993

Part II

Current Trends

5

Local Classifiers as a Method of Analysing and Classifying Signals

Wit Jakuczun

Nencki Institute of Experimental Biology, 3 Pasteur Str., 02-093 Warsaw, Poland
WLOG Solutions, 1A/25 Harfowa Str., 02-389 Warsaw, Poland
W.Jakuczun@wlogsolutions.com

Summary. Biological sciences very often deal with data that are measured in consecutive time periods. Typical examples are EEG, ECG. In mathematical language such data are called signals. Classical methods of analysing and classifying data, like decision trees, are not suitable for signals as they ignore time nature of data.

We propose a novel method, called Local Classifiers, for analysing and classifying signals. The method is a genuine combination of wavelets and Support Vector Machines. On a referential data sets the method proved to be competitive as far as accuracy is concerned to other state-of-the-art methods. We also presented an application of the method to biological data set. The goal of the experiment was to study whether habituated and aroused states can be differentiated in single barrel column of rat's somatosensory cortex by means of analysis of field potentials evoked by stimulation of a single vibrissa. The method proved to be a reliable approach to automatically detection of important parts of local field potentials as far as discrimination between two states of a brain are concerned. The results confirmed previous biological hypothesis.

5.1 Introduction

The continuous process of a technical development results in increasing numbers of available data that are collected by scientists. This increase in the numbers of data creates a need for advanced statistical tools that could be used for analysis while the increase of the computational power of computers opens new opportunities to analysts.

In this chapter, we present a new method for signal analysis and classification. The method combines two approaches to data analysis: 1) quite a new method *lifting scheme* [1], [2] for constructing *wavelets* [3] in *time domain* and 2) *state-of-the-art* classifiers like *Support Vector Machines (SVM)* [4], *Arcing* [5] and its combination. The method is aimed to be one of the *off-the-shelf* methods. This means that the method should be easy to use for non-advanced users while being quite effective.

W. Jakuczun: *Local Classifiers as a Method of Analysing and Classifying Signals*, Studies in Computational Intelligence (SCI) **122**, 105–133 (2008)
www.springerlink.com © Springer-Verlag Berlin Heidelberg 2008

The reader will find both a theoretical basis of the method and examples of the usage. To verify the method a number of experiments were conducted using real and artificial data sets.

5.1.1 Chapter layout

The chapter is divided into six sections. The first section is an introductory section. Sections from the second to the fourth are sections that contain description of the method. Section 2 contains preliminary information with some basic definitions. Section 3 is devoted to description of statistical methods for pattern recognition that were used in the experiments. The description of the method of local classifiers can be found in section 4. The fifth section contains descriptions of the experiments. The last section is devoted to conclusions.

It is not necessary to read the whole chapter to get an idea of the method. We recommend all potentials readers to read sections 1 and 6. If the reader is not interested in the mathematical details of the method they could skip sections from 2 to 4 and concentrate on section 5. Readers familiar with statistical learning can get a quick idea of the method by reading section 4. For the readers that do not feel strong in statistical methods for pattern recognition we recommend reading section 3. To fully understand the method, readers should read the whole chapter.

5.1.2 Other approaches

The method presented in this chapter is not the only method in which the idea of reasoning based on local parts of the signal was used. In literature, we can find a few methods with a similar approach.

The method that is the most similar to the presented one is the method of *interval based literals*. It is described in [6], [7] and in [8]. The method basically creates human readable literals (like "mean of the part of the signal > threshold") for small intervals that the signal is divided into. This method is one of the best methods for classifying signals according to our current knowledge.

Another method that is similar to ours is the method presented in [9]. This method explicitly uses *wavelet* bases for constructing discriminative features. Unfortunately, it has not been thoroughly tested on a variety of data sets.

Among other methods that could be applied for classifying signals we should mention two methods: *TClass* [10] and *Dynamic Time Warping* [11]. The first method is based on features that change in time. This method, similarly to *interval based literals* gives comprehensible classification rules. The second approach belongs to a completely different approach to the classification problem. This approach is based on the construction of a special distance function that measures the similarity between a new example and examples in the training set. Very interesting extensions of the method can be found in [12] and [13].

5.2 Preliminaries

In this section we introduce basic notations and definitions. The definitions and notations are valid throughout the whole chapter unless clearly stated.

First, we define the *multi-channel signal* . This is the type of data the method was constructed for.

Definition 1. *By* multi-channel signal $\mathbf{x}$ *we understand a vector time series* $\mathbf{x} = \{\mathbf{x}_t = (x_{t,1}, \ldots, x_{t,n})^{'} \in \mathbb{R}^n : t \in \mathbb{Z}\} \in \mathbb{R}^{n \times \mathbb{Z}}$. *Index* $t \in \mathbb{Z}$ *is called time index and* $n \in \mathbb{N}$ *defines the dimensionality of the time series. Each component* $\mathbf{x}_i = (\mathbf{x}_{t,i})^{'}_{t \in \mathbb{Z}}$, $i = 1, 2, \ldots, n$, *we call a* channel.

Remark 1. Multi-channel signals very often arise in experimental sciences such as biology. A typical example of the multi-channel signal is *ECG* signal. For each time period, we get a vector of measurements from each electrode. One channel corresponds to the measurements recorded from one electrode.

Remark 2. We use the name *multi-channel signal* for the object that in statistics is known as *multi-dimensional time series*. In the chapter we will use the two names interchangeably.

Remark 3. In practice, we deal only with finite signals. By this we mean both finite length and that the signal was digitalised. We can assume without loss of generality that the time index t belongs to the set $\mathcal{T} = \{1, 2, \ldots, T\}$. This allows us to treat multi-channel signals as matrices of dimension $n \times T$, and each channel as a vector of dimension T.

Remark 4. For reasons of simplicity, we assume that $T = 2^p$ for some $p \in \mathbb{N}$, that is the length of time series is a power of 2. This constraint can be easily relaxed, but the relaxation results in more complicated calculations we would like to avoid in this chapter.

Very often the time series we are given represent different physical objects or different states of the same object. Moreover the data we have represent only a sample of all possible states/objects. The different states/objects are called *decision classes*. Having such data, we would like to extract some general rules that would lead to an accurate description of the new states/objects that we were not given in the original data set.

Definition 2. *By an* example *we understand a pair*

$$\mathbf{z} = (\mathbf{x}, y) \in \mathbb{R}^{n \times T} \times \mathcal{Y} \ ,$$

where $\mathbf{x}$ *is a multi-dimensional time series of the length* T *and the number of channels* $n \in \mathbb{N}$ *and* $y \in \mathcal{Y} = \{1, 2, \ldots, C\}$[1] *is the label that indicates the*

---

[1] In general set $\mathcal{Y}$ can be any finite set, not necessarily containing natural numbers. On the other hand, as it is finite, it can always be represented as a set of a sequence of natural numbers.

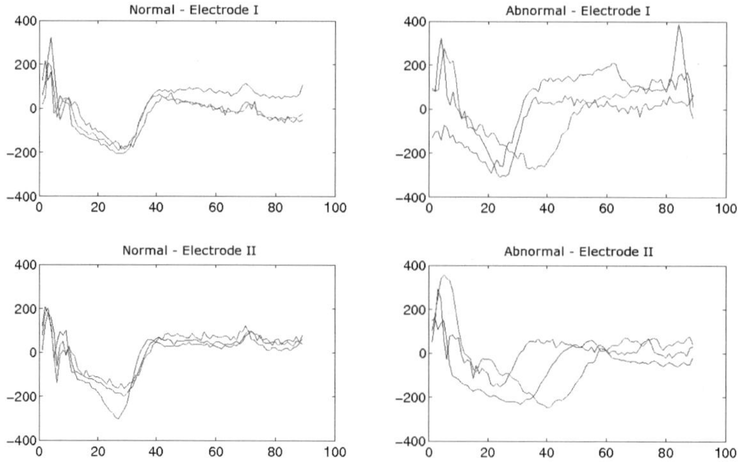

Fig. 5.1. Example of *multi-channel signal*. This is *ECG* signal taken from [14]. There are two electrodes: Electrode I, Electrode II and two decision classes: Normal and Abnormal.

decision class to which the signal **x** *belongs. The set* $\mathcal{Y}$ *is given a priori and its elements encode the decision classes that analysed signals belongs to.*

By $\mathbf{z}_i, i = 1, 2, \ldots, n$ *we will denote a pair*

$$\mathbf{z}_i = (\mathbf{x}_i, y_i) \in \mathbb{R}^T \times \mathcal{Y} \ .$$

The $\mathbf{z}_i$ *consists of only one channel (i-th channel) of the time series.*

Remark 5. If the result of an experiment is a set of time series, then each time series is an *example* according to our notation.

As we have mentioned above the definition of decision classes must be given a priori. In practice, the classes emerge very naturally from the hypothesis we are trying to verify. A typical situation is when we are trying to compare two groups of animals: healthy and unhealthy.

Example 1. In Figure 5.1 we have presented exemplary *multi-channel signals* that were first introduced in [14]. This data set consists of two channel signals that represent two different group of patients: Normal and Abnormal. In our notation, measurements from each electrode is time series **x** and there are two decision classes: Normal and Abnormal that the signals belong to. In practice we usually encode decision classes as integer number, so the set $\mathcal{Y}$ can be defined as $\mathcal{Y} = \{1, 2\}$ where 1 encodes the Normal decision class and 2 encoded the Abnormal decision class.

The aim of this experiment was to create a good discriminating rule/function that would with high accuracy assign a new time series **x** to one of the two decision classes.

As the number of time series belonging to each decision class can be infinite we are only dealing with a finite sample of them. We call the sample we are using for estimating parameters of the method the *training set*. The precise definition is below.

Definition 3. *By a* training set *we understand a set of l examples*

$$\mathcal{Z} = \left\{ \mathbf{z}^k = \left(\mathbf{x}^k, y \right) \in \mathbb{R}^{n \times T} \times \mathcal{Y} \colon k = 1, 2, \ldots, l \right\}$$

By $\mathcal{Z}_i, i = 1, 2, \ldots, n$ we denote a set

$$\mathcal{Z}_i = \left\{ \mathbf{z}_i = \left(\mathbf{x}_i^k, y \right) \in \mathbb{R}^T \times \mathcal{Y} \colon k = 1, 2, \ldots, l \right\}$$

Definition 4. *From set $\mathcal{Z}_i$ we create a* data matrix $\mathbf{X}_i$ *as follows*

$$\mathbf{X}_i = \begin{bmatrix} \left(\mathbf{x}_i^1 \right)' \\ \vdots \\ \left(\mathbf{x}_i^l \right)' \end{bmatrix} \in \mathbb{R}^{l \times T}$$

The data matrix *is a mathematical representation of the data and is used for estimating method parameters (see Section 5.4.3).*

Definition 5. *For any vector $\mathbf{x} \in \mathbb{R}^T$ by $\mathbf{x}_o$ we understand a vector consisting of* odd *indices*

$$\mathbf{x}_o = \left\{ \mathbf{x}_t \colon t \in \{1, \ldots, T\}, \ t \text{ is odd} \right\} \in \mathbb{R}^{\frac{T}{2}}$$

and by $\mathbf{x}_e$ we understand a vector consisting of even *indices*

$$\mathbf{x}_e = \left\{ \mathbf{x}_t \colon t \in \{1, 2, \ldots, T\}, t \text{ is even} \right\} \in \mathbb{R}^{\frac{T}{2}}$$

We can straightforwardly extend this notation for the matrix $\mathbf{X} \in \mathbb{R}^{l \times T}$ to get the matrices $\mathbf{X}_o \in \mathbb{R}^{l \times \frac{T}{2}}$ and $\mathbf{X}_e \in \mathbb{R}^{l \times \frac{T}{2}}$ consisting respectively of odd and even columns of the matrix $\mathbf{X}$.

Definition 6. *For any matrix $\mathbf{X}$ by $\mathbf{X}_{\cdot,m}$ we will denote its m-th column and by $\mathbf{X}_{n,\cdot}$ its n-th row.*

5.3 Statistical methods for pattern recognition

In this section, we present two statistical methods and their combination that will be used in the construction of the local classifiers method. The methods are:

- Proximal Support Vector Machines (PSVM),
- Arcing and Arcing for binary classifiers,
- Arcing for creating *discriminative* features.

5.3.1 Basic definitions

In this subsection we introduce shortly the most important definitions and techniques related to *statistical learning* and *machine learning* [4], [15].

The main assumption of *statistical learning* is that we deal with a finite sample $\mathcal{Z}$ generated randomly, according to some random law Ω, from a set $\mathbb{Z} = \mathbb{X} \times \mathcal{Y}$ of all possible *examples*.

As we mentioned in Section 5.2, each *example* is a pair consisting of a vector of *attributes*[2] describing the physical object and a label that indicates the decision class this object belongs to.

The problem we are dealing in this chapter is called *supervised learning*. It differs from *unsupervised learning* in the fact that in the latter we are not provided with the decision classes. The aim of *supervised learning* is to build a function called *classifier* that calculates the decision class for new examples.

It is very important to understand the fact that we are constructing the *classifier* using only a finite sample $\mathcal{Z}$ called the *training set*. The way we create the classifier is to minimise erratic assignments of the *new* examples to the decision classes. By *new* example we understand examples that were not used in the phase of creation of the *classifier*.

Below we provide mathematical definitions for the concepts mentioned above.

Definition 7. *By* $\mathbb{Z} = \mathbb{X} \times \mathcal{Y}$ *we understand the set of all examples.*

Definition 8. *By* Ω *we understand the probabilistic measure on the set* $\mathbb{X}$.

Definition 9. *By* sample $\mathcal{Z}$ *we understand a finite set randomly generated according to the probabilistic measure* Ω *from the set* $\mathbb{Z}$.

Definition 10. *A* classifier *is any measurable function* $f : \mathbb{X} \rightarrow \mathcal{Y}$.

Definition 11. *The* sample (empirical) error *of the classifier f for the sample* $\mathcal{Z}$ *is the number of misclassified* examples *from the set*

$$\mathrm{err}_{emp}(f, \mathcal{Z}) = \frac{|\{\mathbf{z} = (\mathbf{x}, y) \in \mathcal{Z} : f(\mathbf{x}) \neq y\}|}{|\mathcal{Z}|} .$$

Definition 12. *The* real classification error *of the classifier f is defined as follows*

$$\mathrm{err}_{real}(f) = \mathrm{P}_{\Omega}(\{\mathbf{z} = (\mathbf{x}, y) \in \mathbb{Z} : f(\mathbf{x}) \neq y\}) .$$

---

[2] In the case of multidimensional signals attributes are measurements.

Model selection and validation

Many methods for generating classifiers have parameters[3]. By changing the parameters we get different classifiers, although from the same class. The method of *local classifiers* belongs to this group of methods.

The natural question arises about the best parameters. It is clear that the best parameters are the ones that minimise the *real classification error*. Unfortunately we are unable to calculate the error as we are given only a finite sample of all examples.

We will describe the two most common approaches of *selecting the best model*. One is called *hold-out* method and the second is *cross validation*. For more comprehensive discussion on the subject we refer reader to [15].

Hold-out method

If the training set $\mathcal{Z}$ contains many examples we can divide it randomly (or not) into two subsets: $\mathcal{Z}^{trn}$ and $\mathcal{Z}^{tst}$. For parameters estimation we use only the $\mathcal{Z}^{trn}$ set. We choose the parameters that give the lowest error for examples from set $\mathcal{Z}^{tst}$.

We will denote *Hold-Out* method as $\mathrm{HO}_{det/rnd}\left(\frac{n}{m}\right)$, where n is a cardinality of the set $\mathcal{Z}^{tst}$ and m is the cardinality of the set $\mathcal{Z}^{trn}$[4]. The subscript denotes whether the splitting of the training set was done randomly (rnd) or deterministically (det).

Cross-Validation (CV)

The *method of Cross-Validation*, denoted as $\mathrm{CV}-K$, divides the set $\mathcal{Z}$ into K subsets $\mathcal{Z}^1, \ldots, \mathcal{Z}^K$ of approximately the same number of examples. There is a variant of cross-validation called *stratified crossvalidation* that tries to divide the set $\mathcal{Z}$ in such a way that the fraction of decision classes in each subset $\mathcal{Z}^k, k = 1, \ldots, K$ is approximately the same as in the original set $\mathcal{Z}$.

Having sets $\mathcal{Z}^1, \ldots, \mathcal{Z}^K$ we proceed iteratively. For each $k = 1, \ldots, K$ we estimate classifier's parameters using the examples from the set $\mathcal{Z}^{trn,k} = \mathcal{Z}^1 \cup \ldots \cup \mathcal{Z}^{k-1} \cup \mathcal{Z}^{k+1} \cup \ldots \cup \mathcal{Z}^K$ and evaluate it on the set $\mathcal{Z}^k$. The final error estimation is the average of the K errors for sets $\mathcal{Z}^1, \ldots, \mathcal{Z}^K$.

In both cases we retrain the model on the whole training set with the best model's parameters. Such a model is ready for being used for classifying new examples.

Remark 6. In the case $K = |\mathcal{Z}|$, where $|\cdot|$ is a cardinality of a set, the cross-validation is given a special name *leave on out (LOO)*. This variant gives an *almost* unbiased estimator of *real classification error* and is recommended in the situation the power of the set $\mathcal{Z}$ is small.

---

[3] Please note that method's parameters are not the same as model's parameters. Model's parameters are in fact a function of method's parameters.

[4] It is clear that $m + n = |\mathcal{Z}|$.

Remark 7. It is *highly recommended* to repeat any of the methods a few times and make decisions on the averaged results (using the standard deviation). Such repetition reduces the variance of the results.

5.3.2 Proximal Support Vector Machines

In this subsection we introduce the Proximal Support Vector Machines (PSVM). We start with a general idea behind SVM classifiers and than we present the PSVM method.

Support Vector Machines

The *PSVM* is a variant of widely known, state-of-the-art method of *Support Vector Machines* brought to a wide audience by Vapnik [4]. Both methods are based on the regularisation theory of Tchikhonov and on the theory of *Reproducing Kernel Hilber Spaces (RKHS)* [4].

As before we are given a training set $\mathcal{Z}$, but with the restriction that the $\mathcal{Y} = \{-1, +1\}$. Additionally we are given a function $\phi\colon \mathbb{X} \to \mathbb{R}^s$, where s can be infimum. The function ϕ is called *feature map*. Its role is to transform the original description of the objects into some, possibly infinite, space $\mathbb{R}^s$. This mapping plays a very important role in the *SVM* method as it allows treatment of nonlinear problems as if they were linear. The idea is depicted in Figure 5.2. The assumption that we can separate objects from different decision classes[5] by *hyperspace* is reflected in the definition of the *Support Vector Machine*[6]:

$$f_{SVM}(\mathbf{x}) = \theta\left(h_{SVM}(\mathbf{x})\right) ,$$

where $h_{SVM}(x) = \langle \phi(\mathbf{x}), \mathbf{w} \rangle + b$ is a function that defines a separating hyperspace and $\mathbf{w} \in \mathbb{R}^s$ and b are parameters of the classifier and the function $\theta(\cdot)$ is defined as follows

$$\theta(u) = \begin{cases} -1 \text{ for } u < 0 \\ +1 \text{ for } u \geq 0 \end{cases}.$$

The above definition of classifier f_{SVM} raises two questions:

- What to do when the dimensionality of $\mathbb{R}^s$ is infinite?
- What are good *feature maps*?

The answer to both questions gives the theory of *Reproducing Kernel Hilbert Spaces (RKHS)*. For more details on *RKHS* and its application to *SVM* we refer reader to [4].

Definition 13. *By* Kernel Function *we will understand the function* $k\colon \mathbb{X} \times \mathbb{X} \to \mathbb{R}$ *that has the following properties*

---

[5] Remember that we assume that there are two decision classes
[6] We use $\langle \cdot, \cdot \rangle$ to denote the standard euclidean *inner-product* in $\mathbb{R}^s$

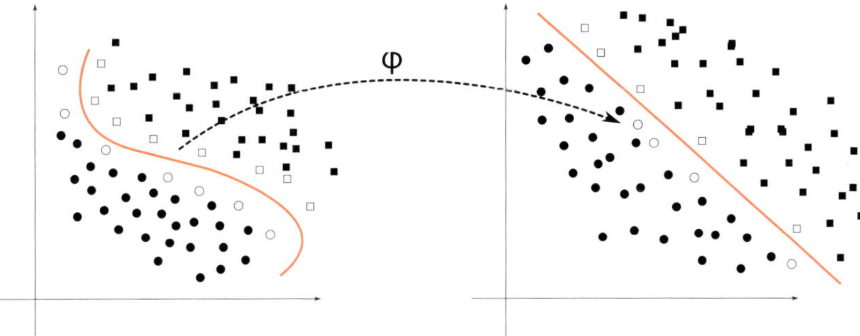

Fig. 5.2. The picture represents the idea of using *feature map* ϕ. The circles and squares represent objects from two different decision classes. The solid line represents the separating curve between the two decision classes. On the left there is the original space, on the right the mapped space. The empty shapes are called supporting vectors. Please note that the separating curve in the original space is the separating hyperspace in the mapped space.

- $\mathrm{K}(\mathbf{x}, \tilde{\mathbf{x}}) \geq 0$ *and the equality takes place only if* $\mathbf{x} = \tilde{\mathbf{x}}$,
- *There exists a feature map* ϕ_{K} *that is connected with the kernel function* k *by the following relation*

$$\mathrm{K}(\mathbf{x}, \tilde{\mathbf{x}}) = \langle \phi(\mathbf{x}), \phi(\tilde{\mathbf{x}}) \rangle_{\mathrm{K}} \ ,$$

where $\langle \cdot, \cdot \rangle_{\mathrm{K}}$ *is an inner product for the RKHS with* K *as the kernel.*

From the above definition it seems that *Kernel Function* is a generalisation of the inner product.

Corollary 1. *The vector* $\mathbf{w}$ *of parameters of* h_{SVM} *is given by the following equation (for the proof see [16])*

$$\mathbf{w} = \sum_{i=1}^{l} \alpha^i \phi \left(\mathbf{x}^i \right) \ ,$$

where $\alpha^i, i = 1, \ldots, l$ *are the coefficients calculated to minimise the training error. The vectors* $\mathbf{x}^i$ *for which* $\alpha^i \neq 0$ *are called* support vectors. *They are denoted as not filled shapes in Figure 5.2.*

The immediate conclusion from the corollary is that the function h_{SVM} can be rewritten in the following form

$$h_{SVM}(\mathbf{x}) = \sum_{i=1}^{l} \alpha^i \left\langle \phi \left(\mathbf{x}^i \right), \phi (\mathbf{x}) \right\rangle + b = \sum_{i=1}^{l} \alpha^i \mathrm{K} \left(\mathbf{x}^i, \mathbf{x} \right) + b \qquad (5.1)$$

The last equation shows the importance of the *kernel function*. Instead of calculating inner products in $\mathbb{R}^s$ with s being very often a huge number it is enough to calculate *kernel function* K which arguments belong to $\mathbb{X}$.

It is not an easy task to find a kernel function K for a given feature map ϕ and vice versa. The most widely used kernels are

- Linear kernel

$$\mathrm{K}_{lin}(\mathbf{x}, \tilde{\mathbf{x}}) = \langle \mathbf{x}, \tilde{\mathbf{x}} \rangle$$

  The feature map connected with this kernel is identity.
- Polynomial kernel

$$\mathrm{K}_{poly}^{c,d}(\mathbf{x}, \tilde{\mathbf{x}}) = (\langle \mathbf{x}, \tilde{\mathbf{x}} \rangle + c)^d \ ,$$

  where $c \geq 0$ and $d \in \mathbb{Z}$ are the method's parameters. The feature map connected with this kernel is the polynomial of degree d.
- Gaussian kernel

$$\mathrm{K}_{gauss}^{\gamma}(\mathbf{x}, \tilde{\mathbf{x}}) = \exp\left(-\gamma \|\mathbf{x} - \tilde{\mathbf{x}}\|_2^2\right) \ ,$$

  where $\gamma \geq 0$ is a method's parameter and the $\| \cdot \|_2$ is the euclidean length of the vector.

Remark 8. Choosing both kernel function and its parameters must be conducted via experiments with the given data set using one of the methods of model selection and validation. It is advised to *always* tune the parameters of the kernels.

Proximal Support Vector Machines (PSVM)

In this subsection we present the PSVM method. The method was first introduced in [17]. The method is similar to the well known method for regression called *ridge regression* [15].

Let $\mathbf{K} = \left[\mathrm{K}(\mathbf{x}^i, \mathbf{x}^j)\right]_{i,j=1,\ldots,l}$ where $\mathbf{x}^i$ and $\mathbf{x}^j$ are *examples* from the training set. The matrix $\mathbf{K} \in \mathbb{R}^{l \times l}$ is called *kernel matrix*. Let $\mathbf{Y} \in \mathbb{R}^{l \times l}$ be a diagonal matrix with i-th diagonal entry equal to ± 1 according to the decision class of the example $\mathbf{x}^i$. Moreover we define vector $\mathbf{e} \in \mathbb{R}^l$ with all entries equal to 1.

The optimal model's parameters $\boldsymbol{\alpha} = [\alpha_1, \ldots, \alpha_l]' \in \mathbb{R}^l$ and $b \in \mathbb{R}$ are defined by the following optimisation problem

$$\min_{\boldsymbol{\alpha} \in \mathbb{R}^l, b \in \mathbb{R}} \frac{1}{2}\left(\|\boldsymbol{\alpha}\|_2^2 + b^2\right) + \frac{\nu}{2}\|\boldsymbol{\psi}\|_2^2 \ , \tag{5.2}$$

subject to

$$\mathbf{Y}\left(\mathbf{K}\boldsymbol{\alpha} + b\mathbf{e}\right) - \boldsymbol{\psi} = \mathbf{e} \ , \tag{5.3}$$

where $\nu \geq 0$ is a *PSVM* method's parameter and $\| \cdot \|_2$ is a standard euclidean norm. Through ν how much the method fits the given training sample can be controlled. The lesser the ν the more fitting is done. The optimal value for ν should be estimated using one of the methods described in Section 5.3.1.

Let us define matrix $\mathbf{H} = \mathbf{Y}\,[\mathbf{K}\ \mathbf{e}]$ and vector $\tilde{\boldsymbol{\alpha}} = \begin{bmatrix} \boldsymbol{\alpha} \\ b \end{bmatrix} \in \mathbb{R}^{l+1}$. Now, the problem (5.2) and (5.3) can be defined as follows

$$\min_{\tilde{\boldsymbol{\alpha}}=[\boldsymbol{\alpha}\ b]'\in\mathbb{R}^{l+1}} \|\boldsymbol{\alpha}\|_2^2 + \frac{\nu}{2}\|\mathbf{H}\tilde{\boldsymbol{\alpha}} - \mathbf{e}\|_2^2 \ . \tag{5.4}$$

The reader familiar with statistical methods should recognise that (5.4) is equivalent to *ridge regression* applied to the regression problem with data matrix $\mathbf{H}$ and vector $\mathbf{e}$ as a right hand side.

In Figure 5.3 we present the idea of the *PSVM* method. The method tries to approximate examples from two decision classes with two hyperspaces while keeping the hyperspaces as far from each other as possible.

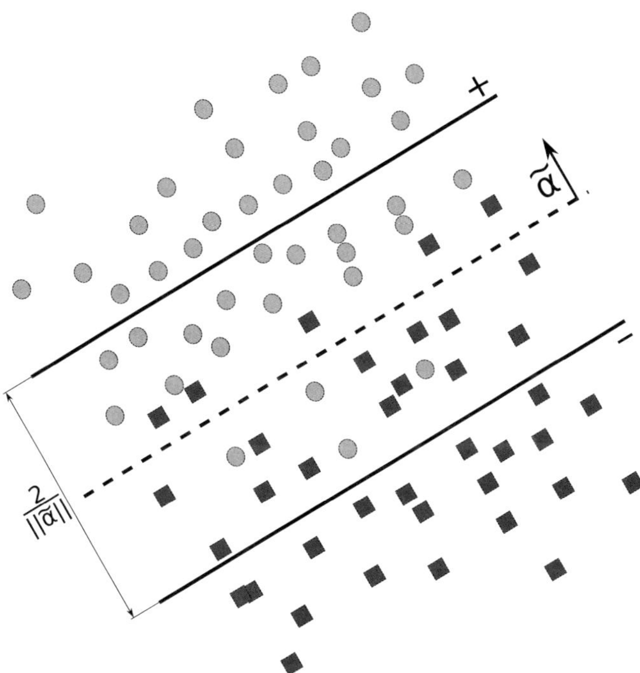

Fig. 5.3. The idea of PSVM method. The lines denoted as + and - are the regression lines for two classes (denoted by squares and circles). The dotted line defines the decision rule. This picture presents the situation for $\mathbb{R}^2$ and linear kernel.

The vector $\tilde{\boldsymbol{\alpha}}$ that is a solution to (5.4) is given by the following formula:

$$\tilde{\boldsymbol{\alpha}} = \left(\frac{1}{\nu}\mathbf{I} + \mathbf{H}'\mathbf{H}\right)^{-1}\mathbf{H}'\mathbf{b} \tag{5.5}$$

From (5.5) it is clear that PSVM is very simple and easy to implement. Other variants of SVM require solving either quadratic or linear programming problems. PSVM needs only solving system of linear equations. This was the main reason for choosing this particular variant of SVM for our method of local classifiers.

SVM for non-binary problems

So far we have assumed that the number of decision classes is equal to 2. Unfortunately in practice it is a rare case. The good thing is that having binary classifiers such problems cen be dealt with by creating a set of binary sub-problems.

There are two most popular schemes: *One-Versus-All* and *All-Versus-All*. In both schemes a number of binary classifiers is created for a modified training set.

In the case of *One-Versus-All* we build $|\mathcal{Y}|$ binary classifiers. Each classifier is given the original data with transformed decision classes. One class represents selected decision class and the second class represents other decision classes. To classify a new example we choose the class that corresponds to the classifier with the strongest response. By response, we understand the value of the function h_{SVM}.

Scheme *All-Versus-All* is very similar to the scheme presented above. The difference is that we construct $\frac{|\mathcal{Y}|(|\mathcal{Y}|-1)}{2}$ classifiers. Each classifier is given a subset of training sample consisting of examples belonging only to two decision classes. Classification of a new example is based on a voting scheme. The decision class that is assigned to the new example is the one that was most frequently pointed by all classifiers.

5.3.3 Arcing

This subsection is devoted to the description of one of the most interesting and effective methods of creating *ensemble of classifiers* called *Arcing*[7]. The method was first introduced in [5].

We assume that we are given an algorithm $\mathcal{A}$ that can produce classifiers using a training sample $\mathcal{Z} = \{\mathbf{z}^1, \ldots \mathbf{z}^l\}$, the natural number J and a positive number μ. The Arcing method is as follows

1. Let $\mathbf{m} \in \mathbb{R}^l$ be a vector of l zeros.
2. For $j = 1, 2, \ldots J$ do the following
   a) Let $\mathbf{w}_i^j = (1 + \mathbf{m}_i)^\mu$ for $i = 1, 2, \ldots, l$. Normalise vector $\mathbf{w}^j$ so that $\|\mathbf{w}^j\|_1 = \sum_{i=1}^l \mathbf{w}_i^j = 1$.
   b) Generate a copy $\mathcal{Z}^j$ of the training sample. The copy is generated by sampling the set $\mathcal{Z}$ with replacement according to the weight vector $\mathbf{w}^j$.
   c) Build a classifier f^j using the algorithm $\mathcal{A}$ and the set $\mathcal{Z}^j$.

---

[7] *Arcing* is acronym from **A**daptively **R**esample and **C**ombine

   d) For each $i = 1, \ldots, l$ increase $\mathbf{m}_i$ by one if i-th example $\mathbf{z}^i$ is wrongly classified by the classifier f^j.

3. The output of the method is a set of classifiers $\mathcal{F} = \{f^1, f^2, \ldots, f^J\}$.

Having the set $\mathcal{F}$ we classify a new example $\mathbf{z} = (\mathbf{x}, ?)$ using the classical voting approach. In this approach outputs of classifiers $f^j(\mathbf{x})$ are aggregated and the decision class that was most often indicated is assigned to the example $\mathbf{z}$.

Arcing for *binary* classifiers

The classical Arcing as presented above was primarily created for classifiers that can deal with more than two decision classes. To use this approach for classifiers that are limited only to two decision class problems, like SVM and its variants, we needed to modify it.[8]

The modified algorithm is as follows

1. Let $\mathbf{M} \in \mathbb{R}^{l \times |\mathcal{Y}|}$ be a matrix with elements equal to 0.
2. For $j = 1, 2, \ldots J$ do the following

   a) Let $\mathbf{w}_i^j = \left(1 + \sum_{c=1}^{|\mathcal{Y}|} \mathbf{M}_{i,c}\right)^{\mu}$ for $i = 1, 2, \ldots, l$. Normalise vector $\mathbf{w}^j$ so that $\|\mathbf{w}^j\|_1 = \sum_{i=1}^{l} \mathbf{w}_i^j = 1$.

   b) Generate a copy $\mathcal{Z}^j$ of the training sample. The copy is generated by sampling the set $\mathcal{Z}$ with replacement according to the weight vector $\mathbf{w}^j$.

   c) Create a decision class mapping function $\kappa^j : \mathcal{Y} \rightarrow \{-1, 1\}$.

   d) Build a classifier f^j using the algorithm $\mathcal{A}$ and the set $\mathcal{Z}^j$ with class labels transformed with function κ^j.

   e) For each $i = 1, \ldots, l$ if i-th example $\mathbf{z}^i = (\mathbf{x}^i, y^i)$ with transformed class label by the function κ^j is wrongly classified by f^j which is expressed by the condition

$$f^j(\mathbf{x}^i) \neq \kappa^j(y^i)$$

   then increase $\mathbf{M}_{i,c}$ for all $c = 1, \ldots, |\mathcal{Y}|$ such that, $\kappa^j(c) = \kappa^j(y^i)$.

3. The output of the method is a set of binary classifiers $\mathcal{F} = \{f^1, f^2, \ldots, f^J\}$. Each classifier gives output ± 1 that corresponds to a subset of $\mathcal{Y}$ given respectively by $\left(\kappa^j\right)^{-1}(-1)$ and $\left(\kappa^j\right)^{-1}(+1)$ for $j = 1, 2, \ldots, J$.

To classify a new example $\mathbf{z} = (\mathbf{x}, ?)$ we collect votes for each class from the binary classifiers $f^1, f^2, \ldots, f^J$. Each classifier votes for a subset of $\mathcal{Y}$ and the votes for all classes from the subset are equal. The example $\mathbf{z}$ is assigned to the decision class with the greatest number of votes.

---

[8] One of the schemes presented in 5.3.2 could also be used.

Remark 9. The idea of the presented modification of the *Arcing* algorithm for handling non-binary problems with binary classifiers is based on a similar modification for another algorithm *Boosting* [15]. The modification is presented in [18].

Remark 10. In our experiments as a function κ^j we used a function that randomly divided the set $\mathcal{Y}$ into two non-coinciding subsets of approximately the same cardinality. Our observations that such function is good enough from practical point of view supported claims stated in [18].

5.3.4 Combining Arcing and other classifiers

The classical Arcing algorithm is based on a voting scheme as it was presented in the previous subsection. In this subsection we show how we could replace the voting scheme by any classifier.

Let $\mathcal{F} = \{f^1, f^2, \ldots, f^J\}$ be a binary classifier ensemble created by the Arcing algorithm as described in the previous subsection. Furthermore, let[9]

$$f^j(\mathbf{x}) = \theta\left(\hat{f}^j(\mathbf{x})\right) ,$$

where

$$\theta(u) = \begin{cases} -1 & u \le 0 \\ +1 & u > 0 \end{cases} ,$$

and $\hat{f}^j(\mathbf{x}) \colon \mathbb{R}^n \to \mathbb{R}$.

For each example $\mathbf{z}^i = (\mathbf{x}^i, y^i)$ from the training set $\mathcal{Z}$ we create a new example

$$\hat{\mathbf{z}}^i = (\hat{\mathbf{x}}^i, y^i) ,$$

where $\hat{\mathbf{x}}^i \in \mathbb{R}^J$ is a vector defined as follows

$$\hat{\mathbf{x}}^i = \left(\hat{f}^1(\mathbf{x}^i), \ldots, \hat{f}^J(\mathbf{x}^i)\right) .$$

The training set $\hat{\mathcal{Z}}$ built from the examples $\hat{\mathbf{z}}^1 = \left(\hat{\mathbf{x}}^1, y^1\right), \ldots, \hat{\mathbf{z}}^l = \left(\hat{\mathbf{x}}^l, y^l\right)$ is used as an input for a method for constructing classifiers. As a result we get a classifier $f \colon \mathbb{R}^J \to \mathcal{Y}$.

To classify the new example $\mathbf{z} = (\mathbf{x}, ?)$ we first calculate the vector $\hat{\mathbf{z}} = (\hat{\mathbf{x}}, ?)$ using the set of functions $\hat{f}^1, \ldots, \hat{f}^J$ and than we estimate decision class by applying the classifier f.

Remark 11. The presented idea of combining *Arcing* and other classifiers is similar to the idea of *Stacking* introduced in [19]. It can also be viewed as a method for creating *discriminative features*.

---

[9] This is the case for f^j being SVM classifiers with $\hat{f}^j = h^j_{SVM}$.

5.4 Method of *local classifiers*

In this section we present the *method of local classifiers* . First we present the *update-first* version of the *lifting-scheme.* Then we introduce the idea of *local classifier.* Next we show how to create good classifiers using *local classifiers.* We also show how to create ensembles of local classifiers and how to use local classifier for constructing good discriminative features for other classifiers.

Throughout the section we assume that we are dealing with the l multi-channel signals with M channels. For notation please refer to Section 5.2.

5.4.1 Update-first lifting Scheme

In this section we describe a *update-first lifting scheme* [2]. The *lifting scheme* [1] is a flexible method for constructing *wavelets* [3]. Its main virtue compared to classical algorithms is that it allows the creation of wavelets in terms of only time (space) scale. The classical algorithms are based on Fourier analysis and therefore are rather cumbersome for more complicated situations like wavelets on intervals.

The *lifting scheme* is an iterative procedure. Each iteration denoted by upper script $j = 1, \ldots, J$, consists of three steps:

- **SPLIT** From the data matrix $\mathbf{X}_i^j \in \mathbb{R}^{l \times T^j}$ we create two matrices: $\mathbf{X}_{i,e}^j \in \mathbb{R}^{l \times \frac{T^j}{2}}$ and $\mathbf{X}_{i,o}^j \in \mathbb{R}^{l \times \frac{T^j}{2}}$,

- **UPDATE** From matrices $\mathbf{X}_{i,e}^j$ and $\mathbf{X}_{i,o}^j$ we calculate the matrix $\mathbf{C}^j$ as follows

$$\mathbf{C}_i^j = \frac{1}{2}\left(\mathbf{X}_{i,e}^j + \mathbf{X}_{i,o}^j\right) = \begin{bmatrix} \left(\mathbf{c}^{1,i,j}\right)' \\ \vdots \\ \left(\mathbf{c}^{l,i,j}\right)' \end{bmatrix}.$$

 The matrix $\mathbf{C}_i^j$ is called a *coarse approximation* of the matrix $\mathbf{X}_i^j$. It has the property that its rows consist of low-pass filtered and down-sampled rows of $\mathbf{X}_i^j$.

- **PREDICT** In this step we calculate *wavelet-like coefficients* using the matrix $\mathbf{C}_i^j$ and columns of the matrix $\mathbf{X}_{i,e}^j$. The result of this step is the matrix $\mathbf{D}_i^j \in \mathbb{R}^{l \times \frac{T^j}{2}}$. Each column $m = 1, \ldots, \frac{T^j}{2}$ of the matrix $\mathbf{D}_i^j$ is calculated using the subset of columns of the matrix $\mathbf{C}_i^j$ and the m-th column of matrix $\mathbf{X}_{i,e}^j$. Using only a subset of columns of $\mathbf{C}_i^j$ makes the method *local* in the sense that only partial information from the data matrix is used to calculate each column of the matrix $\mathbf{D}_i^j$.

 If we go back from matrices to time series this means that we are calculating a *wavelet-like coefficients* using only a small part of the time series. This is a key property of the method that allows it to concentrate only on the important, from the classification point of view, parts of the analysed signals.

The procedure starts with $\mathbf{X}_i^1 = \mathbf{X}_i$ and $T^1 = T$. For the next iteration we take the matrix $\mathbf{C}_i^j$ as the matrix $\mathbf{X}_i^{j+1}$. In Figure 5.6 we present one iteration of the algorithm. A careful reader must have noticed that the number of iterations J of the algorithm is bounded from above by $\lfloor \log_2(T) \rfloor$. Moreover the number of columns of consecutive matrices $\mathbf{D}_i^j$ for $j = 1, 2, \ldots, J$ is given by the formula $\lfloor \frac{T}{2^j} \rfloor$. In the last iteration J we get two matrices: $\mathbf{C}_i^J$ and $\mathbf{D}_i^J$. In Figure 5.4 the output of the *update-first lifting scheme* for some matrix $\mathbf{X}_i$ is schematically shown.

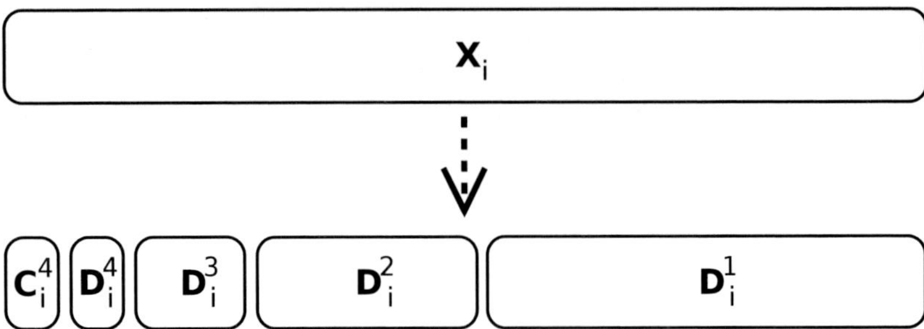

Fig. 5.4. Diagram illustrating what the output of the *update-first lifting scheme* is after 4 iterations. The matrix $\mathbf{X}_i$ is transformed into 4 matrices $\mathbf{D}_i^1, \ldots, \mathbf{D}_i^4$ and matrix $\mathbf{C}_i^4$. The number of columns of i-th matrix is approximately twice the number of columns of matrix $i+1$-st.

5.4.2 PREDICT step in details

Assume that we are at the step j. To calculate the matrix $\mathbf{D}_i^j$ we proceed as follows.

- For each $m = 1, 2, \ldots, \frac{T^j}{2}$ we define
    - A natural even number $L_i^{j,m} \geq 2$ through which we control the time resolution of the method. The bigger the number the more columns of the matrix $\mathbf{C}_i^j$ are used to calculate the m-th column of the matrix $\mathbf{D}_i^j$.
    - Special operator

$$\mathrm{PREDICT}_i^{j,m} : \mathbb{R}^{(1+L_i^{j,m}) \times l} \to \mathbb{R}^l$$

- Subset of $L_i^{j,m}$ columns of the matrix $\mathbf{C}_i^j$ that will be used to calculate the m-th column of the matrix $\mathbf{D}_i^j$. We denote the selected columns by $\mathbf{C}_i^{j,m}$.

Let $\mathbf{D}_i^{j,m}$ be the m-th column of the matrix $\mathbf{D}_i^j$ and $\mathbf{X}_{i,e}^{j,m}$ be m-th column of the matrix $\mathbf{X}_{i,e}^j$. Having defined the necessary parameters we calculate the m-th column of the matrix $\mathbf{D}_i^j$ using the following formula:

$$\mathbf{D}_i^{j,m} = \text{PREDICT}_i^{j,m} \left(\left[\mathbf{X}_{i,e}^{j,m} \ , \ \mathbf{C}_i^{j,m} \right] \right)$$

We will denote the matrix $\left[\mathbf{X}_{i,e}^{j,m} \ , \ \mathbf{C}_i^{j,m} \right]$ as $\mathbf{A}_i^{j,m}$.

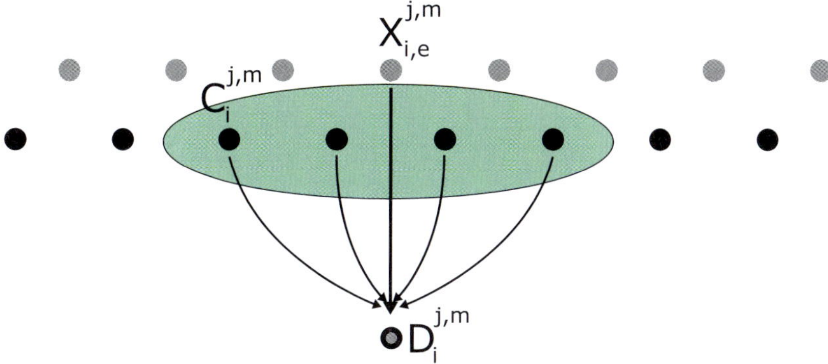

Fig. 5.5. Diagram illustrating the idea of the *locality property* of the operator PREDICT$_i^{j,m}$. The black dots in the light ellipse indicate columns of matrix $\mathbf{C}_i^{j,m}$. Gray dots indicate columns of matrix $\mathbf{X}_{i,e}^{j,m}$. The gray dot with black frame is m-th column of matrix $\mathbf{D}_i^j$

Remark 12. The most important parameters here are $L_i^{j,m}$ and $\mathbf{C}_i^{j,m}$. The first controls the *locality* of the method. The bigger $L_i^{j,m}$ the bigger part of the signal is used to calculate *wavelet-like* coefficients in the matrix $\mathbf{D}_i^j$. By controlling which columns constitute the matrix $\mathbf{C}_i^{j,m} \in \mathbb{R}^{l \times L_i^{j,m}}$ we can model transient structure in the signal. In most situations the selected columns are the columns $k - L_i^{j,m}/2 + 1, \ldots, k - 1, k, k + 1, \ldots, L_i^{j,m}/2$ but the method allows any selection of $L_i^{j,m}$ columns. In Figure 5.5 we have presented the idea of how the columns of matrices $\mathbf{C}_i^j$ and $\mathbf{X}_{i,e}^j$ are combined to calculate the m-th column of the matrix $\mathbf{D}_i^j$.

In practice we used one value of $L^{j,m}$ for all j and m. The best value was selected according to one of the validation method (see Section 5.3.1).

Remark 13. It is worth mentioning that in the case of PREDICT$_i^{j,m}$ for all m and j being linear both the *update-first lifting scheme* and the original *lifting scheme* algorithm implicitly calculate coefficients of expansion of each row of the matrix $\mathbf{X}_i$ in some bi-orthogonal base. Properties of the base are controlled through the operator PREDICT. For more details the reader can refer to [1] and [2].

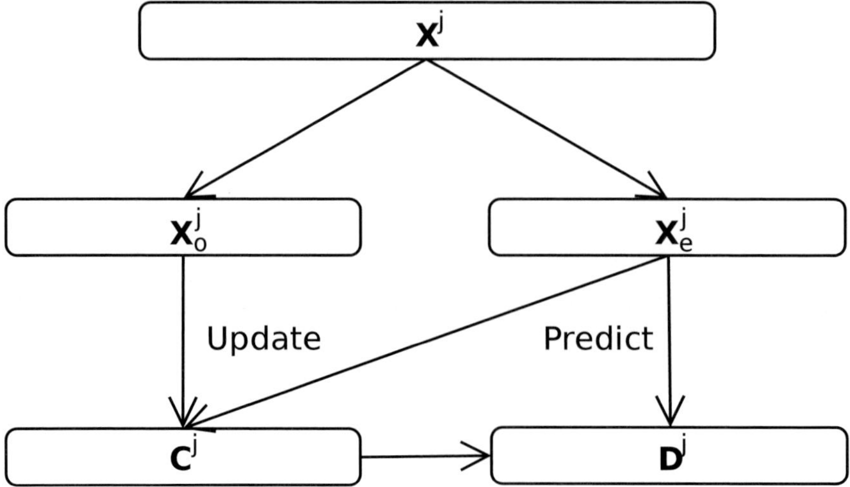

Fig. 5.6. Diagram illustrating one iteration of *update-first lifting scheme* algorithm

5.4.3 Local classifiers

In this subsection we show how to create good, from the classification point of view, *operators* $\text{PREDICT}_i^{j,m}$ using *Proximal Support Vector Machines (PSVM)* [17].

Suppose we are given a set of parameters $\nu_i^{j,m}$ for all j, m and i. Having constructed the matrix $\mathbf{A}_i^{j,m} \in \mathbb{R}^{l \times (1 + L_i^{j,m})}$ as described in Section 5.4.2 we can use it as an input for the *PSVM* algorithm. We proceed as follows:

- The training set $\mathcal{Z}_i^{j,m}$ we build from the l rows of the matrix $\mathbf{A}_i^{j,m}$ and the class labels associated with the signals, that is

$$\mathcal{Z}_i^{j,m} = \left\{ (\mathbf{x}_i^{j,m,s}, y^s) : s = 1, \ldots, l \right\} \,,$$

  where $\mathbf{x}_i^{j,m,s}$ is the s-th row of the matrix $\mathbf{A}_i^{j,m}$ and y^s is a decision class label for the s-th signal $\mathbf{x}^s$.

- We build a classifier $f_i^{j,m} : \mathbb{R}^{1+(L_i^{j,m})} \to \mathcal{Y}$ using the PSVM algorithm and the training set $\mathcal{Z}_i^{j,m}$. The classifier is of the form

$$f_i^{j,m}(\cdot) = \theta(h_{SVM,i}^{j,m}(\cdot)) \,,$$

  where $h_{SVM,i}^{j,m}$ is defined as in (5.1) with the parameters estimated according to the equation (5.5).

- The operator PREDICT_m^j is defined as follows

$$\text{PREDICT}_m^j(\cdot) = h_{SVM,i}^{j,m}(\cdot) \,.$$

The values of this operator for signals from different classes group on different sides of some threshold.

Definition 14. *By* a local classifier *we understand a function*

$$f_i^{j,m}(\cdot) = \theta(h_{SVM,i}^{j,m}(\cdot)) \ .$$

By local discriminative features *we understand functions*

$$h_{SVM,i}^{j,m}(\cdot) \ .$$

Remark 14. In the case when the set $\mathcal{Y}$ contains more than 2 elements we cannot strictly apply the above method. For such problems we need to use one of the schemes *One-Versus-All* or *All-Versus-All* (see Section 5.3.2).

5.4.4 Ensemble of local classifiers with *Arcing*

In this section we describe how to build ensembles of local classifiers. We use Arcing for binary classifiers scheme as an ensemble construction algorithm (see Section 5.3.3).

To use Arcing scheme for constructing an ensemble of local classifiers we have to change only one step of the algorithm in which we create a classifier. This step is constructed as follows:

1. Having a random copy $\mathcal{Z}^j$ of the training set $\mathcal{Z}$ we create local classifiers for each channel $m = 1, 2, \ldots, M$ using the data $\mathcal{Z}^{j,m}$.
2. From all the local classifiers we select a classifier with the best accuracy for the training set. This classifier is included in the ensemble.

The constructed ensemble consists of a mixture of local classifiers that are based on the data from different channels and different parts of given signals. By analysing the ensemble we could conclude which parts of the signals and which channels are the most important from the classification point of view. The results of such analysis could lead to a better understanding of the data.

5.4.5 Local discriminative features

Using the method described in Section 5.3.4 we can use any classifier for the problem of classifying signals. The method first uses Arcing scheme for construction of local discriminative features and then using any method creates the final classifier.

This approach could be very useful for a situation in which we would like to use classifiers that are not very good for classifying signals but have other virtues. A good example are decision trees [15] that are well known for producing comprehensive classification rules that are fairly easily understood.

Another advantage of using local discriminative features is the possible increase in classification accuracy. We have used *PSVM* method with local discriminative features and it gave very good results for a number of reference data sets.

5.5 Experiments

In this section we present results of experiments we conducted to verify presented method of local classifiers. The experiments are divided into two parts. The first part contains results for a number of reference data sets that gives an overall impression as to the accuracy of the method. The second part is a description of the experiment we conducted with cooperation with *the Nencki Institute of Experimental Biology of the Polish Academy of Science.* The goal was to verify physiological hypothesis about functionality of an awake rat's brain.

5.5.1 Accuracy of the method of local classifiers

In this section we show the results of experiments whose aim was to verify the accuracy of the presented method. The description of the results is divided into two parts. In the first part we present the used data sets in short, the second part consists of the results themselves.

Description of the reference data sets

The reference data sets represent a wide range of signals both real and artificial. We claim that this collection of data sets are very good for verifying usefulness of methods for signals classification.

1. *Waveform* - this well known artificial data set was first introduced by Leo Breiman [20]. We used a version of these data proposed in [9]. The data set consists of one channel signals of the length of 32 samples. There are 3 decision classes.
2. *CBF* - artificial data sets. *CBF* was introduced in [9]. It consists of one channel signals of the length of 64 samples belonging to one of 3 decision classes.
3. *Control-Charts* - another well known artificial data set. It was first presented in [21]. There are 6 decision classes.
4. *2-Patterns* - this artificial data set was first introduced in [22]. There are 4 decision classes and the length of the signals is 128 samples.
5. *Trace* - this is simplified version of the artificial data set proposed in [23]. The data set was used for comparison of signal classification methods for signals from control systems in industry.
6. *Gunx* - a real data set introduced in [12]. The data set originates from an automatic surveillance system. It consists of two decision classes of one channel signals of the length 150 of samples.
7. *Auslan Clean and Flock* - a very interesting real data set first described in [10]. The data set consists of multichannel signals that represent 94 words in Australian Sign Language (Auslan). Each word is a decision class and the signals were collected with a special glove. The version Flock of

the data set is the original data set and the version Clean is simplified version of the data set. The second was introduced in [6].

8. *Japanese Vowels* - another real data set. It was first introduced in [24]. The data set consists of multichannel signals that belong to 9 decision classes. It was collected in such a way that 9 male speakers were to say a japanese vowel /ae/.

9. *Pendigits* - another quite interesting real data set presented in [25]. The data set represents 10 handwritten digits and was collected via a tablet. Signals in this data set represent positions of a pen recorded during the writing process.

10. *ECG* - the real data set introduced in [14]. The data set is a subset of a bigger data set that is available from `http://www.physionet.org`. This data set contains ECG recordings that belong to one of two classes.

11. *Wafer* - another real data set introduced in [14]. This data set origins are in the computer industry and it contains measurements of parameters collected from the special chamber in which wafers are crafted. There are two decision classes.

12. *USPS* - the most famous real data set first introduced in [26]. This data set consists of grayscale bitmaps of handwritten digits. To apply our method we first transformed bitmaps into time series using space-filling Hilbert's curve.

Notation

The following notation was used in all tables:

ARCING	denotes the Arcing algorithm (see Section 5.3.3)
SVM+ARCING	denotes the combination of Arcing algorithm and PSVM algorithm (see Section 5.3.4)
DT+ARCING	denotes the combination of Arcing algorithm and Decision Tree algorithm [20] (see Section 5.3.4)

Presentation of the results

In this section we present the results we have obtained for the reference data sets described in the previous section. The results are divided into three parts:

- In Table 5.2 are results for artificial data sets.
- In Table 5.3 are results for real data sets.
- Table 5.4 contains comparison of our method with best known result from literature

Moreover in Table 5.1 the information of the verification method used for each data set is presented. The verification methods are the same as in the literature so we can make a direct comparison.

From Tables 5.2 and 5.3 we can conclude that best results were obtained for SVM+ARCING. The second is ARCING and the last is DT+ARCING. As

Table 5.1. Verification methods for used data sets.

Data set	Number of decision classes	Number of channels	Method of verification
Tracedata	4	1	CV−10
Waveform	3	1	CV−10
CBF	3	1	CV−10
Control	6	1	CV−10
2-Patterns	4	1	$HO_{det}\left(\frac{1000}{4000}\right)$
Gunx	2	1	CV−10
Auslan(Flock)	95	22	CV−5
Auslan(Clean)	10	8	CV−5
Japanese Vowels	10	12	$HO_{det}\left(\frac{640}{370}\right)$
Pendigits	10	2	$HO_{det}\left(\frac{7494}{3498}\right)$
ECG	2	2	CV−10
WAFER	6	2	CV−10
USPS	10	1	$HO_{det}\left(\frac{7291}{2007}\right)$

Table 5.2. Percentage errors of the method ($\pm$ standard deviation) for artificial data sets.

Data set	ARCING	SVM+ARCING	DT+ARCING
Tracedata	2.00 ± 0.75	$\mathbf{1.34 \pm 1.58}$	3.70 ± 1.12
Waveform	14.03 ± 1.30	$\mathbf{13.30 \pm 1.02}$	16.73 ± 1.37
CBF	1.43 ± 0.59	$\mathbf{0.83 \pm 0.52}$	1.53 ± 0.69
Control	$\mathbf{0.43 \pm 0.18}$	$\mathbf{0.41 \pm 0.23}$	4.00 ± 0.48
2-Patterns	5.52 ± 0.64	$\mathbf{2.89 \pm 0.53}$	15.35 ± 1.92

Table 5.3. Percentage errors of the method ($\pm$ standard deviation) for real data sets.

Data set	ARCING	SVM+ARCING	DT+ARCING
Gunx	$\mathbf{0.85 \pm 0.62}$	1.05 ± 0.50	1.75 ± 0.68
Auslan (Flock)	−	$\mathbf{2.20 \pm 0.26}$	16.16 ± 0.96
Auslan (clean)	7.05 ± 1.38	$\mathbf{2.20 \pm 0.84}$	20.15 ± 3.15
Japanese Vowels	6.30 ± 1.34	$\mathbf{0.97 \pm 0.14}$	17.84 ± 1.35
Pendigits	19.33	$\mathbf{2.6}$	8.6
Ecg	13.21 ± 0.84	$\mathbf{10.82 \pm 1.94}$	15.87 ± 2.78
Wafer	0.69 ± 0.12	$\mathbf{0.49 \pm 0.11}$	0.54 ± 0.18
USPS	10.61	$\mathbf{5.33}$	14.85

Table 5.4. Comparison of the method and the best known result from the literature. All errors are in percentages ($\pm$ standard deviation).

Data set	Local classifiers	Best known result
Tracedata	1.35 ± 1.58	**0.18 ± 0.26** [7]
Waveform	**13.30 ± 1.02**	14.60 [6]
CBF	0.83 ± 0.52	**0.0** [10]
Control	0.41 ± 0.23	**0.17** [8]
2-Patterns	2.89 ± 0.53	**0.59 ± 0.08** [7]
Gunx	0.85 ± 0.62	0.5 [8]
Auslan (Flock)	2.20 ± 0.26	**1.28** [8]
Auslan (clean)	2.20 ± 0.84	**2.00** [6]
Japanese Vowels	**0.97 ± 0.14**	1.41 [8]
Pendigits	2.6	**1.59** [8]
Ecg	**10.82 ± 1.94**	15.50 [27]
USPS	5.33	**2.0^{*}**

far as comprehensibility is concerned the order is quite different. The most comprehensible classifier is DT+ARCING as the size of the trees is small. The other two methods does not give comprehensible classification rules. Nevertheless we can analyse the grown ensemble and make reasoning based on its elements.

We should mention that combining ARCING with other classification methods could also yield good results. We used DECISION TREES and SVM because they represent two completely different approaches to constructing classifiers.

From the comparison in Table 5.4 we can say that the presented method gives comparable results to the best known results in the literature. The comparison is not fair for the method of local classifiers as it is compared to a variety of other methods that are not necessarily of the same class as the method described in this chapter.

Moreover some of the presented best results were obtained with only one run of the verification method. Such results are not reliable. For example all results from [8] were obtained for only one run of the $CV-K$ verification method. We present averaged results from 10 runs of the $CV-K$ method. For some data sets we could find runs that gave better or similar results to those presented in [8].

5.5.2 Analysis of local field potentials

In this section we present the results of the analysis of local field potentials recorded within barrel cortex of an awake rat. The work was done with cooperation with *the Nencki Institute of Experimental Biology of the Polish Academy of Science*. The presentation of the experiments is not going to be very detailed due to space limitations and we refer reader to [28] for more detailed description.

The problem description

In Figure 5.7 we present schematically the experiment. One session of the experiment lasted for a few days. An awake rat was placed in a special device which prevented it from moving its head. A stimulator was fastened to one of his vibrissa. The electrode was placed in a barrel. By moving vibrissa regularly the local field potential was generated and recorded via an electrode. The experiment was divided into two parts: Control and Conditioning. In the first part only the stimulator was active. In the second after the stimulation, the rat was given an electric shock in his ear. By this classical conditioning paradigm we wanted to check whether habituated and aroused states of an awake rat's brain can be differentiated by means of analysis of local field potentials.

The aim of the experiment was to analyse the information flow in the rat' barrel cortex. It has been shown in [29] and [30] that the experiment makes it possible to claim that there are two different states of the cortex: Active and Inactive. In [31] a hypothesis of the information flow in the rat's barrel cortex was formulated as a result of previous experiments.

The previous approaches were based on PCA [32], ICA and rough-sets [33] and wavelets [30]. All the methods roughly confirmed the hypothesis. Unfortunately by treating the potentials as a whole the methods could not answer the question about which part of the signal was the most important. Our method was able to give such answer.

Applying local classifiers to the analysis of local field potentials

The data we dealt with consisted of recordings of local field potentials from a group of five rats. For each rat we collected a few recordings from both the Control and the Conditioning part. We used the Control part as an approximation of the Inactive state and the Conditioning part as an approximation of the Active state.

For each rat we selected local classifiers using specially crafted permutation test[10]. The test controlled whether the selected classifiers were reliable. By reliable classifier we understand a classifier that has significantly better training accuracy for original data comparing to data with shuffled labels. After selection we compared indicated classifiers for all rats. This gave us the information of the part of the signal that was used by local classifiers from all rats. The result of this selection is shown in Figure 5.8.

The indicated parts of the local field potentials are roughly near 13 ms mark from the stimulation. This part of the signal has a strict physiological interpretation and fully supports the hypothesis stated in [31].

This result can be regarded as quite a success as the method was used without any *a priori* knowledge about analysed signals. Moreover it is repetitive and clear.

---

[10] The detailed description of the permutation test can be found in [28].

We claim that the method of local classifiers supported with permutation test is a reliable method for analysis of local field potentials. It is also easy to use by experimenters.

Another question is about its classification power. Unfortunately this subject needs more research and the results presented in [28] should be revised and improved.

5.6 Conclusions

The presented method is based on the already used paradigm of a local classifier. From the conducted experiments, we can claim that its accuracy is comparable to the best known methods (see Section 5.5.1). It has also proved to be useful for analysing biological signals (see Section 5.5.2).

5.6.1 Choosing right method

It is hard to say which method should be used by the researcher. The decision should be based on a particular problem and perhaps the intuition of the researcher. If only accuracy is needed, then the method from this chapter might be a choice, especially scheme SVM+ARCING but also method of *interval based literals* or *dynamic time warping* should be used as an alternatives. If the comprehensibility of classification rules is a must then DT+ARCING should be used. The alternatives could be *TClass* or *interval based literals*.

5.6.2 Future research

We believe that the method could be extended in the following directions:

- treating multi dimensional signals (like bitmaps) more directly by using specialised version of the lifting scheme [34],
- replacing ARCING with methods that construct ensembles more suited to SVM classifiers [35],
- usage of the idea of *random forests* [36] for constructing random forest of local classifiers,
- replacing the UPDATE operator with more sophisticated smoothing methods like *moving average*.

References

1. Sweldens W (1998) The lifting scheme: A construction of second generation wavelets, SIAM Journal on Mathematical Analysis 29(2):511–546
2. R Claypoole RN R Baraniuk (1998) Adaptive wavelet transforms via lifting, In: Transactions of the International Conference on Acoustics, Speech and Signal Processing, 1513–1516

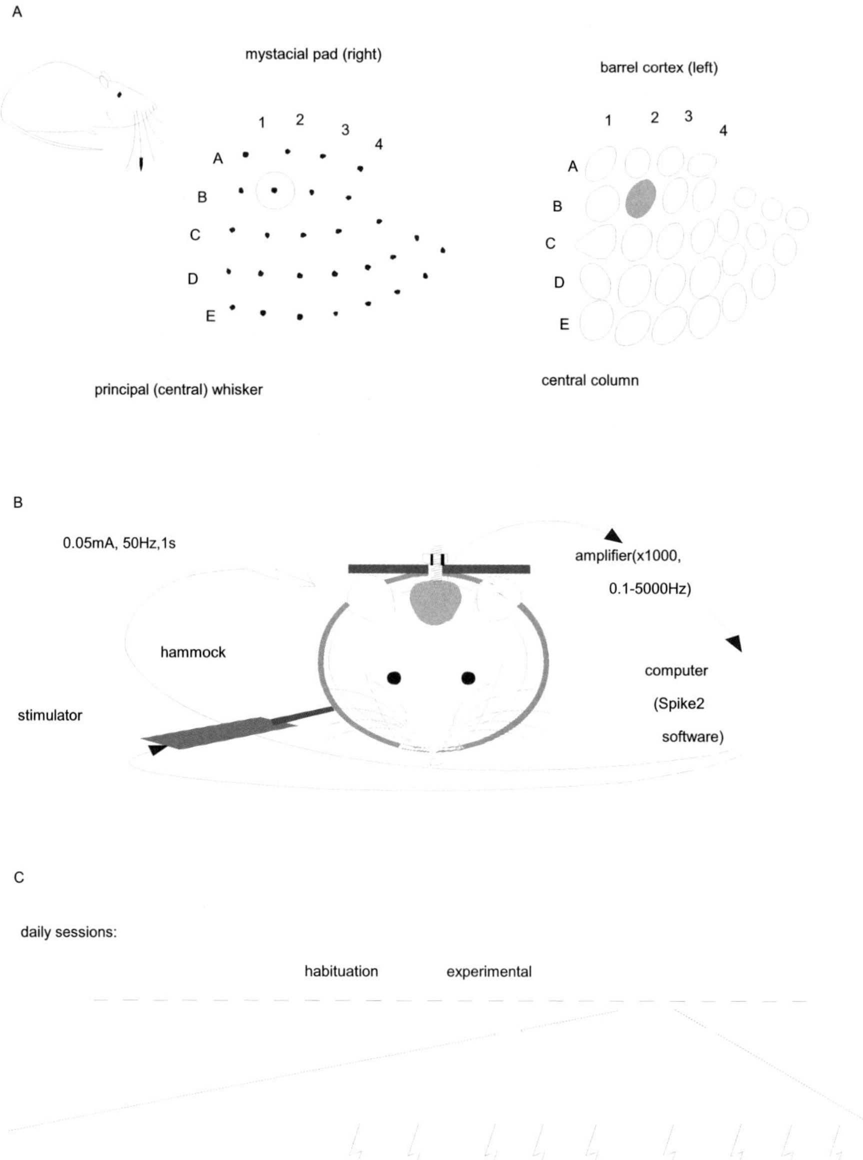

Fig. 5.7. (A) The scheme presenting connection between vibrissae and the barrel cortex. (B) Scheme of the experiment. (C) Exemplary session. (This picture was created by Ewa Kublik [28]).

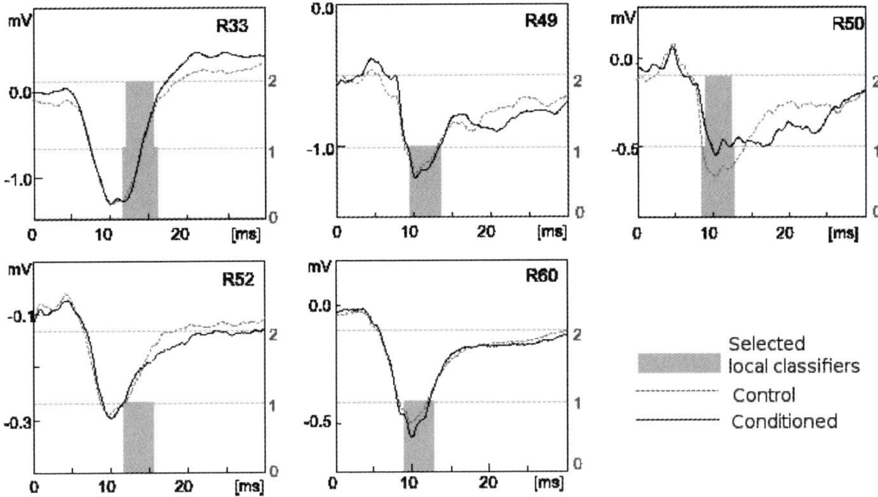

Fig. 5.8. Selected local classfiers with averaged evoked potentials from both groups: *Control* and *Conditioned* for the analysed rats. The grey boxes represents part of the evoked potentials that were indicated by the method. The height of the bar is equal to the number of local classifiers that were used for classification.

3. Daubechies I (1992) Ten Lectures on Wavelets. Society for Industrial and Applied Mathematics, Philadelphia, PA, USA

4. Vapnik VN (1995) The nature of statistical learning theory. Springer-Verlag New York, Inc., New York, NY, USA

5. Breiman L (1998) Arcing classifiers, The Annals of Statistics 26(3):801–849

6. Gonzalez CA, Diez JJR (2000) Time series classification by boosting interval based literals, Inteligencia Artificial, Revista Iberoamericana de Inteligencia Artificial 11:2–11

7. Rodríguez JJ, Alonso CJ (2004) Interval and dynamic time warping-based decision trees, In: SAC '04: Proceedings of the 2004 ACM symposium on Applied computing, 548–552. ACM Press, New York, NY, USA

8. Rodríguez JJ, Alonso CJ, Maestro JA (2005) Support vector machines of interval-based features for time series classification., Knowl-Based Syst 18(4-5):171–178

9. Saito N (1994) Local Feature Extraction and Its Application Using a Library of Bases. Ph.D. Thesis, Yale University

10. Kadous MW (2002) Temporal Classification: Extending the Classification Paradigm to Multivariate Time Series. Ph.D. Thesis, School of Computer Science & Engineering, University of New South Wales

11. Berndt D, Clifford J (1994) Using dynamic time warping to find patterns in time series, In: AAAI Workshop on Knowledge Discovery in Databases, 229–248

12. Ratanamahatana C, Keogh EJ (2004) Making Time-Series Classification More Accurate Using Learned Constraints., In: Berry MW, Dayal U, Kamath C, Skillicorn DB (eds.) SDM. SIAM

13. Keogh EJ, Pazzani MJ (1999) Scaling up Dynamic Time Warping to Massive Dataset., In: PKDD, vol. 1704 of *Lecture Notes in Computer Science*, 1–11. Springer
14. Olszewski RT (2001) Generalized Feature Extraction for Structural Pattern Recognition in Time-Series Data. Ph.D. Thesis, Carnegie Mellon University
15. Hastie T, Tibshirani R, Friedman JH (2001) The Elements of Statistical Learning. Springer
16. Vapnik VN (1998) Statistical Learning Theory. John Wiley & Sons
17. Fung G, Mangasarian OL (2001) Proximal support vector machine classifiers, In: Knowledge Discovery and Data Mining, 77–86
18. Schapire RE (1997) Using output codes to boost multiclass learning problems, In: Proc. 14th International Conference on Machine Learning, 313–321. Morgan Kaufmann
19. Wolpert DH (1990) Stacked generalization. Tech. Rep. LA-UR-90-3460, Los Alamos, NM
20. Breiman L, et al. (1984) Classification and Regression Trees. Chapman & Hall, New York
21. Alcock RJ, Manolopoulos Y (1999) Time-Series Similarity Queries Employing a Feature-Based Approach, In: Proceedings 7th Panhellenic Conference in Informatics (PCI'99), III.1–9
22. Geurts P (2002) Contributions to decision tree induction: bias/variance tradeoff and time series classification. Ph.D. Thesis, University of Liège, Belgium
23. Roverso D (2000). Multivariate temporal classification by windowed wavelet decomposition and recurrent neural networks
24. Kudo M, Toyama J, Shimbo M (1999). Multidimensional curve classification using passing-through regions
25. Alimoglu F (1996) Combining Multiple Classifiers for Pen-Based Handwritten Digit Recognition. MA Thesis, Institute of Graduate Studies in Science and Engineering, Bogazici University
26. Cun YL, Boser B, Denker JS, Henderson D, Howard RE, Howard W, Jackel LD (1990) In: Advances in Neural Information Processing Systems II, 396–404. Morgan Kaufmann, San Mateo, CA
27. Geurts P, Wehenkel L (2005) Segment and combine approach for non-parametric time-series classification, In: Proceedings of the 9th European Conference on Principles and Practice of Knowledge Discovery in Databases
28. Jakuczun W, Wrobel A, Wojcik D, Kublik E (2005) Classifying evoked potentials with local classifiers, Acta Neurobiologiae Experimentalis
29. Musial P, Kublik E, Wrobel A (1998) Spontaneous variability reveals principal components in cortical evoked potentials, NeuroReport 9:2627–2631
30. Wypych M, Kublik E, Wojdyllo P, Wrobel A (2003) Sorting functional classes of evoked potentials by wavelets, Neuroinformatic
31. Wrobel A, Kublik E, Musial P (1998) Gating of the sensory activity within barrel cortex of the awake rat., Experimental Brain Research
32. Kublik E, Musial P, Wrobel A (2001) Identification of principal components in cortical evoked potentials by brief suface cooling, Clinical Neuropshysiology
33. Smolinski TG, Boratyn GM, Milanova M, Zurada JM, Wrobel A (2002) Evolutionary Algorithms and Rough Sets-based Hybrid Approach to Classificatory Decomposition of Cortical Evoked Potentials, In: Alpigini JJ, Peters JF, Skowron A, Zhong N (eds.) Rough Sets and Current Trends in Computing,

Third International Conference, RSCTC 2002, no. 2475 in Lecture Notes in Artificial Intelligence, 621–628. Springer-Verlag
34. Kovacevic J, Swelden W (1997). Wavelet families of increasing order in arbitrary dimensions
35. Valentini G, Dietterich TG (2004) Bias-variance analysis of support vector machines for the development of svm-based ensemble methods., Journal of Machine Learning Research 5:725–775
36. Breiman L (2001) Random forests, Machine Learning 45(1):5–32

Using Neural Models for Evaluation of Biological Activity of Selected Chemical Compounds

Ryszard Tadeusiewicz

AGH University of Science and Technology,
30 Mickiewicza Ave., 30-059 Krakow, Poland
rtad@agh.edu.pl

Summary. The chapter shows how we can predict and evaluate the biological activity of particular chemical compounds using neural networks models. The purpose of the work was to verify the usefulness of various types and different structures of neural networks as well as various techniques of teaching the networks to predict the properties of defined chemical compounds, prior to studying them using laboratory methods. The huge number and variety of chemical compounds, which can be synthesized makes the prediction of any of their properties by computer modeling a very attractive alternative to costly experimental studies. The method described in this chapter may be useful for forecasting various properties of different groups of chemical compounds. The purpose of this chapter is to present the studied problem (and obtained solutions) from the point of view of the technique of neural networks and optimization of neural computations.

The usefulness and wide-ranging applicability of neural networks have already been shown in hundreds of tasks concerning different and often very distant fields. Nevertheless, the majority of investigators tend to attain particular pragmatic ends, treating the used neural models purely as tools to get solutions: some particular network is arbitrarily chosen, results are obtained and presented, omitting or greatly limiting the discussion on which neural network was used, why it has been chosen and what could have been achieved if another network (or other non-neural methods, like regressive ones) had been applied.

In this situation, every researcher undertaking any similar problem once more faces the serious methodological question: which network to select, how to train it and how to present the data in order to obtain the best results. This chapter will present the results of the investigations, in which, to the same (difficult) problem of predicting the chemical activity of quite a large group of chemical compounds, various networks were applied and different results were obtained. Basing ourselves on the results, we will draw conclusions showing which networks and methods of learning are better and which are worse in solving the considered problem. These conclusions cannot just be mechanically generalized because every question on the application of neural networks has its own unique specificity, but the authors of this chapter hope that their wide and precisely documented studies will appear useful

R. Tadeusiewicz: *Using Neural Models for Evaluation of Biological Activity of Selected Chemical Compounds*, Studies in Computational Intelligence (SCI) **122**, 135–159 (2008)
www.springerlink.com   © Springer-Verlag Berlin Heidelberg 2008

for persons wanting to apply neural networks and considering which model to use as a starting one.

6.1 Introduction

The method described here may be useful for forecasting **various** properties of **different** groups of chemical compounds. However, the focus of the work is to predict the chemical activity of alkylaromatic and alkylhetero-cyclic compounds in the enzymatic reaction catalyzed by ethylbenezene dehydrogenase [3]. EBDH, the key enzyme in metabolism of ethylbenzene and propylbenzene in anaerobic denitrifying bacteria, catalyses the first known oxygen-independent enantioselective oxidation of hydrocarbon substrates providing a completely novel type of reaction in biochemistry.

The practical aim of the research was to investigate the architecture of an optimal neural model, which could be used for screening of EBDH activity with new compounds without the need for expensive experimental tests. There was also hope for getting insight into the mechanistic aspects of the reaction catalyzed by the enzyme. Furthermore, in order to be able to describe Electric and structural features of all studied compounds the compounds descriptors were derived from quantum chemical calculations on DFT (Density Functional Theory) level. Such an approach avoids limitation encountered in the QSAR (quantitative structure activity relationship) correlation analysis based on standard fragment descriptors (such as Hammett or hydrophobic constants), where one is limited to a common structural core of compound group. Moreover, some DFT parameters used, such as energies of the frontier orbitals (HOMO and LUMO) provide feasible chemical interpretation as ionization potential and electron affinity, from which one can estimate the absolute hardness and the electro-negativity of a particular compound. Summarizing, the problem addressed in the study was to build a model connecting a range of various theoretical parameters describing chemical and geometrical features (input variables) of EBDH substrates and inhibitors with their biological activity described by their relative reaction rate (output variable). However, up to date there is no common scale that allows description of both substrates (which differ in reaction rate) and inhibitors, which influence either reaction rate of a reference substrate (ethylbenzene) and/or the equilibrium constant of enzyme-substrate complex formation. To the best of our knowledge, such a description of both substrate and inhibitors is a truly innovative approach.

This chapter presents solutions of the problem of predicting chemical activity of quite a large group of chemical compounds where various networks were used. Based on this we draw conclusions and show which of the utilized learning methods are better in solving the considered problem. Although these conclusions cannot be fully generalized, as every application has its own unique specificity, we think that our studies will be useful to those using neural networks for modeling in the fields of biology, bioinformatics, biomedicine, but also many others.

6.2 Problem statement from the biochemical point of view

The example of the structure of the chemical compound studied in this work (*e.g.*, alkylaromatic molecule) is shown in Fig. 6.1.

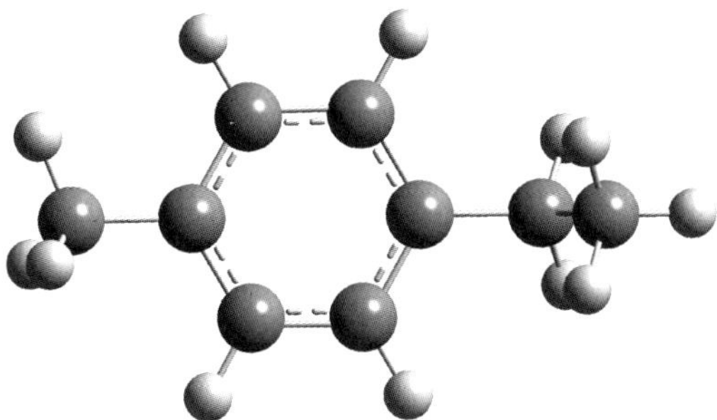

Fig. 6.1. *Para*-ethyltolunene - an example of the structure of molecules studied in this work.

Nevertheless, the chemical aspects of the problem are not in the scope of this chapter and only the following aspects seem to be important:

- The task is based on a large number of premises (22 in the maximum set, see Fig. 6.2.). They include the data on the topology of considered chemical molecules (3 quantitative data and 1 descriptive datum, determining the location of the substituent in relation to the active center of the molecule in 3 categories: para, meta, and ortho), the data describing the substituent's electronic parameters (15 quantitative data resulting from the quantum mechanics computations) and the parameters describing its size (3 quantitative data, including 2 from the quantum mechanics computations). Thus the input vector consists of maximum 24 parameters (24 input neurons). The input descriptor determining the location of the substituent in the aromatic ring is encoded with one-of-N method and needs use of 3 neurons for the 3 possible categories in the input layer (see Fig. 6.2).
- Because of various estimates of the usefulness (importance) of different input data, the number of input signals has been limited in some experiments. Therefore, several of the networks have smaller number of inputs than it would appear from Fig. 6.2.

- The reduction of the number of input data (data dimensionality) is not only the starting point in building the network but often also an important result of its work. The relative estimate of the importance of various inputs has been possible ex post based on the analysis of the parameters of the trained network, or based on the analysis of the sensitivity of the neural model to variations of its particular inputs. The comparison of the functioning of the network at various numbers of inputs data is one of the results discussed in this chapter.

- The model always had only one output signal, but depending on the formulation of the problem, this output could have the interpretation of a variable of either quantitative type or qualitative type. The fact that the considered network always has only one output, enables, during the choice of the architecture and way of the functioning of the network, application of structures that are designed exclusively to the models of one output only (for instance GRNN - Generalized Regression Neural Network). Moreover, due to the fact that the evaluation of the activity could have been considered at one time as quantitative data (when the experimentally obtained data on the activity are used) and at another time quantified in several arbitrary categories, *e.g.*, as a negative (inhibition), small, average, or large, it was possible to compare the efficiency of the functioning of the studied networks in either regressive or classifying mode.

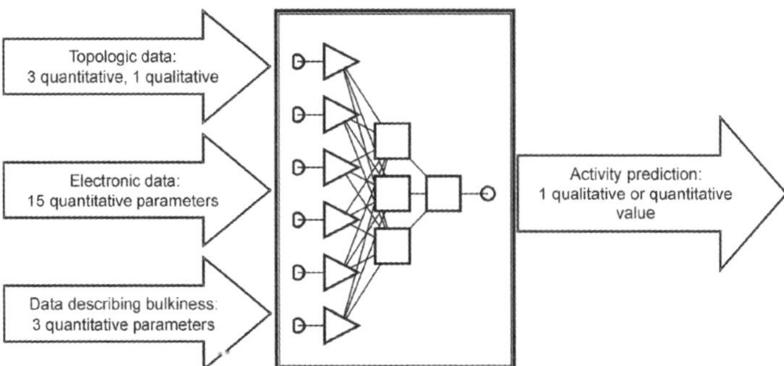

Fig. 6.2. General outline of the neural model studied in this work.

Various aspects of the networks were investigated, as described later. The results of our studies show how neural networks solve particular problems of modeling complex, non-linear relations. Such results not only enrich the particular domain for which they are created, but also form the general methodology of creating and exploiting neural networks.

6.3 Neural networks used for biological activity prediction

The following neural network types were taken into account: LIN (linear type network), MLP (Multi-Layer Perceptron network - regressive), MLP/LIN (regressive Multi-Layer Perceptron network with linear element in the output layer), MLP/CL (Multi-Layer Perceptron network with the discrete output value - network performing classification task instead of regressive evaluation), and GRNN (Generalized Regression Neural Network). For each type many network architectures were applied, optimized, trained, tested and evaluated for elaborating the general assessment. All networks were created, trained and analyzed using the Statistica Neural Networks 7.1 software. For each network type the Intelligent Problem Solver (IPS) option was used and usually at least 1000 models of the network of a chosen type were created and investigated. The optimization procedure utilized uniform training methodology (for example 100 epochs of back propagation algorithm, 25 epochs of conjunct gradient and 25 epochs of conjunct gradient with momentum) for every network while varying topology of the input and hidden layers. The optimization of the architecture and the input vector by the Intelligent Problem Solver was performed on the basis of validation error minimization. Models which achieved the smallest validation errors were selected from the whole network population (usually more than 1000) yielding a set of 50 best networks. All selected models were subjected to external cross-validation on test cases, which were not used for network development. From the set of optimized neural networks only one, giving the best results in prediction for both validation and test cases, was used in further analysis.

Below, many types and configurations of networks, characterized by different number of neurons on particular layers will be compared. Also the dimension of input vector, which was optimized in the study, includes the input layer into the optimization process. For short and effective description of the studied and compared networks we introduce following notation, based on the schemes used for programming languages syntax description:

<*network type*> <*number of input signals*> : <*number of neurons in input layer*> [- <*number of neuron in first hidden layer*>] [- <*number of neuron in second hidden layer*>] - <*number of neuron in output layer*> : <*number of output signals*>

Elements presented above in < and > parenthesis will be replaced by particular names and numbers respectively, elements located in [and] brackets are optional. In this chapter we take into account only the following values of the <*network type*> terms:

<*network type*> = LIN which denotes linear type network

<network type> = MLP which denotes typical Multi-Layer Perceptron network (regressive)

<network type> = MLP/LIN which denotes Multi-Layer Perceptron network with linear element in the output layer

<network type> = MLP/CL which denotes Multi-Layer Perceptron network with the discrete output value (network performing classification task instead of regressive evaluation)

<network type> = GRNN which denotes Generalized Regression Neural Network

6.4 Learning set

The learning set is very important element in every investigation, when neural networks are used. For the full characterization of the investigated problem, we are listing studied chemical compounds dataset together with their relative activities in Table 6.1.

In the investigations with the classification networks, the above data were converted in such a way that four classes of the activity were created, denoted with the following codes:

-1 – inhibitors,

0 – activity up to 50%,
1 – activity of 51 - 150%,
2 – activity above 150%.

The classes contained the following number of chemical compounds:

"-1" class – 8 compounds,

"0" class – 9 compounds,
"1" class – 5 compounds,
"2" class – 4 compounds.

During the investigations, the dataset was divided (randomly) into three parts: learning (L), validation (V) and testing (T) dataset. The first (L) was used for network learning, another (V) for ascertaining the time at which the training procedure failed in order to avoid the effect of **loosing** the ability of generalization, and the last (T) for the final result control, in order to check whether any accidental coincidence of the learning and validation data did take place. Such an accidental event leads to falsification of the results of the network trainieng and performance. The proportions of the division of the dataset (Table 2) were as follows: 14:2:2 (L:V:T) for the networks predicting exclusively the activity of the substrates (18 chemical compounds) and 18:4:4 (L:V:T) in the case of the MLP and GRNN networks predicting both the activity of the substrates and of the inhibitors (26 chemical compounds).

Table 6.1. The collection of the dataset available. Relative activity (reaction rate, k_{cat}) given in %, taking ethylbenzene activity as 100% and classes of activity used in classification neural network.

Compound name	Relative activity	Classes
1,2-diethylbenzene	0	-1
1,4-diethylbenzene	35	0
1-ethylnaphtalene	0	-1
2-ethylaniline	94.53	1
2-ethylfuran	133.92	1
2-ethylnaphtalene	9.3	0
2-ethylphenol	56.14	1
2-ethylpyridine	0	-1
2-ethylpyrrol	234.63	2
2-ethylthiophene	242.92	2
2-ethyltoluene	3.8	0
2-methylfuran	0	-1
2-methylpyrrole	0	-1
2-methylthiophene	0	-1
3-ethylphenol	24.31	0
3-ethylpyridine	16.94	0
3-ethyltoluene	10	0
4-ethylphenol	259.01	2
4-ethylpyridine	0	-1
4-ethyltoluene	28	0
4-fluorethylbenzene	15	0
ethylbenzene	100	1
n-propylbenzene	14	0
toluene	0	-1
4-ethylaniline	134	1
4-propylphenol	180	2

Table 6.2. Division of data set into learning, validation and test cases for networks predicting substrate activities (LIN, 18 cases) and both substrate and inhibitor activities (rest of networks, 26 cases).

Type of network	Learning cases	Validation cases	Test cases
LIN	14	2	2
MLP	18	4	4
MLP/LIN	18	4	4
MLP/CL	18	4	4
GRNN	18	4	4

In order to evaluate and compare the overall ability of selected models to deal with our modeling problem, the training was repeated 20 times yielding a set of 20 networks for each optimized network type. The network model was determined by the network architecture, selection of particular elements for the input vector, cases partitioning into learning, validation and test groups as well as the training procedure. For each network, the weights were randomly determined and the learning cases were randomly distributed among L,V,T subsets. As a result of such an approach, for each category of networks a distribution (not a single value!) of learning, validation, and test errors was obtained. This allows to take into account the changeability of the performance results, which leads to objective comparison of the obtained model types along with analysis of the best examples from the group.

6.5 Results obtained for particular neural networks

Now, we show and compare the results obtained for every particular neural network type used in this study.

6.5.1 Linear network (LIN)

A linear network was constructed and studied in order to obtain a full understanding on how various types of the networks behave, and to enable a comparison of the results given by the neural networks with the solutions provided by the QSAR method (commonly used in chemistry). QSAR models routinely used in chemistry and drug design are usually based on multiple linear regression (MLR) and provide a linear relationship between range of parameters and chemical or biological activity (*i.e.*, reaction rate, equilibrium constant, toxicity, etc.). Therefore, from the mathematical point of view, linear QSAR equations are identical with linear networks.

The linear neural network, similarly to all others considered in this chapter, was formed by the automatic optimization procedure and the non-iterative algorithm of learning by pseudo-inversion of the matrix was applied. It is known that in the linear network both the structure and its parameters are fully determined by the learning set. For this reason, optimization is limited exclusively to the optimal input dataset. Namely, in the selected network instead of 24 neurons 20 neurons were taken, as a result of the automatic optimization. The network selected for evaluation has the following structure:

LIN 20:20-1:1

The created network was working as the regression network, *i.e.*, the values, which may be treated as the values of the activities of the compounds, predicted by the network were expected in the output of the network. These values are directly compared with the experimental data for the investigated chemical compounds. The obtained results were astonishingly good. Fig. 6.3

shows a very high degree of correlation between the empirically determined values of the activities and those obtained by computing with the use of the linear network. Despite the fact that the discrepancy between computed and experimental data was very large for a few compounds (they are shown in Fig. 6.3), the mean values of the errors, given in Table 6.3 are quite good.

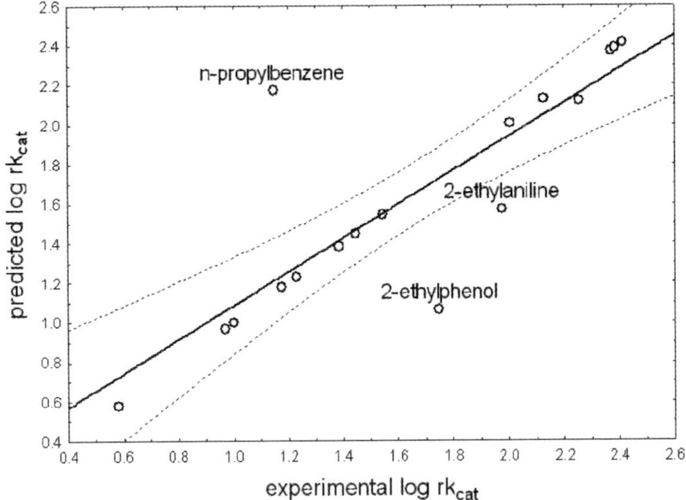

Fig. 6.3. The comparison of the values of the relative activity of the studied chemical compounds obtained experimentally (horizontal axis) and those determined by the linear network (vertical axis): r=0.8467 and p=9 $* 10^{-7}$.

Table 6.3. Mean values of the errors made by the linear network and averaged errors of 20 LIN networks along with $\pm$ 0.95 confidence ranges (mean value is shown with a measure of its statistical distribution).

	Learning error	Validation error	Test error
Best network	$1.31 * 10^{-15}$	0.399	0.308
Averaged set	$1.84 * 10^{-15}$ $\pm 7.5 * 10^{-16}$	2.41 ± 2	2.58 ± 1.9

One measure of the quality of the neural network performance to the empirical data is the correlation coefficient of the data predicted by the network and the empirical results. This coefficient was equal to r=0.8467 for the best linear network and gave the statistical significance of these values at the level of p=9 $* 10^{-7}$. It is interesting to compare the result obtained by the linear network with the result of the forward stepwise MLR. Stepwise regression is more widely known and commonly used in creation the optimal possible MLR models. When applied to the same set of data, the equation with four variables was formulated. Obviously, the necessary precondition for including particular variables into the MLR equation was confirmation that its correlation with log rk_{cat} is statistically significant. The activity predicted by the regression model correlates with the experimental data at the level of r = 0.8704 (which corresponds to the statistical significance of p = 0.00059).

The multiple network validation procedure provided additional information about the characteristics of the linear architecture (Table 6.3). The average learning error $(1.84 * 10^{-15})$ is 15 orders of magnitude lower than the average error for validation and test group (2.41 and 2.58, respectively). Moreover, the deviation of test and validation errors is far much greater than that observed for training cases (Fig. 6.4). This is a clear indication, that linear network is properly modeling only learning cases.

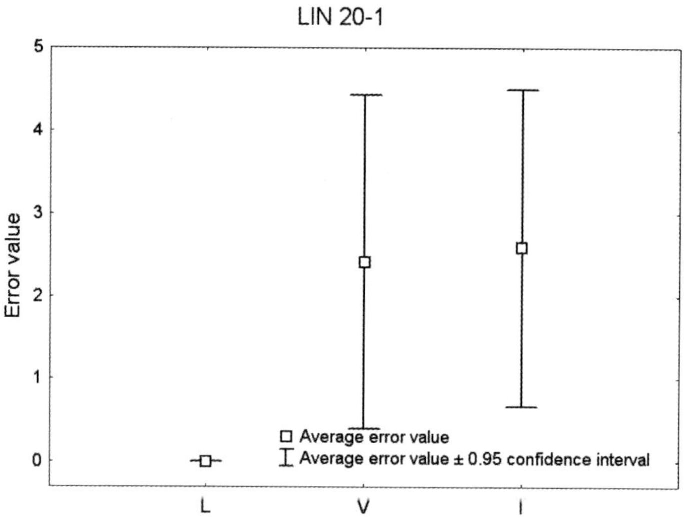

Fig. 6.4. Box-whisker plot of average errors (with its distribution) for learning (L), validation (V), and test (T) datasets obtained in a group of the LIN 20-1 networks. The range indicators provide 95% confidence interval for each set.

6.5.2 Non-linear MLP network

Taking the above-mentioned values as the reference, we searched for op-timal non-linear network - in the beginning limiting ourselves to the MLP network (Multi-Layer Perceptron) [19]. We analyzed the performance of about two thousand networks, in which the input signals and the number of the hidden neurons were variable. It was assumed that the optimized net-work would have only one hidden layer. The networks with two hidden layers were also studied, but they did not give good results.

After roughly two thousand attempts, the optimal network topology was found to be:

MLP 20:20-8-1:1

This network has a limited number of input elements (the optimization procedure had chosen the best topology containing 20 from 24 input data) and a relatively small number of hidden neurons (8). The network was first trained with the quick back propagation algorithm and then with the quasi-Newton algorithm with the momentum factor of 0.3. Moreover, Gaussian noise of the amplitude of 0.05 was added to the input signals, to avoid get-ting stuck in the local minima.

The procedure led to satisfactory results. The quality of the work of the network presented in this chapter is shown in Table 6.4.

Table 6.4. Mean values of the errors made by the non-linear MLP 20:20-8-1:1 network and averaged errors of 20 MLP 20:20-8-1:1 networks along with $\pm$ 0.95 confidence ranges (mean value is shown with a measure of its statistical distribution).

	Learning error	Validation error	Test error
Best network	0.028075	0.026944	0.049206
Averaged set	0.13 ± 0.037	0.15 ± 0.043	0.26 ± 0.08

The errors of the trained MLP network are at least one order of magnitude smaller than the respective errors made by the linear network. This concerns the error of the validation and testing sets, because only these errors are conclusive (may give a good estimation of the errors made by the network during a typical, working exploitation, *i.e.*, for finding the properties of new, totally unknown or not studied chemical compounds). It can be seen that in this respect the quality of the best MLP network is **considerably** better than the quality of the best linear network.

The learning error made by the network is worth discussing. It has an extremely little value (practically equal to zero) in the linear network. It is the result of using the precise optimization algorithm for the linear network learning, instead of the iterative adaptive technique, as in MLP. Such a situation is commonly considered as undesirable, because of the loss (or serious limitation) of the network's ability to generalize the results of the learning procedure.

The good quality of the selected MLP network is also confirmed by Fig. 6.5 (similar to Fig. 6.3), showing the relationship between the predicted and

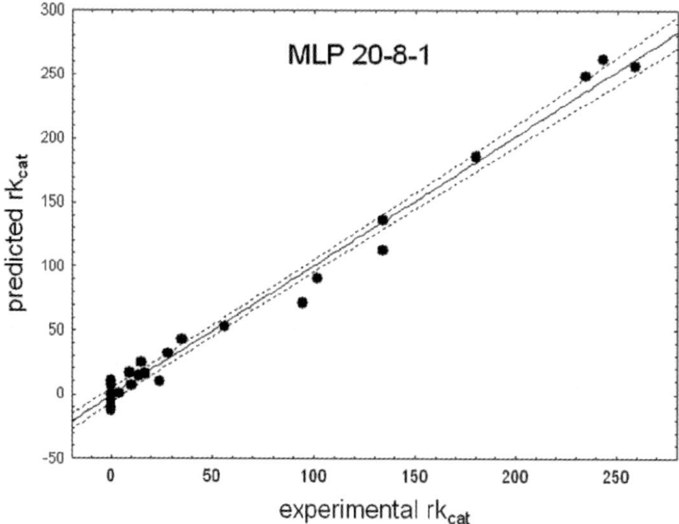

Fig. 6.5. Correlation plot of experimental data and the data expected by the selected best MLP 20:20-8-1:1 network (r=0.9925 and p=$1.95*10^{-23}$).

empirically found activity values for the considered group of standards. The narrow range of the discrepancy (denoted with the dashed lines) accessible with 95% confidence interval is worth noting. The analogical area in Fig. 3 is several times larger. It means that the relationship between the real data and the data obtained using the forecast of the chemical activity of the investigated chemical compounds using MLP 20:20-8-1:1 neural network is very strong and well determined. This is also confirmed by the values of correlation coefficient and its significance, obtained for this net-work: r=0.9925 and p=$1.95*10^{-23}$.

The multiple network validation procedure provides additional information about the errors distribution in the set comprising of neural networks of identical architecture as MLP 20-8-1 (see Fig. 6.6). The average error for learning dataset was 0.129, for validation dataset 0.151 and test dataset 0.261. The error distribution was presented as 95% confidence interval. Even the errors in a test group exhibit twice as wide distribution range as for learning and validation cases all averaged mean errors are generally of comparable value. Such a behavior indicates that fully non-linear MLP can be a reliable tool with high generalization capabilities.

6.5.3 The variant of the MLP type network with a linear output neuron

MLP networks are considered in literature in two variants. The first one, used in the experiments presented in the previous section concerns the network with

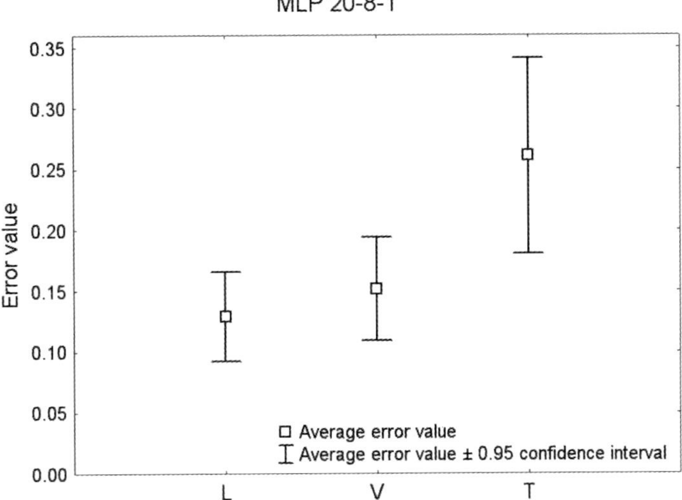

Fig. 6.6. Box-whisker plot of average errors for learning (L), validation (V), and test (T) datasets obtained in a group of the MLP20-8-1 networks. The range indicators provide 95% confidence interval for each set.

all non-linear neurons (most often of a sigmoid character, although other non-linear neurons such as hyperbolic tangent are also discussed). On the other hand, in some literature reports it is suggested that in regression tasks, in which the network output is a **value**, not a **decision** predicted by the model, **linear** neurons may be used as the neurons of the **output layer**. Sometimes it is expressed that linear neurons are **recommended** in this type of network, because it gives more freedom for the choice of the behavior of a neuron model (the output is not restricted by the saturation of the sigmoid at the zero and one level).

Using the well-collected input data for the considered problem of the analysis of activity of chemical compounds and the efficient tool of the optimization of the network structure it was decided to check the behavior of the MLP type network with a linear neuron in output layer.

For the purpose of preliminary investigations, the automatic optimization procedure was used and it yielded the network of the following structure:

MLP/LIN 20:20-9-1:1

It can be easily observed that this structure is similar to the one that was found (in separate and independent studies) for a typical, fully non-linear MLP type network, described in the previous section. It should be stressed, however, that the input dataset, selected by optimization, differed in some elements from that used in the fully non-linear network. The only visible structural difference is a somewhat larger hidden layer, than the one chosen in the preliminary trials by the IPS for the fully non-linear network (MLP 20:20-8-1:1). This suggests that the correct training of the network with a linear element is a **more difficult** task than the training of a fully non-linear network. Therefore more indirect data (signals produced by a hidden layer) are needed to achieve a good functioning of the network.

Completely disappointing were the results obtained with this type of the network. The first warning signal was obtained while analyzing the correlation plot of the experimental and predicted data, shown in Fig. 6.7. It emerged that the correlation in this case was worse than for the fully non-linear MLP network (r=0.9717; p=$1.5*10^{-6}$), which proves that the quality of the model is unsatisfactory.

Even more alarming results were found while observing the variations of values of the error made by the network for the learning set (continuous line) and for the validating set (dashed line), see Fig. 6.8.

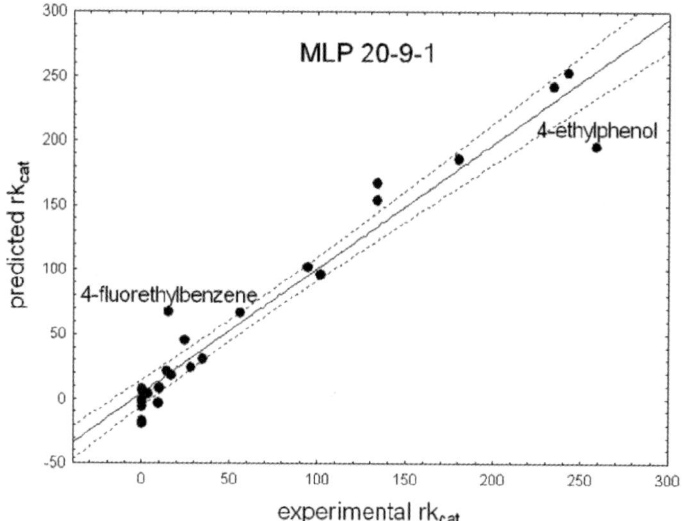

Fig. 6.7. Correlation plot of the experimental and predicted data by the best MLP network with a linear output element (r=0.9717; p=$1.5*10^{-6}$). Two significant outliners were labeled.

As is clearly seen in this figure, the investigated network is learning quite efficiently but it has significant difficulties with the generalization of the results of the training procedure, because the validation error curve proceeds much above the line of errors obtained for the learning set. On the contrary, it is worth looking at the analogical diagrams obtained for the non-linear network, described in section 6.5.2 (Fig. 6.9).

In this case, it is clear that the improvement of the quality of the net-work performance for the learning set was closely correlated with that for

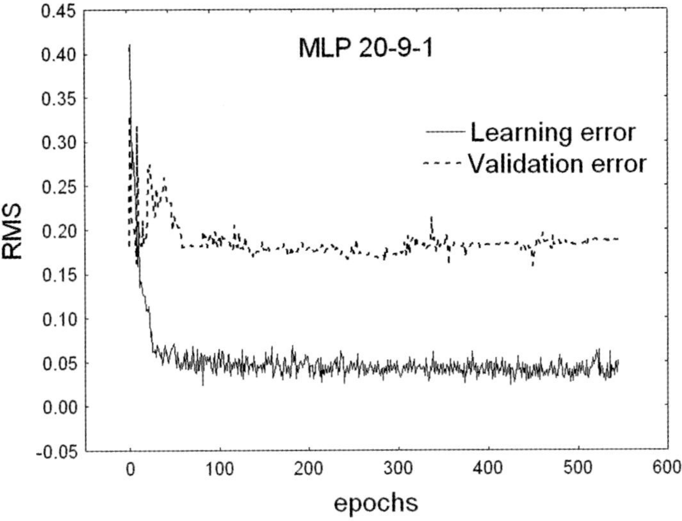

Fig. 6.8. The training curve of the optimal MLP network with a linear output element.

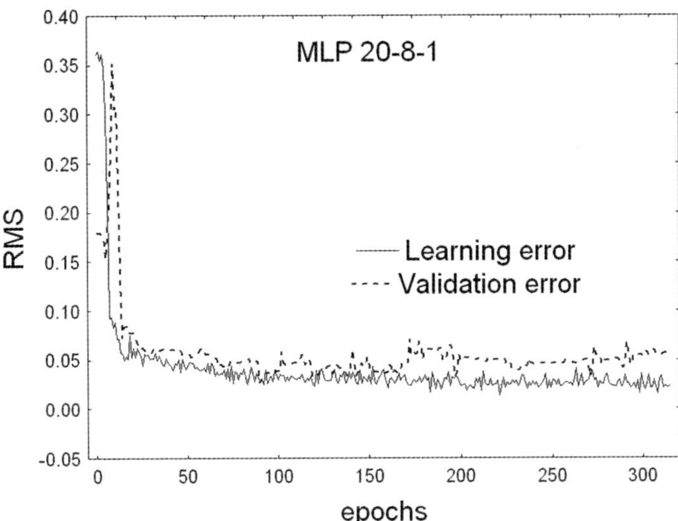

Fig. 6.9. The training curve the optimal full non-linear MLP network.

the validation set. Moreover, a detailed analysis of the diagram as in Fig. 9 might indicate that around the 100th epoch the learning process entered very smoothly the phase of "over-training." Further reduction of an average trend of the error line for the learning set was accompanied by visible to experts deterioration of the network performance in activity estimation for validation cases. The effect of the over-training, although subtle, was in this case discovered by automatic algorithm controlling the quality of the network. Therefore, after an arbitrary stopping of the learning procedure, the network with the weights from the 89th epochs was regained (15 epochs of quick propagation and 74 epochs of the quasi-Newton algorithm).

Turning back to the question of the MLP network with a linear output element it should be stated that the course of learning mentioned in Fig. 6.8 indicates clearly that the network gained significantly better results in predicting the chemical activity of particular chemical compounds for these molecules, which were supplied during the learning procedure, than for those on which we were testing the quality of the functioning of the network (in the validation and control sets). This is confirmed by the values of the appropriate errors listed in Table 6.5 (worth comparing with Table 6.4).

Table 6.5. Mean values of the errors made by the non-linear MLP type network with a linear output element and averaged errors of 20 MLP/LIN 20:20-9-1:1 networks along with $\pm$ 0.95 confidence ranges (mean value is shown with a measure of its statistical distribution).

	Learning error	Validation error	Test error
Best network	0.032292	0.158581	0.114655
Averaged set	0.12 ± 0.046	0.17 ± 0.070	0.27 ± 0.066

Multiple network validation procedure (Table 6.5, Fig. 6.10) yielded following average errors: 0.12, 0.17, and 0.27 for learning, validation and test data set. The average error values are comparable to those obtained for fully non-linear MLPs (with test errors slightly higher than validation and learning errors). This, again, proves that MPL architecture (regardless of the type of output neuron) is of generally good and reliable quality.

Summing up this group of experiments, we may state that the hypothesis which postulates an advantageous effect of the linear characteristic of the output neuron in the MLP network **has not been confirmed**. The reason for this may lie in the fact that the IPS chose one hidden layer comprising of 9 neurons as the optimal solution for the MLP/LIN, while the optimization for the fully non-linear network ended with the structure with 8 hidden neurons. Sometimes such adding of **one** neuron to the hidden layer of the network may "convert" the network from the mode of the learning with the generalization to the mode of learning through memorization. This is particularly true when

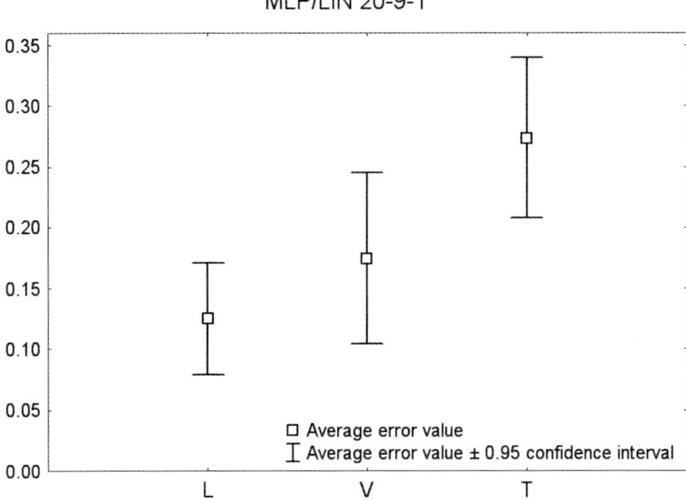

Fig. 6.10. Box-whisker plot of average errors for learning (L), validation (V), and test (T) datasets obtained in a group of the MLP/LIN 20-9-1 networks. The range indicators provide 95% confidence interval for each set.

the network has a relatively big number of inputs (here - 20) and is trained on a rather small learning set (in this case there were only 18 learning cases, 4 validation cases and 4 test cases). Maybe the reason for obtaining such results should be sought in slightly different output vector (although it has been selected for this type of the network experimentally from thousands other input sets). Nevertheless, it seems well worthwhile to present and discuss the result observed in these studies.

6.5.4 Non-linear networks of the GRNN type

The process of building the model in the GRNN network is divided into two stages. In the first step, groups of similar cases are localized in the space of the input signals. This stage is realized using the radial layer of the GRNN network. In the second stage, the regression approximation of the searched relationship is formed. Based on the earlier input space division by radial layer and the degree of similarity of the considered input signal to particular class the decision is made and the result is obtained.

Even this short and schematic description of the functioning the GRNN indicates that a network of this kind should have, in our case, a particularly high degree of efficiency. This is because GRNN performance goes along the lines of reasoning of an expert (a man) during the attempt of predicting the properties of an unknown chemical compound. The researcher usually collects knowledge through experimental trials for a group of compounds, classifies them based on similarities and then compares them to an unknown substance looking for analogy - this is precisely what the specific model of GRNN networks offers. For this reason, we hoped that it would be possible to apply the GRNN as the neural model designed to predict chemical activity of unknown chemical compounds.

An additional premise for focusing ourselves on GRNN networks was the fact that in all networks described here we had one output (only), which is the necessary condition when applying GRNN networks.

For this reason we carried out the preliminary investigations (using the IPS) for roughly 2000 variants of the GRNN. For further detailed studies, not one, but three structures of this type of network were chosen. The selected networks had the structures presented in Table 6.6.

Table 6.6. Structures of the GRNN networks selected for further studies.

Number of the network	Structure
GRNN40	GRNN 24:245-14-2-1:1
GRNN48	GRNN 23:23-14-2-1:1
GRNN58	GRNN 22:22-14-2-1:1

As can be seen, the IPS had chosen in a consistent way as the most advantageous the structure of the network containing 14 neurons in the radial layer and 2 neurons in the regression layer. The differentiation of the net-works selected as those most advantageous was done in the selection of the subsets of the input data, taken into account during the neural model formation.

The networks listed in Table 6.6 were subjected to the learning procedure, giving the results listed in Table 6.7.

Table 6.7. List of the best GRNN networks, together with the errors for the learning, validation, and test sets.

Type of the network	Learning error	Validation error	Test error
GRNN 24:24-14-2-1:1	0.857	0.739	1.189
GRNN 23:23-14-2-1:1	0.964	0.589	1.213
GRNN 22:22-14-2-1:1	1.037	0.481	1.137

The analysis of table 6.7 and its comparison with the data given earlier in similar tables for MLP network, or even for the linear network does not leave any doubts: the GRNN networks in the considered example are not "competitive" to any of the earlier considered networks.

6.5.5 Non-linear networks of the GRNN type with MLP 20-8-1 input vector

The apparent defeat of GRNN networks was so surprising, that we decided to investigate the matter more deeply. Could it be that the reason for not achieving good performance lied in the IPS optimization procedure for GRNN and not network architecture itself? In order to check this, the previously used input vector for MLP 20-8-1 was used. 200 GRNNs were generated by IPS and top five networks were saved. All of them had similar quality (r around 0.98) and the same architecture: GRNN 20:20-18-2-1:1. The Fig. 6.11 presents the correlation plot for the best networks while network errors are presented in Table 6.8.

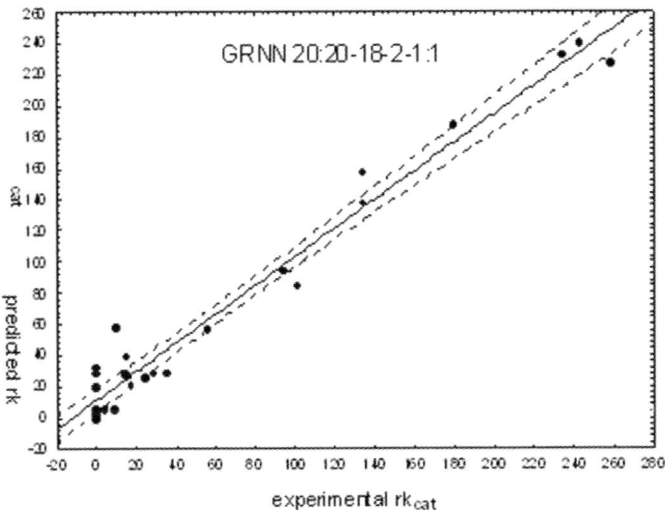

Fig. 6.11. Correlation plot of the experimental and predicted data by the best GRNN network element with input vector from MLP 20-8-1 (r=0.9827; p=4.36 ∗ 10^{-19}).

Apparently, for some reason the automatic optimization procedure could not achieve optimal input vector (*i.e.*, eliminate redundant information). When the input vector that was optimal for the regression jobs was provided, it was possible to obtain a model comparable with fully non-linear MLP. The analysis of errors obtained for the best network shows good quality of obtained model, as all errors (for training, validation and test group) are of the same magnitude. Interestingly, Fig. 6.12 shows that on average these types of networks tend to exhibit approximately 10 fold lower training error then the validation and the test one. This may suggest slight over-training tendency.

Table 6.8. Mean values of the errors made by the GRRN type network with input vector from MLP 20-8-1 and averaged errors of 20 GRNN networks along with ± 0.95 confidence ranges (mean value is shown with a measure of its statistical distribution).

	Learning error	Validation error	Test error
Best network	$1.29 * 10^{-3}$	$2.57 * 10^{-3}$	$3.30 * 10^{-3}$
Averaged set	$2.44 * 10^{-4}$	$1.34 * 10^{-2}$	$1.54 * 10^{-2}$
	$\pm 1.42 * 10^{-4}$	$\pm 4.23 * 10^{-3}$	$\pm 4.97 * 10^{-3}$

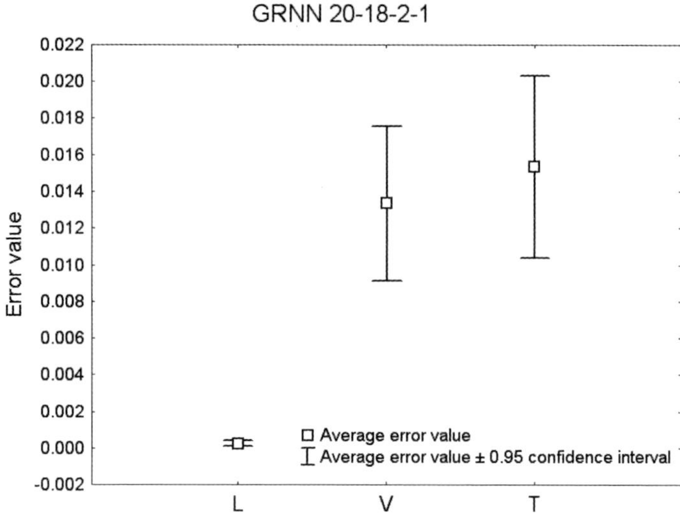

Fig. 6.12. Box-whisker plot of average errors for learning (L), validation (V), and test (T) datasets obtained in a group of the GRNN 20-18-2-1 networks with input vector from MLP 20-8-1. The range indicators provide 95% confidence interval for each set.

6.5.6 MLP network of classification type

After completing the studies on the structure and size of the network and solving the studied problem, a new a new attempt was made to try to achieve even better results. This time the influence of representation of the network's output results on the network robustness was checked. As was already shown an activity can be expressed quantitatively and compared with the experimental results. However, in practice, it may be also described as the classification of the selected compound by the range of its activity. When the **classes** of activity are taken instead of the **values** of the activity, the task of the network is changed from regression to classification. We have attempted to formulate and solve such a classification task within the context of deliberated problems, introducing one **quality** variable (instead of quantity) of one of the four categories (-1 – inhibitors, 0 – substrate with activity up to 50%, 1 – 51 - 150%, 2 – above 150%) described already in the section with presentation of the learning set.

Based on such assumptions and taking the typical form of representation of such networks, given in the form of one-of-N representation again the IPS was used to obtain the optimum structure of the network. 1000 networks studied by the IPS were trained with the same algorithm, *i.e.*, first with back propagation algorithm, then conjugate gradients and finally epochs of conjugate gradients with the momentum factor algorithm. The selected best network had structure presented below:

MLP/CL 13:13-10-4:1

It is worth noting that the work of the IPS resulted this time in a significant reduction of the input dataset (only 13 variables from 24 possible were qualified for use in the application of the network).

The above-mentioned neural architecture was afterwards manually retrained. The learning algorithm comprised of back propagation (100 epochs) followed by 19 epochs of the quasi-Newton algorithm with the momentum factor equal to 0.3. Similarly as in the case of learning the regression MLP networks the Gaussian noise of the amplitude of 0.05 was added to the signals supplied to the network, to avoid the stopping of the training process in local minima.

The results surpassed our highest expectations. The formally calculated values of the errors shown in Table 6.9 are definitely the best of all those achieved in the described (and others omitted for the lack of space) experiments. Moreover, these results are fully consistent with the classification of the same data carried out by an expert. This refers to the learning, the validation, and the test data. This result does not require any further comments.

Table 6.9. Mean values of errors made by the non-linear MLP network with classifying output elements and averaged errors of 20 MLP/CL networks along with ± 0.95 confidence ranges (mean value is shown with a measure of its statistical distribution).

	Learning error	Validation error	Test error
Best network	0.107042	0.009889	0.000605
Averaged set	0.36 ± 0.18	0.32 ± 0.24	1.74 ± 0.84

The multiple network validation procedure (Table 6.9 and Fig. 6.13) shows that the obtained MLP/CL 13-10-4 network is really exceptional in its quality when compared to the group. Its validation and test errors are lower than learning errors. On the other hand, in a group of 20 MLP/CL 13-10-4 the

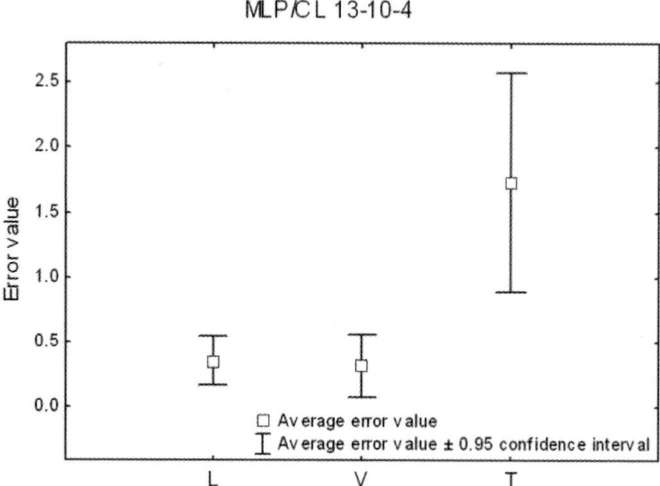

Fig. 6.13. Box-whisker plot of average errors for learning (L), validation (V), and test (T) datasets obtained in a group of the MLP/CL 13-10-4 networks. The range indicators provide 95% confidence interval for each set.

averaged mean errors for learning and validation cases are of the same magnitude, while test errors are an order of magnitude higher. This indicates that on average MLP/CL may be less predictive for new cases then regressive MLPs. However, as depicted by the best network, one can achieve exceptionally good results with classification approach 100% of correct classifications. Such a behavior suggests that training the MLP/CL 13-10-4 depends strongly on initial weights setup and cases partitioning in-to L,V,T groups.

6.6 Comparison of all networks used

Combining together the results of error evaluations during the learning, validation, and testing processes, one can draw conclusions on general abilities of particular networks. Taking into account only the errors encountered during the learning process (see Fig. 6.14) one can see better results for the GRNN networks than for all MLP networks. The worst result obtained for the MLP/CL network is a little surprising, but it can be due to different definition of the error value for regression and for classification networks. The very good result obtained for linear network is off course the artifact produced by the pseudoinverse method applied in the "training" of such models. In fact, this method produces optimal solution for every learning set, but its weakness is the lack of generalization capabilities.

During the validation process (Fig. 6.15) GRNN type network occurred to be the best one as well, but the other results were very close to the theoretical

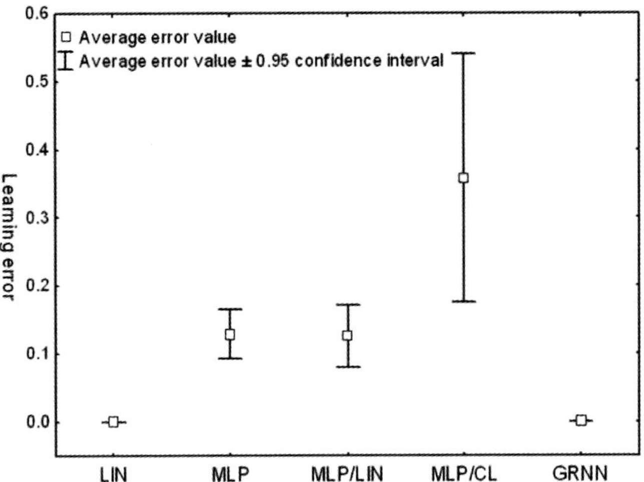

Fig. 6.14. Box-whisker plot of average errors for learning datasets for all types of networks. The range indicators provide 95% confidence interval for each set.

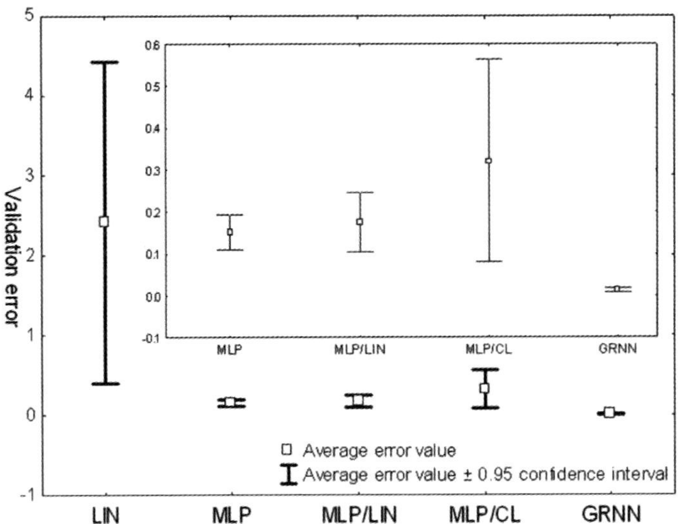

Fig. 6.15. Box-whisker plot of average errors for validation datasets for all types of networks. The range indicators provide 95% confidence interval for each set.

expectations. The huge errors for linear networks were observed (with also big measure of its statistical distribution) and a quite good behavior of all MLP networks. Control of the network behavior performed by means of separate test data (Fig. 6.16) confirm the general observations made on the base of validation set. It means that a good (fully random) segmentation of the known data between learning, validation, and test datasets was achieved and results of the investigations can be found as statistically correct.

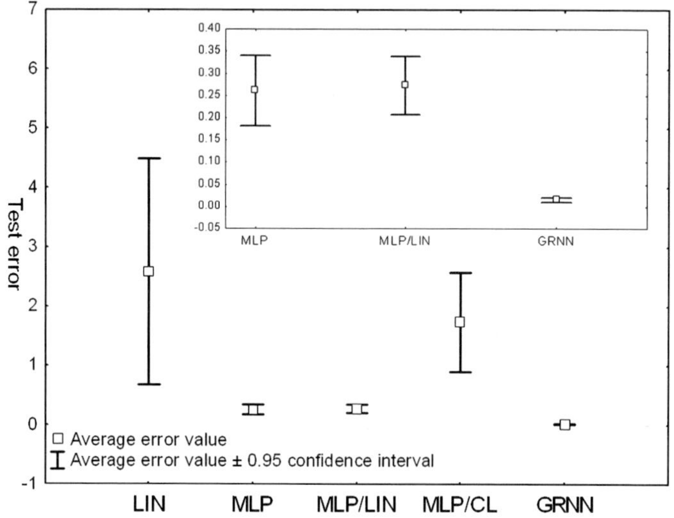

Fig. 6.16. Box-whisker plot of average errors for test datasets for all types of networks. The range indicators provide 95% confidence interval for each set.

6.7 Conclusion

The chapter presents many examples of how neural networks can be used for evaluation and prediction of biological activity of selected chemical compounds. Of course every problem must be solved individually, but there are no doubts that by using neural networks models, we can predict biological activity of new enzymes and drugs. This way of selection of promising compounds is much easier and much cheaper than making synthesis of such compounds and testing them in the lab. Proposed methodology can be specially interesting in the case of chemical compounds which must be tested on animals (*in vivo*). Modeling of the compounds for biological purposes with neural network models, instead of *in vivo* tests, is not only cheaper and easier, but should also be preferred from the deontological point of view.

References

1. Haas O, Burnham K (eds.) (2007) Intelligent and Adaptive Systems in Medicine, CRC Press
2. Horzyk A, Tadeusiewicz R (2005) Comparison of plasticity of self-optimizing neural networks and natural neural networks. In: Mira J, Alvarez JT (eds.) Mechanism, Symbols, and Models Underlying Cognition, Lecture Notes in Computer Science, vol. 3561, Springer-Verlag, Berlin - Heidelberg - New York, Part I, pp. 156–165
3. Szaleniec M, Goclon J, Witko M, Tadeusiewicz R (2006) Application of artificial neural networks and DFT-based parameters for prediction of reaction kinetics of ethylbenzene dehydrogenase. Journal of Computer-Aided Molecular De-sign, 20(3):145–157
4. Tadeusiewicz R (1993) Neural Networks (in Polish). Handbook printed by AOW, Warsaw, Poland
5. Tadeusiewicz R, Izworski A, Bulka J, Wochlik I (2005) Unusual effects of "artificial dreams" encountered during learning in neural networks. In: Yeung DS, Wang X, Zhang L, Huang J (eds.) Proceedings of 2005 International Conference on Machine Learning and Cybernetics, IEEE Press (IEEE catalog number 05EX1059), vol. 7, pp. 4205–4209

7

Using Machine Vision to Detect Distinctive Behavioral Phenotypes of Thread-shape Microscopic Organism

Bai-Tao Zhou and Joong-Hwan Baek

School of Electronics, Telecommunication, and Computer Engineering, Korea Aerospace University, Koyang City, South Korea
{zhou,jhbaek}@kau.ac.kr

Summary. Distinctive Behavioral Phenotypes are defined as those exclusive behaviors that a species has yet others do not. It has facilitated the genetic analysis of many important aspects of nervous system function and development in the nematode such as *Caenorhabditis elegans* (*C.elegans*). Although a large number of distinctive phenotypes can be distinguished by an experienced observer, automatic and precise detection methods have not been systemically addressed. Here we describe a behavioral phenotype detection system for microscopic thread-shape organisms, which is based on automatically-acquired image data. The first part of this chapter introduces an animal auto-tracking and imaging system capable of following an individual animal for a long time by saving a time-indexed image sequences representing its locomotion and body postures. Then we present a series of image processing procedures for gradually shrinking the thread-shape representation into a Bend Angle Series expression (BASe), which later is the foundation of n-order-difference calculation for static and locomotion pattern extraction. Finally, for mining distinctive behaviors, the Hierarchical Density Shaving (HDS) [1] clustering method is applied for compacting, ranking and identifying unique static and locomotion patterns, which combined represent distinctive behavioral phenotypes for a specific species. The behavioral phenotypes detected by this system are partially consistent with animal behaviors described in the literature materials.

7.1 Introduction

Thread-shape animals such as nematodes are the most numerous multicellar animals on the earth. A handful of soil contains thousands of the microscopic worms, many of them are parasites of insects, plants or animals. There are nearly 20,000 described species classified in the *Phylum Nemata*.

The reasons why we study these animals are for computerizing survey records and developing computer-aided identification expert system. In *Phylum Nemata*, some species are threatening the development of farming and forestry all over the world. Therefore, there are requirements for designing

B.-T. Zhou and J.-H. Baek: *Using Machine Vision to Detect Distinctive Behavioral Phenotypes of Thread-shape Microscopic Organism*, Studies in Computational Intelligence (SCI) **122**, 161–182 (2008)
www.springerlink.com

a non-specialist nematode identification system for custom quick inspection and quarantine. Another more important reason is for facilitating phenotype-based genetic interaction analysis and neuronal system modeling. For this purpose, *C.elegans* are such ideal model organism for extensive integrated analysis of the genetics of development. And behavior phenotype analysis to these animals represents a fundamental challenge in this field.

Although a large number of behavioral phenotypes can be distinguished by experienced observers, automatic and precise detection methods have not been systematically addressed. Therefore, researchers are either hindered by the complex computations or turned to other ways. In the field of genetic interaction analysis, some researchers bypassed the formidable phenotype ob-servation and analysis with statistic ways. Recently, Zhong et al [2-5] ex-perimentally predicted interactions for two human disease-related genes and identified 14 new modifiers by statistically combining functionally linked or-thologous genes. They employed a statistic way instead of the modifier screen methods, which rely on easily detectable phenotypes.

The genetic interaction analysis depends on the availability of reliable as-says to detect distinctive abnormal behavior phenotypes of specific species. However, distinctive abnormal behaviors among different species, particularly in some more complex behaviors, such as locomotion, are often subtle and complex, and therefore long-time naked-eye observation is imprecise and sub-jective. For enhancing the reliability in great extent, it is essential to develop an automated recording and analysis systems, and it would be great help for practitioners if there exits automatic phenotype detection tools.

The last two decades have seen the dramatic developments in such disci-plines as computer control, image/video processing, and data mining, which made it possible to conquer those above issues by designing an automatic an-imal tracking, imaging and data analyzing system. The system should have the capabilities of tracking and recording of individual animals for a long time periods, and then followed by implementing a series of image processing and video summarization schemes for recording and abstracting animal's morpho-metric and morpho-anatomical characters. Finally those characterized data can be fed to data mining tools for species identification/classification and knowledge querying purposes.

In this chapter, based on previous experiments [6, 7, 8-10], we introduce an automatic behaviorial phenotype detection system for thread-shape or-ganisms by applying image-sequence processing procedures, simplified data representations, and a series data mining methods. Considering the morpho-logical properties of thread-shape organisms, an orientation-impervious-shape representation is devised for characterizing postures and reducing complexity of geometric transformations and computations as well. In order to find static and locomotion patterns, n-order difference operation are employed. Finally, for mining the distinctive behaviors, typically, only a small number of func-tionally related postures or locomotion actions are clustered into one or more groups for characterizing each species, and the rest need to be pruned. So, the Hierarchical Density Shaving clustering method is chosen for compacting and ranking unique image sequences, which are representing behavioral phenotypes for the current essay sample. In the experimental results, the detected behavioral phenotypes are partially consistent with animal behaviors, which are described in the literature materials.

This chapter is organized as following. After briefly introducing the research background, animal photographing and image processing are given in Section 3 and Section 4 respectively. Then Section 5 discusses data preparation and phenotype detection methods. Further, experiment results with *Caenorhabditis elegans* are given in Section 6. Finally, Section 7 concludes the work.

7.2 Background

Two school of researchers have been dedicating their works to computerized tool design for nematode identification. One group focuses on the designing a taxonomy system by using morphometric and morpho-anatomical measurements. The other group concentrates on designing an automatic morphological and locomotion measuring system for nematode identification and its behavioral phenotype detection.

On the field of taxonomy and identification, some computer systems had been proposed to help for the species identification in nematode genera. To replace the dichotomous keys, Fortuner et al [11-14] proposed computer programs in the identification of species in *HelicotylelzchusSteiner*. They calculated the general coefficient of similarity of Gower upon morpho-anatomical measurements, non-variable characters and variable characters. Jim et al [15] published an article that summarized the characters from all published descriptions and a set of integrated identification tools. Those identification systems are believed to be more reliable and convenient than a traditional dichotomous key because:

(1) new species can easily be added to the nematode database;

(2) the intraspecific and interspecific variability of measurements and morphological characters are represented quantitatively;

(3) the results are presented in a simple manner which allows specific identifications by non-specialists;

(4) the versatility of the program should make it acceptable to all scientists. Worldwide access to the program is assured through public data networks.

For vision-based automatic phenotype identification, in the last two decades, researchers designed new computational algorithms and methodologies for quantitatively analyzing and identifying behavioral phenotypes using morphological and locomotion features. Silva et al [16] proposed a system to detect the existence of nematode in microscopic images. Baek et al [6] designed a machine-vision based *C.elegans* tracking, analysis and classification system, which can distinguish 6 types of animals by feeding 94 morphological and locomotion features. Geng et al [7] classified 16 (15 mutant-types, one wild type) types of worm using a Random Forests (RF) [17] classifier with average out-of-bag error rate of 9.1%. Our previous work [8, 9] proposed a performance-enhanced system, especially for the mutant-type worm identifications, which employed representative shape features by applying orientation-impervious shape representation. The orientation-impervious shape representation forms the basis for behavioral phenotype detection discussed in this chapter.

The above systems are all domain/problem-oriented, and their typical output is a classifier supported by a subset of features and their calibrated measurements, which combined represents unique phenotypes/identifications for each species. The systems' architectures of [6-9] can be described as Figure 7.1.

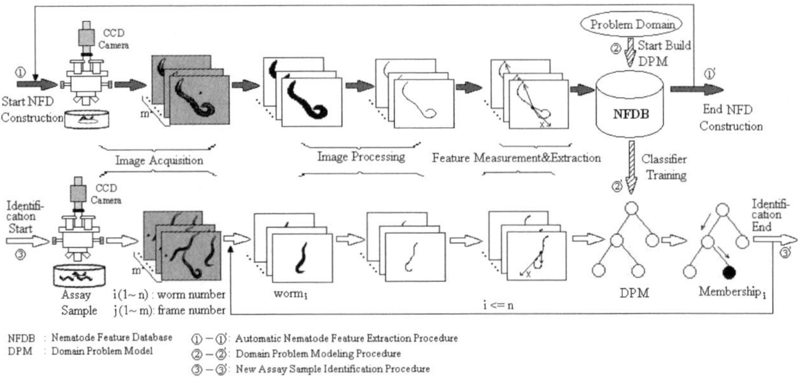

Fig. 7.1. Architecture of Nematode Identification Systems.

However, those features and measurements are more suitable to be biological keys for taxonomy than behavioral phenotypes for neurobiology. In those systems, the described features, like "Maximal tail thickness=6.5", "Local head and tail movement ratio=2.3", and "Average tail brightness=64", are more fitted for narrowing down categories to achieve a species ID. But it's hard for a neurologist to apply those features as a whole to discriminate and interpret abnormal behavior phenotypes caused by uncoordinated mutants and drug inhibitions.

Therefore, higher-level intuitive phenotype representations are necessary. In this subject, Christopher et al [18] proposed an automated system for measuring nematode sinusoidal movements with such parameters as body bend extent, amplitude and wavelength. Unique behavior phenotypes are demonstrated by comparing of those movement parameters, which need further explanations and interpretations. In this chapter, instead of representing behavioral phenotype as some scattered measurements, our new system's output is posture-image sequences (video clips), which can replay the summarized the video clips of a worm's distinctive behaviors, and therefore give much better intuitions to the observers.

7.3 Animal Tracking and Imaging for Slow Moving Microscopic Organism

Although some model organisms are feasible for long time indoor study due to their small sizes and ease of manipulations, the animal movements and microscope magnified effects make long time naked-eyes observations weary and fatigue job. Especially for some active animals such as *C.elegans*, their adults are able to crawl about 25 times their body-length per minute. So, in order to make long-time automatic observation, recording and measurement, it is essential to design a vision-based control system for keeping the animals in the center of field of view and imaging system under a constant illumination as well.

During a recording, first a CCD camera captures and digitizes images at a preset frequency. Next, a tracking program identified the animal in the field of view and saved a grayscale image of the worm, the stage position, the position of the worm in the field of view, and the time of capture. Finally, the tracking software estimate drives the motored stage for counteracting worm locomotion and keeping worm in the center of view. Thus, the system can generate a time-coded sequence of images that represented a nearly complete record of the animal's body movements over an indefinitely long time period. An automatic Animal Tracking and Imaging system is schematically shown in Figure 7.2.

In our previous experiment [6], *C.elegans* locomotion was tracked with a Zeiss Stemi 2000-C Stereomicroscope mounted CCD video camera. And a computer-controlled tracker (Parker Automation, SMC-1N) was employed to put the worms in the center of the optical field of the stereomicroscope. The stereomicroscope was fixed to its largest magnification ($50\times$) during operation. Each single animal was snapped every 0.5 second for at least five minutes, which forms a video file unit containing around 600 images to record the animal shapes and behaviors. For each species, the recording procedure was repeated one hundred times on different animals, which produced around 60,000 frames for each species and totally 60,000$\times$5 worm frames for entire assay samples.

Among those image pixels with values less than or equal to the average value minus three times the standard deviation, the largest connected component was found. The image was then trimmed to the smallest axis-aligned rectangle, and saved as eight-bit grayscale data. The dimension of each image and the coordinates of the center of mass of the worm in the tracker field were also saved simultaneously as the references for the location of an animal in the tracker field at the corresponding time point when the images are snapped.

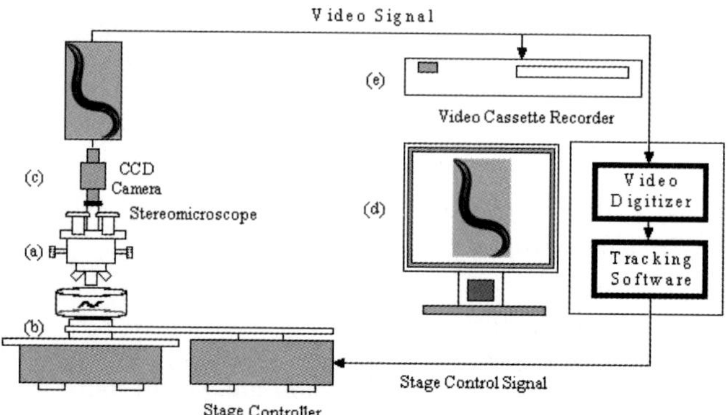

Fig. 7.2. Animal Tracking and Imaging system. (a) Stereomicroscope; (b) Motorized stage connecting to PC serial port; (c) Monochrome analog CCD camera and its view; (d) PC with a video acquisition board showing the trimmed image; (e) VCR cross verification and behavioral tracking.

Depending on the type and the posture of a worm, the number of pixels per image frame varied although the number of pixels per millimeter was fixed at 312.5 pixel/mm for all worms.

7.4 Image Processing Methods for Thread-shape Microscopic Organism

To facilitate behavioral phenotype analysis, the grayscale images are subjected to a series image processing procedures for achieving simplified representations of animal shapes and postures, which are shown in Figure 7.3, and the processing results are shown in Figure 7.4.

First, the gray scale image is converted into binary image based on the distribution of the gray value. Secondly, the spots inside the worm body are removed with a morphological closing operator. Thirdly, the multiple objects are segmented with the sequential algorithm [19] for component labeling, and the areas containing worms are processed separately. The areas containing isolated objects, such as eggs, are removed by setting off pixel. At last, a morphological skeleton can be obtained by applying a skeletonizing algorithm [20] through iterating dilation operation along the binary image.

Another aspect need to mention here is to distinguish the looped holes and noise spots inside the worm body. If we use the operation mentioned in the second step, some holes may be filled as the spots inside the worm body. To remove the spots inside the worm body and to keep the holes simultaneously, we measure thickness of these two regions with 16 vectors. The details are given in 7.4.2.

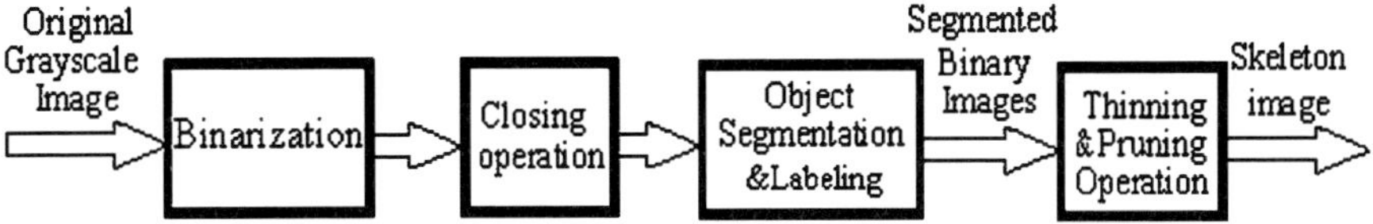

Fig. 7.3. Image processing procedures of thread-shape microscopic animal image.

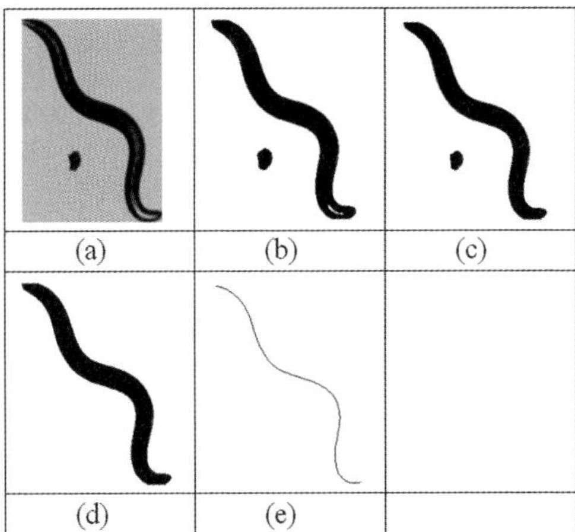

Fig. 7.4. Image processing results. (a) Original gray level image; (b) Binary image after thresholding operation; (c) Binary image following closing operation; (d) Final clean binary image after removal of isolated object; (e) Skeleton obtained through thinning and pruning.

7.4.1 Binarization

First, background intensity level (b) can be found by taking the maximum value of the four corner points of the trimmed image because the background intensity has a relatively constant value for the microscope controllable illuminance, and at least one of the four corners are not occupied by the worm's body. To decide upon the threshold, a 5×5 moving window is used for scanning over the experimental image, and the mean (m) and standard deviation (σ) of the 25 pixels inside the window are computed at every pixel position. If the mean was less than $0.7b$ or the (σ) is larger than $0.3m$, then the pixel is decided as a pixel of the worm body and assigned 1. Otherwise it is marked as foreground value 0. The binarization process is defined as Equation 7.1.

$$g(x,y) = \begin{cases} 0 & \text{if } m(x,y) < 0.7b \text{ or } \sigma(x,y) > 0.3m \\ 1 & \text{otherwise} \end{cases} \tag{7.1}$$

where

$$m(x,y) = \frac{1}{5 \times 5} \sum_{i=-2}^{2} \sum_{j=-2}^{2} f(x+i, y+j)$$

$$\sigma(x,y) = \sqrt{\frac{1}{5 \times 5} \sum_{i=-2}^{2} \sum_{j=-2}^{2} [f(x+i, y+j) - m(x,y)]^2}$$

After binarization, median filtering is performed. A median filter [21] has a superior effect for removing impulse noise. Median filtering can preserve small sized holes and remove impulse noise, which are caused by light reflection from the worm body. In this paper, we use a 9×9 window for median filtering to remove impulse noise in the binary worm image.

7.4.2 Morphological Operations with Hole Detection

A morphological closing operator (binary dilations followed by erosions) [21] erases noise and spots on and inside the worm body. In order to avoid occasional false cleaning up holes formed by head-tail touching, as shown in Figure 7.5, a reference binary image is also generated on parallel by detecting, labeling, and filling holes.

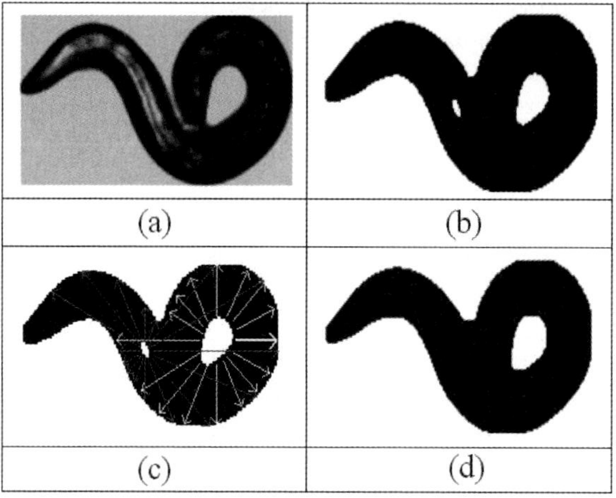

Fig. 7.5. Hole detection. (a) Original grayscale image. Image contains a coiling hole. (b) Binary image. Image is obtained after binarization and median filtering. (c) Test each labeled area except worm body using 16 vectors. (d) Binary image after hole detection and noise removal.

To eliminate the remaining noise and keep important holes, a procedure for distinguishing between hole and noise is applied. First, image inversion is performed for the binary image, which converts black pixels to white, and white to black. Then, object labeling is applied to the noise or hole regions, except worm's body and the background. And the coordinates of the centroid of each region are computed. Therefore, the determination of whether the region is a hole or noise can be made by measuring the total thickness of the body enclosing the region. In order to measure the thickness, 16 vectors are defined. Each vector traverses and counts the number of pixels from the centroid of a region in its direction until it reaches the background. Then thickness is calculated by multiplying the magnitude by the number of pixels traversed. The total thickness is the sum of two opposite direction thicknesses. So, among the 8 total thicknesses, if the minimum is less than a threshold, the region is considered as noise. If a region is determined as noise, the region will be filled with body pixels, otherwise, preserved the region as a hole. These procedures are repeated for all of the labeled regions.

7.4.3 Object Segmentation and Labeling

Even though the closing operation is performed, it is possible that other objects could exist apart from the worm body and hole. A worm's crawling tracks or eggs could cause the unwanted objects. In order to remove those small objects, sequential algorithm [18] for component labeling will be performed by scanning the image in x and y directions sequentially, and then the largest connected object is found as the worm's body while other labeled objects are removed. After removing the isolated objects, the hole region will be restored onto the worm's body.

7.4.4 Thinning and Pruning

After isolating unwanted objects by object labeling, a morphological skeleton was obtained by applying a skeletonizing algorithm [20]. Redundant pixels on the skeleton were eliminated by thinning, which is shown in Figure 7.4(e). To avoid branches on the ends of skeletons, the skeleton was first shrunk from all its end points simultaneously until only two end points were left. These two end points represent the longest end-to-end path on the skeleton. A clean skeleton can then be obtained by growing out these two remaining end points along the unpruned skeleton by repeating a dilation operation.

7.5 Distinctive Behavioral Phenotype Detection

7.5.1 Head and Tail Recognition for Each Frame

Since animal's head and tail are always presenting different behaviors, it is essential to design an automatic identifying process, which forms the base step for further behavioral phenotype analysis. Although the mass center of each animal can be obtained during the tracking and imaging step, it was not a proper time for detecting head and tail due to the manipulating inconvenience of the graylevel and binary images.

After getting skeleton images, the two end points on each skeleton image are used to represent animal's head and tail, but by far the correspondences are not clear. So an automatic identification scheme is designed for grouping and marking head and tail ends of each skeleton image within each video unit. The head-tail identification is based on two clues that come from field observations: 1) the head moves more frequently than the tail; 2) head-head and tail-tail distances between two consecutive frames are relatively much smaller than the distances of head-tail and tail-head in the same two frames. Figure 7.6 shows the statistic results of one video unit.

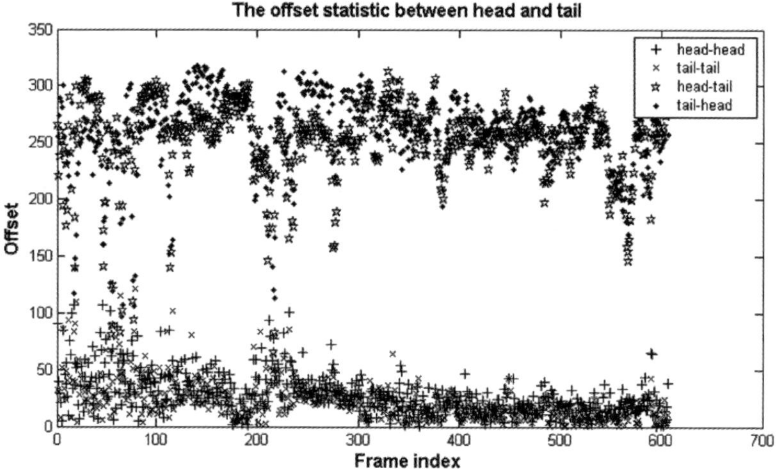

Fig. 7.6. The offsets of head-head, tail1-tail, head-tail, and tail-head from every two consecutive frames of 610 frames in a video file. The head-head and tail-tail offsets is much smaller then the ones of tail-head and head-tail.

For head and tail recognition, first, those two end points will be divided into two groups (*group1* and *group2*) according to the position similarity between two consecutive frames, same method can also be found in [7]. Let $Endpt_m(t) = (endpt_{mx}(t),\ endpt_{my}(t))$ denote the mth endpoint in frame t, and $Endpt_n(t+1) = (endpt_{nx}(t+1),\ endpt_{ny}(t+1))$ as the nth endpoint in frame $t+1$. The distance between endpoint n in frame $t+1$ and the endpoint m in frame t is d(m, n). We initialize the group1 as $(endpt_{1x},\ endpt_{1y})$ and group2 as $(endpt_{2x},\ endpt_{2y})$.

For each following frame, if the head and tail are not being accidentally flipped, an endpoint will be assigned into the same group in which its location is closest to the endpoint in previous frame, under the condition that the distance between the other point in current frame and the other point in the previous frame is not the maximum. So the the kth endpoint in frame $(t+1)$ $Endpt_k(t+1)$ will be assigned as group g according to the Equation 7.2 provided that the other point which was assumed into the other group, and the distance between the other point and the endpoint that belongs to same group in frame t is not the maximum at the same time. Here $g(v)$ is the group to which the vth endpoint in frame t belongs. This grouping rule actually is implemented according to the coordinate of the end points in adjacent two frames.

$$(k,g) = arg(k, g(v))\{d(Endpt_k(t+1), Endpt_v(t)) == min\ \{d(m,n)\}\}$$
$$\text{where } d(Endpt_{\bar{k}}(t+1), Endpt_{\bar{v}}(t)) \neq max\{d(m,n)\},$$
$$v, k \in \{0,1\},\ m \in \{Endpt_0(t), Endpt_1(t)\},$$
$$n \in \{Endpt_0(t+1), Endpt_1(t+1)\} \tag{7.2}$$

The groups with bigger offset will be marked as the head group ($Headg$), and the one with smaller offset as the tail group ($Tailg$). The experiment result can be seen in Figure 7.7, in which, the head is marked with a solid circle. This result is obtained by feeding the sample points to the wormtools2p1 of Caltech Nematode Movement Analysis System, which is an amazing software provided by Sternberg Lab [18].

7.5.2 Orientation-impervious Shape Representation

Even though a skeleton image (a pixel sequence) can describe animal shape or posture with much less data comparing with the original graylevel images and binary images, it is still not efficient and convenient for shape representation and comparison. The animal's orientation arbitrariness and length variability make skeleton-based shape computation complex and impracticable. In addition, it is burdensome work to mining 5 species data of 5×100 files, each of which contains around 600 skeletons represented by 130 pixels approximately. Therefore, it is essential to find a more efficient way of shape representation and further reducing the amount of data.

In orientation-impervious-shape representation scheme [22] - a Bend Angle Series expression (BASe) is devised and applied to maintain adequate information for characterizing shape/postures, and to reduce computation complexity of geometric transformations and data mining computations as well. In BASe, an animal's posture/shape is represented by a $n-2$ bend-angle series(vector): $B_\alpha = \{\alpha_1, \alpha_2, ..., \alpha_{n-2}\}$, where α_i is an included angle formed by lines l_i and l_{i+1}. l_i is connected by points p_i and p_{i+1}, which are evenly sampled from head to tail in the skeleton pixel sequence. The point p_i can be expressed as (x_i, y_i); and line l_i can be $(x_i, y_i, x_{i+1}, y_{i+1})$. All the n sampled points consist of $2\times n$ matrix P, from which all the $(n-1)$ lines consist of $(n-1)\times 4$ matrix L, and from these lines, all $(n-2)$ angles construct $1\times(n-2)$ matrix α, as following:

$$P = \begin{bmatrix} x_1 & x_2 & \cdots & x_n \\ y_1 & y_2 & \cdots & y_n \end{bmatrix}$$

$$L = \begin{bmatrix} x_1 & y_1 & x_2 & y_2 \\ x_2 & y_2 & x_3 & y_3 \\ \cdots \\ x_{n-1} & y_{n-1} & x_n & y_n \end{bmatrix}$$

$$\alpha_i = \arccos \frac{(x_{i+1}-x_i, y_{i+1}-y_i)\cdot(x_{i+2}-x_{i+1}, y_{i+2}-y_{i+1})}{\parallel(x_{i+1}-x_i, y_{i+1}-y_i)\parallel\parallel(x_{i+2}-x_{i+1}, y_{i+2}-y_{i+1})\parallel} \tag{7.3}$$

BASe obviously has its advantages. It shrinks the shape representation from two dimensions (skeleton points) to one dimension (angle sequences), which is a data reduction process on the condition of keeping enough shape information. Representing shape as a series angle sequence, the BASe not only provides more flexible and convenient ways for shape comparison and detection, but also removes the disadvantage led by animal orientation arbitrariness (crawling towards any direction). In addition to the orientation-invariant property, evenly sampled BASe also has the property of scale invariance: different animals' length within one species do not affect the behavior analysis. In our experiment, each $C.elegan$ animal is represented by 19 points, evenly sampled on its skeleton from head to tail, forming 18 segments by lining each two adjacent points, which approximately represent a segment of the skeleton. And the worm's shape is abstractly pictured by using the 17 bend angles between these 18 lines, each of which is formed by two adjacent line segments. And these 17 angles are represented in radians using a vector $B_\alpha = [\alpha_1, \alpha_2, ..., \alpha_{17}]$. Figure 7.7 shows those shapes with different orientations, which can be grouped together by the BASe.

7.5.3 Habitual Behavior Pattern Detection

A habitual behavior of an animal is described as a series of time-indexed frames (a video clip) that observed in a video file, and the clip can also be found in other video files with a certain frequency. The habitual behaviors are detected and collected to form a video clip set, which combined are used to characterize a species's behaviors and locomotion.

As described before, an animal's habitual behavior or time-index postures can be described by those vectors of bend angle series. An alternative way of describing a habitual behavior is using (0, 1 ,..., n)-order differences. Shown in Equation 7.4 and Figure 7.8, an example of (k+1)-frame habitual behavior, which starts from $b(i)$, can be represented by (0 , 1,..., k)-order differences of $b(i)$: $\Delta^0[b(i)]$, $\Delta^1[b(i)]$, $\Delta^2[b(i)]$, ..., $\Delta^k[b(i)]$. 0-order differences depict the

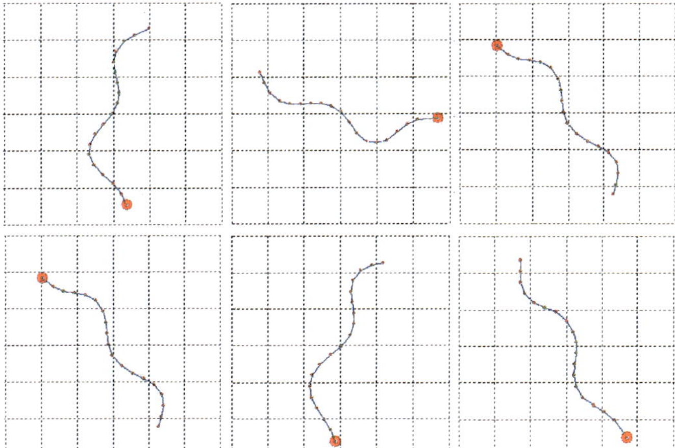

Fig. 7.7. One of the shape patterns of *Unc36* with 1.5 absolute distances. 6 similar shapes represented by BASe can be detected and grouped together even though with different orientations.

dissimilarity between *b(i)* and *b(j)*. 1-order differences describe changing tendency between each two continuous frames, and 2-order differences are representing the changing tendency between each three neighbor frames. Similarly, (k-1)-order difference depict the two changing tendencies of the first k frames (b(i), ..., b(i+k-1)) and the second k frames (b(i+1), ..., b(i+k)), and the k-order difference depicts the changing tendency of all the k+1 frames, or the global variances.

A (k+1)-frame habitual behavior pattern can be detected and decided by comparing (0 , 1,..., k)-order differences between source and target behavior. The detection process is illustrated in Figure 7.8. Assume one source behavior composed of two frames starts from *b(i)* frame in a video file. Whether the behavior is a pattern or not depends on whether the same behaviors are frequently and evenly appeared in other video files or not. To make this judgement, frame comparisons should be carried between *b(i)* and any other forthcoming frame, for instance, b(j), and also between *b(i+1)* and *b(j+1)*. In practice, instead of directly doing the 4 frame comparisons, 1-order differences ($\Delta^1[b(i)]$ and $\Delta^1[b(j)]$), which representing changing tendencies of those two two-frame patterns, are compared first. If the global variances are within a tolerance value ε, then secondly, the corresponding starting frames (*b(i)* and *b(j)*) will be compared to determine whether they are initializing with the same shape or not. If the starting frames are similar, then the two-frame behavior started from *b(i)* may potentially be a behavior pattern and it needs further confirmation by finding the same behavior in other video files; Otherwise, those two two-frames have the similar changing tendency, but not the same starting point - therefore they are different behaviors. In general, in order to find a *(k+1)*-frame behavior pattern, firstly the k-order differences of the source and target behaviors will be compared to measure the global similarity of all k+1 frames. If match, then do *(k-1)-order* differences comparison for sub-global similarity measuring. In this way, the process continues until the 2-order and finally the 1-order difference comparison, then a potential *(k+1)* frame behavior is found. If in the middle of difference comparison chain, any chain fails the comparison, the process stops and claim the source and target's *(k+1)* frames do not belong to the same pattern.

$$\Delta^k[b(i)] = \sum_{m=0}^{k} \binom{m}{k}(-1)^{k-m}b(i+m) \tag{7.4}$$

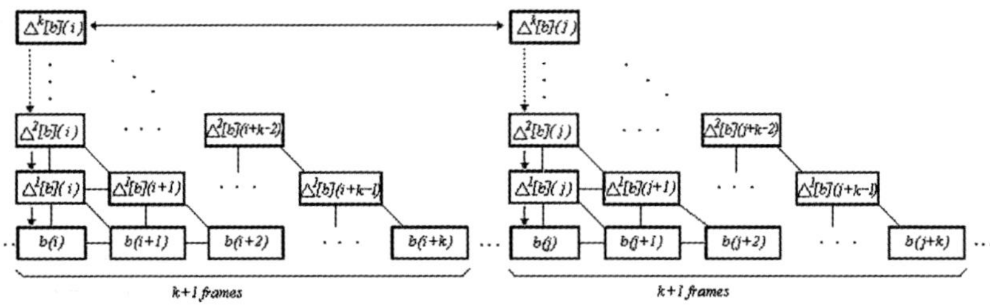

Fig. **7.8.** Habitual behavior description and detection by using k-order difference.

In this experiment, M_α, a BAS matrix, is employed to store and represent those video frames. And 1-order and 2-order differences of the M_α are used to represent changing tendencies of very two and three consecutive BASs. For shape detection, we will compare the difference between any two rows from M_α, and the 1-order and 2-order differences of M_α are employed for locomotion pattern discovery.

$$M_\alpha = \begin{pmatrix} \alpha_{1,01} & \alpha_{1,02} & \cdots & \alpha_{1,17} \\ \alpha_{2,01} & \alpha_{2,02} & \cdots & \alpha_{2,17} \\ \alpha_{3,01} & \alpha_{3,02} & \cdots & \alpha_{3,17} \\ & \cdots & & \\ \alpha_{n,01} & \alpha_{n,02} & \cdots & \alpha_{n,17} \end{pmatrix} \tag{7.5}$$

As shown in Figure 7.9, Three-frame A and B belong to a habitual behavior pattern, which are extracted using (1, 2)-order differences. A and B come from different video recordings of two *n2* worms, one is from frame 105 to 107, and the other is from 300 to 302. Figure 7.10 demonstrates why those two three-frame patterns belong to the same behavior pattern. A1, A2,and A3 show the BAS comparisons of corresponding three pairs (105 to 300, 106 to 301, 107 to 302), their similarities lead to the corresponding 1-order differences (shown

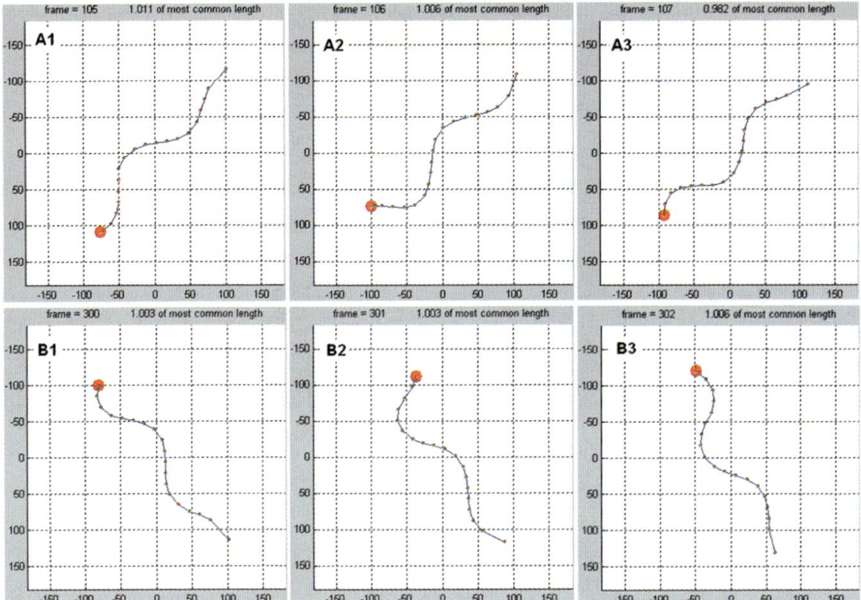

Fig. 7.9. A habitual behavior pattern shown in A and B. A and B come from two worms within a species. The behaviors measured by the BAS (0-order), 1-order and 2-order difference: static epl=0.2, locomotion ep2=0.25.

in B1 and B2) drawing in the same plot, and therefore the 2-order difference in B3. From here, Figure 7.9 and Figure 7.10 show the BASe and n-order differences are efficient and reasonable for extracting locomotion pattern.

7.5.4 Distinctive Behavioral Phenotype Extraction

In general, behavior pattern contains two classes: static patterns and locomotion patterns. Static pattern is a posture behavior, in which the animal keeps stillness or tiny movement during a specific time span. Locomotion pattern contains several different postures, which depict a series of posture changes. Different strains may have various occurrences of specific static pattern and locomotion patterns.

Distinctive Behavioral Phenotypes are defined as those exclusive behaviors that a species has yet others do not. The habitual behavior patterns can be extracted for each type, as discussed in Section 7.5.2. For 3-frame patterns, the static pattern has smaller 1-order difference and 2-order difference. The static patterns can be explored although the corresponding frames or animals have different shape as shown in Figure 7.11, depending on the difference operation, which ignores the single frame shape.

Locomotion patterns are the second aspect to describe behaviors of animals. Same changing tendencies can be found by comparing the corresponding

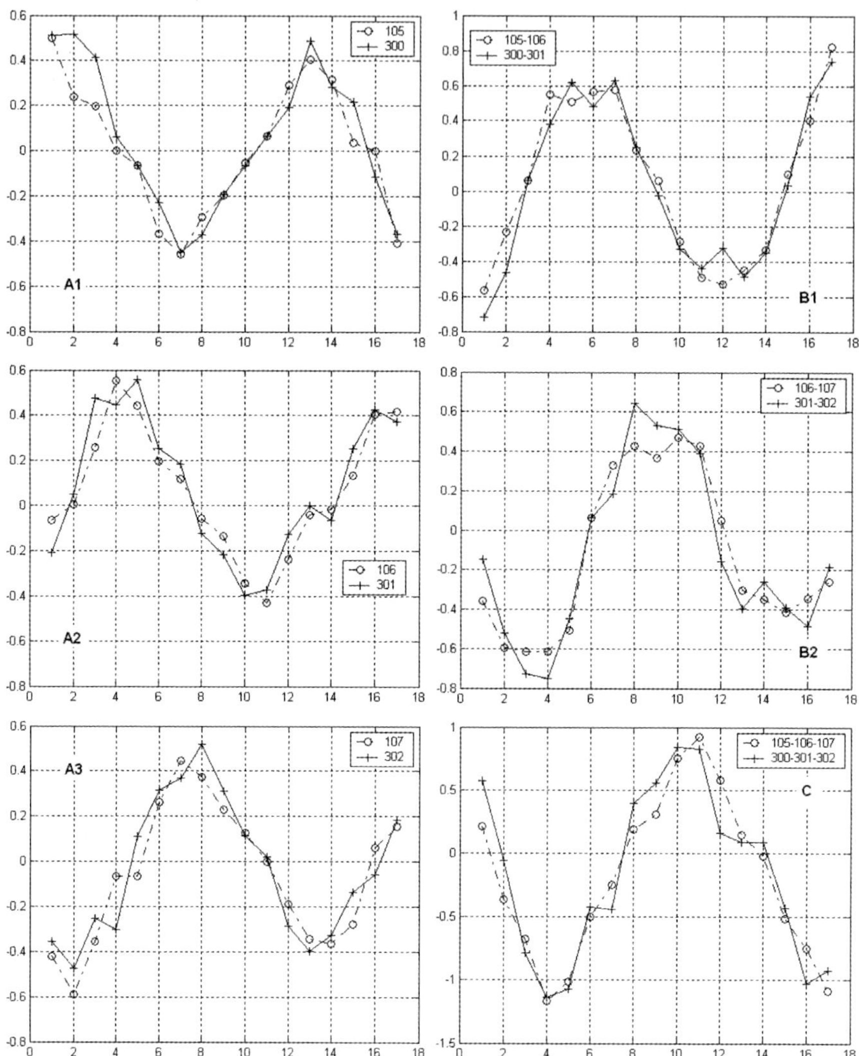

Fig. 7.10. The BAS(0-order), 1-order, and 2-order differences representation of the habitual behavior pattern shown in Fig.8. A1 to A3 are the BAS; B1 and B2 show the 1-order differences; C is the 2-order differences.

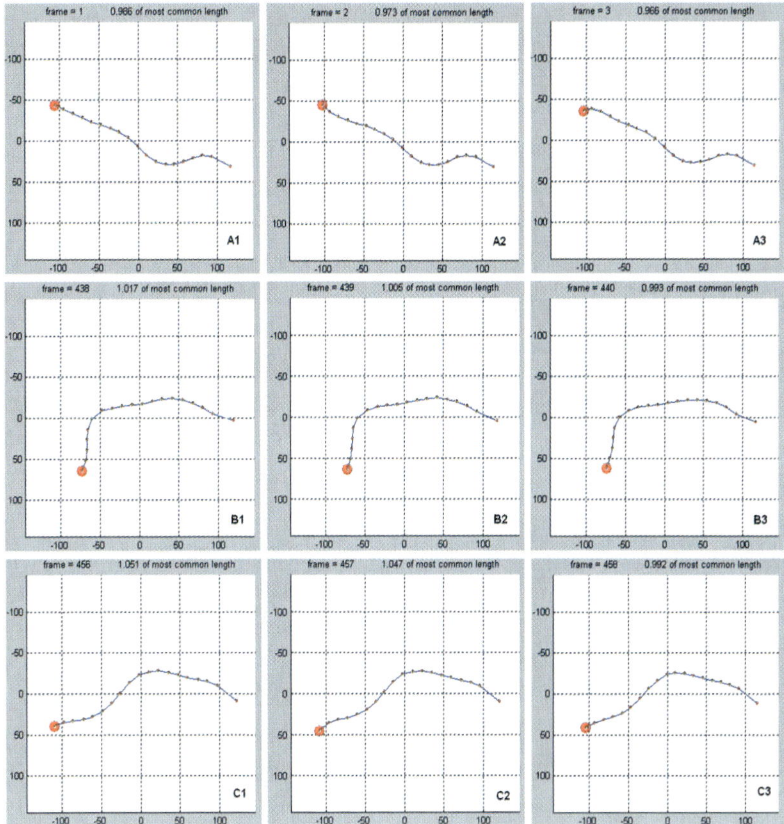

Fig. 7.11. Static pattern examples found in *nic-1* animals based on the BASe, with 1-order and 2-order difference. Even though four animals with different posture, they have similar changing tendencies, keeping still or move little distance

2-order difference and 1-order difference. Same or similar changing tendencies in 3 continuous frames mean same or similar 2-order difference and 1-order difference, which is indicated as Fig. 10 (C, B1 and B2). These two *n2* animals moved severely with same locomotion pattern. Their BASes of the continuous three original skeletons are as Fig. 10 (A1-A3); the Fig.10 (B1 and B2) dictates the offset of the bend angle from frame 1 to frame 2 and from frame 2 to 3 respectively. The Fig.10 (C) reflects the global locomotion trend among these 3 frames.

After obtaining the locomotion patterns, the distinctive and represented patterns will be selected for each type. Density shaving is suitable for this requirement, which is employed to extract a subset of data clustering well. Gunjan et al [23] proposed the Automated Hierarchical Density Shaving (Auto-HDS) in 2006, "a framework that consists of a fast, hierarchical, density-based clustering algorithm and an unsupervised model selection strategy".

The selected data corresponding to the locomotion patterns are fed into the Auto-HDS software - Gene Density Interactive Visual ExplorER (DIVER) to check the distribution of those locomotion patterns from different species. The clustering finds the stand-alone areas which are not or less overlapping by other strains. The locomotion patterns located in those areas are seen as the distinctive locomotion phenotypes.

7.6 Experiment Results with Caenorhabditis elegans

7.6.1 Strains of Animals

In this experiment, there are 5 strains (1 wild type *n2*, 4 mutant-types as *nic-1, unc29, unc36, unc38*), which are described in [7] and shown in Table 7.1. For each strain, there are two groups of data sets (Data10, Data20) consisting of 10 (20) 5-min recordings, in which images are captured at the frequency of 2Hz. Each recording forms a video unit containing around 600 images representing the animal shapes and behavior. The phenotype data is extracted through binarization, skeleton, sampling and until the final BASe. The bend angle matrixes consisting of these continuous bend angle vectors are the input data for extracting the distinctive phenotypes, including shape phenotype in one frame, static phenotype and locomotion phenotype in a specific duration.

Table 7.1. The Description of Mutant Types

Strains	Defective molecule	Description	Source
n2	N/A	Normal	Mendel et al., Science 267, 1652 (1995); Segalat, et al., Science 267, 1648 (1995)
nic-1	Type1 glycosyltransferase	Hyperactive, defective male mating	Mendel et al., Science 267, 1652 (2995)
unc29	Nicotinic receptor Beta subunit	Weak kinker, head region stiff, moves better in reverse	Fleming et al., Neurosci. 17, 5843 (1997)
unc36	Voltage-gated calcium channel (VGCC) alpha2 /theta subunit	Very slow, thin loopy at rest	Brenner, Genetics 77, 77 (1974)
unc38	Nicotinic receptor alpha subunit	Weak kinker, sluggish, slightly dumpyish	Fleming et al., J. Neurosci. 17, 5843 (1997)

7.6.2 Data Preprocessing

Each column of the BAS matrixes has small values; The difference between each two columns is big sometimes. And also each column, which represents one of the 17 angle features, has different data distribution. To avoid the dependence on some of those columns, the BAS matrixes are first normalized by using z-score.

In addition, the density shaving are only applied to locomotion phenotypes because the Auto-HDS's shaving operations will be misled by those static phenotypes, which has relatively smaller dissimilarity than locomotion ones. So, The static phenotypes will be filtered out before clustering, and the density shaving processes are only for locomotion phenotypes.

7.6.3 Extracting Results

The proposed approach was tested for the $wild(n2)$, $nic - 1$, $unc29$, $unc36$ and $unc38$. The static behavior phenotype is calculated with the occurrence ratio. We checked the occurrence of static pattern in different size of datasets (Data10, Data20) respectively. In addition, the locomotion behavior phenotype is represented as cluster numbers, the experiment results from those two dataset are listed here: Table 7.2 is the static pattern result for 5 types of animals, using Data10 and Data20; Table 7.3 and Table 7.4 are for the locomotion pattern results. For static patterns, the static pattern occurrence number and the ratio of this number are calculated as measurement.

Table 7.2. The number of occurrence of static pattern for each assay species

Dataset	Measurement	*n2*	*nic-1*	*unc29*	*unc36*	*unc38*
Data10	# Occurrence	9	317	160	132	69
Data10	Occurrence Ratio	0.0016	0.0534	0.0345	0.0255	0.0148
Data20	# Occurrence	25	617	249	531	226
Data20	Occurrence Ratio	0.0023	0.0511	0.0268	0.049	0.0241

Locomotion Pattern Extract Result from Data10:

The top 37 patterns are regarded as the final clustering result, among them $n2$ has 12 distinctive patterns; $nic - 1$ has 3 distinctive patterns; $unc29$ doesn't have distinctive patterns; $unc36$ has 11 distinctive patterns; $unc38$ doesn't have distinctive patterns. Besides, there are 11 patterns which are shared by at least two types of worms as Table 7.2.

Locomotion Pattern Extract Result from Data20:

Top 100 patterns are selected as the representative patterns for these 5 types of animal. Similar to the results from 10 files in each type, $n2$ and

Table 7.3. The number of distinctive patterns without overlapped and with less overlapped for each assay species from Data10

	n2	nic-1	unc29	unc36	unc38
# of Distinct Patterns	12	3	0	11	0
# of Shared Patterns	0	6	1	2	1
Ratio of Shared Patterns	0	0.66667	0.33333	0	0
	0	0.66667	0.33333	0	0
	0	0.66667	0	0	0.33333
	0.00729	0.59142	0.09489	0.08029	0.22628
	0	0.66667	0	0	0.33333
	0	0.59091	0.022727	0.13636	0.25
	0	0	0.66667	0	0.33333
	0	0	0.33333	0.66667	0
	0	0.33333	0	0.66667	0
	0	0.33333	0	0	0.66667

unc36 have obvious distinctive patterns; *nic-1*, *unc29* and *unc38* shared several patterns, which mean they have similar locomotion patterns, as Table 7.3.

Table 7.4. The number of distinctive patterns without overlapped and with less overlapped for each assay species from Data20

	n2	nic-1	unc29	unc36	unc38
# of Distinct Patterns	43	4	0	25	0
# of Shared Patterns	0	6	4	11	1
Ratio of Shared Patterns	0	0.6667	0	0	0.3333
	0	0	0.6667	0.3333	0
	0	0	0.6667	0.3333	0
	0	0.3333	0.6667	0	0
	0	0.2500	0	0.7500	0
	0	0	0.3333	0.6667	0
	0	0	0	0.8000	0.2000
	0	0	0	0.7500	0.2500
	0	0	0	0.6667	0.3333
	0.00631	0.2162	0.0541	0.6486	0.0811
	0	0.3333	0	0	0.6667

From the above experiment results, we can see *n2* and *unc36* have obvious locomotion patterns. There exists many shared patterns among *nic-1*, *unc36* and *unc38*, which match the description as Table 7.4. from [7] to some extent. *nic-1* has a few distinctive patterns and obvious high ratio of statistic patterns, the combination of these two measurements can be its distinctive features. Besides, *unc38* also has higher static locomotion patterns than *unc29*.

7.7 Conclusion and Future Work

The study represents the first step toward developing computer vision methods for detecting distinctive behavioral phenotypes of thread-shape microscopic animals. The output is a high-level phenotype representations, a summarized posture-image sequences (video clips), which gives better intuitions. In our system, even though photographing frequency is low and three out of the five types are moving poorly, which should affect the performance greatly, the system can still find distinctive locomotion phenotypes for each species by combining static and locomotion patterns.

Future work can be carried and observed on hyperactive animals with bigger video data of higher frame-rate. Besides, worm body contour-bend-angle-series and higher-order difference comparisons may give finer description of animal's posture and locomotion.

References

1. Gupta GK (2006) Robust Methods for Locating Multiple Dense Regions in Complex Datasets. Ph.D. thesis, The University of Texas at Austin
2. Zhong W, Sternberg PW (2006) Genome-wide prediction of C. elegans genetic interactions. Sci 311: 1481–1484
3. Troyanskaya OG, Dolinski K, Owen AB, Altman RB, Botstein D (2003) A Bayesian framework for combining heterogeneous data sources for gene function prediction (in Saccharomyces cerevisiae). Natl. Acad. Sci. 100:8348–8353
4. Jansen R, et al. (2003) A Bayesian networks approach for predicting protein-protein interactions from genomic data. Sci. 302:449–453
5. Lee I, Date SV, Adai AT, Marcotte EM (2004) A probabilistic functional network of yeast genes. Sci. 306:1555–1558
6. Baek J, Cosman P, Feng Z, Silver J, Schafer WR (2002) Using machine vision to analyze and classify C.elegans behavioral phenotypes quantitatively. J. Neurosci. Methods 118:9–21
7. Geng W, Cosman P, Berry CC, Feng Z, Schafer WR (2004) Automatic tracking, feature extraction and classification of C. elegans phenotypes. IEEE Trans. Biomedical Engineering. 51:1181–1820
8. Zhou B, Nah W, Lee K, and Baek J (2005) A general image based nematode identification system design. Lecture Notes in Artificial Intelligence, Springer-Verlag Berlin Heidelberg 3802:899–904
9. Zhou B and Baek J (2006) An automatic nematode identification methods based on locomotion patterns and representative shape features. Lecture Notes in Bioinformatics, Springer-Verlag Berlin Heideberg 4115:372–380
10. Zhou B and Baek J (2006) A nematode identification system based on locomotion features and representative shape patterns. Proc. of the 2006 Korean Signal Processing Conference (KSPC06), Korea 79–83
11. Fortuner R (1970) On the morphology of Aphelenchoides besseyi Christie, 1942 and A. siddiqii n. sp. (Nematoda, Aphelenchoidea). J. Helminth. 44:141–152

12. Fortuner R and Wong Y (1984) Review of the genus Helicotylenchus Steiner, 1945. 1 : A computer program for identification of the species. Revue Nématol. 7:385–392
13. Fortuner R (1988) Nematode Identification and Expert System Technology. New York, NY, USA, and London, UK, Plenum Press
14. Fortuner R (1998) Computer assisted semi-automatic identification of Helicotylenchus species. The program NEMAIDCalif: Pl. Pest and Dis. Rep. 2:45–48
15. Diederich J, Fortuner R, Milton J (2000) Genisys and computer-assisted identification of nematodes. J. Nematology. Earth and Environmental Sci. 2.1:17–30
16. Silva CA, Magalhaes KMC, Dória Neto AD (2001) An intelligent system for detection of nematodes in digital images. Proc. of the International Joint Conference on Neural Networks, Vol. 1 (2003) 20–24
17. Breiman L (2001) Random forests. Machine Learning 45(1):5–32
18. Cronin CJ , Mendel JE, Mukhtar S, Kim Y-M, Stirbl RC, Bruck J, Sternberg PW (2005) An automated system for measuring parameters of nematode sinusoidal movement. BMC Genetics 6(1):5
19. Jain R, Kasturi R, Schunck BG (1995) Machine Vision. McGraw-Hill Inc.
20. Zhang TY, Suen CY (1984) A fast parallel algorithm for thinning digital patterns. Comm. ACM 27(3):236–239
21. Gonzalez R, Woods R (2002) Digital Image Processing, 2nd ed. Englewood Cliffs, NJ: Prentice Hall
22. Golland P, Grimson WEL, Kikinis R (1999) Statistical Shape Analysis Using Fixed Topology Skeletons: Corpus Callosum Study. Proc. of 16th International Conference on Information Processing and Medical Imaging, Leture Notes in Computer Science 1613:328–387
23. Gupta G, Ghosh J (2006) Bregman Bubble Clustering: A robust, scalable framework for locating multiple, dense regions in data. Proc. of the International Conf. on Data Mining (ICDM'06) 232–243

Contour Matching for Fish Species Recognition and Migration Monitoring

Dah-Jye Lee[1], James K. Archibald[1], Robert B. Schoenberger[2], Aaron W. Dennis[1], and Dennis K. Shiozawa[3]

[1] Department of Electrical and Computer Engineering, Brigham Young University, Provo, UT 84602, USA
   djlee@ee.byu.edu, jka@byu.edu, aaron_dennis@byu.net
[2] Symmetron, LLC a div. of ManTech International Corp. Fairfax, VA 22033, USA
   Robert.Schoenberger@mantech-ist.com
[3] Department of Biology, Brigham Young University, Provo, UT 84602, USA
   dennis_shiozawa@byu.edu

Summary. A variety of matching and classification techniques have been employed in applications requiring pattern recognition. In this chapter we introduce a simple and accurate real-time contour matching technique specifically for applications involving fish species recognition and migration monitoring. We describe FishID, a prototype vision system that employs a software implementation of our newly developed contour matching algorithms. We discuss the challenges involved in the design of this system, both hardware and software, and we present results from a field test of the system at Prosser Dam in Prosser, Washington. In tests with up to four distinct species, the algorithm correctly determines the species with greater than 90 percent accuracy.

8.1 Introduction

Measurements of the abundance, distribution, and movement of fish are critical to fishery management. On rivers in the western United States, migrating fish frequently encounter a variety of man-made barriers, including hydroelectric and diversion dams and associated reservoirs. The detrimental impact on fish stocks are substantial: an estimated 5 to 20 percent of juvenile salmonids are killed in passing each dam in the Columbia River basin [12], and dams are widely viewed as major contributors to the decline of the salmon in the Pacific Northwest [29].

D.-J. Lee et al.: *Contour Matching for Fish Species Recognition and Migration Monitoring*, Studies in Computational Intelligence (SCI) **122**, 183–207 (2008)
www.springerlink.com

Most dams have special passageways that make it possible for fish to bypass the structure. Figure 8.1(a) shows a typical diversion dam, and Fig. 8.1(b) shows the dam's fish ladder, a narrow passage through which fish can swim upstream. Fish ladders diminish, but do not eliminate, the negative impact of dams on migrating fish. Because they constrain fish movement to small corridors, fish ladders are convenient locations to collect measurements of passing fish.

Reliable observation data obtained by monitoring fish movement can provide critical information about population counts, periodicity of movement, and passage survival of the observed fish. From compiled databases of observations over time, projections of the strengths of runs can be made, long term trends in populations can be compared, and even the relative success of various mitigation measures can be appraised. Observation data can also be used to determine the influence of abiotic factors on fish growth, to relate migration timing to regional sea surface temperatures, and to investigate changes in relative species composition relative to upstream habitat changes. Even mortalities as fish move progressively upstream can be estimated [8].

(a)   (b)

Fig. 8.1. (a) Prosser Dam on the Yakima River in Prosser, Washington and (b) its fish ladder.

At present, almost all fish monitoring is performed by humans, either as on-site observers or as viewers of recorded video or images. The accuracy of recorded observations is therefore constrained by human error, local labor resources, and available funding. The expenses incurred can be significant — fish are counted and monitored at over 50 percent of the facilities run by the US Bureau of Reclamation (USBR) and the US Army Corps of Engineers. To reduce the cost and overhead of manual fish observation, some facilities record passing fish on video, but this generates lengthy recordings that must be reviewed manually to collect accurate data. Because the image quality in the recorded video is typically poor, it is sometimes difficult for biologists to identify species from the video. Equipment maintenance for video recording is also a concern, as analog video recording technology has become obsolete. According to biologists of the USBR, state governments, universities, and Native American Tribes (who manage tribal lands and associated natural resources), an automated fish recognition and monitoring system is urgently needed.

The manual observation of wildlife is labor-intensive, and various technologies have been employed to assist with data collection. In most cases, the benefits have resulted from improved sensing or sensor deployment (aerial thermal imaging, for example), rather than from automated processing of the observation data. Huettmann discussed the feasibility of using automated software for the generation of an accurate animal census [14]. Later, he explored the use of automated software based on artificial intelligence to count roe deer and red fox [15]. Other researchers developed an automated counting program to process aerial photographs to detect black brant [7]. In similar work, Laliberte and Ripple investigated the processing of aerial photographs and high-resolution satellite images to produce counts of geese and caribou using a public-domain image-analysis program [17].

Other researchers have addressed the monitoring and measuring of fish for various applications. Most commonly, previous work has focused on monitoring the size of a particular species in a fish farm, or controlling a cutting machine for fish processing. Chan et al. used a stereo imaging system to relate salmon morphology to mass [5]. The system allows salmon farmers to make informed decisions about feeding, grading and harvesting strategies. Fuzzy C-Mean (FCM) was used for fish recognition in a fishery; the approach requires a priori knowledge of the analyzed data [30]. Naiberg and Little built a stereo vision system to measure fish size to monitor fish growth, to determine feed and medication, and to decide when to harvest [32]. In [33], Nogita and colleagues describe an imaging system built to analyze fish behavior and movement patterns; the information is used to monitor the presence of acute toxicants in water. An image processing system was built by Gamage and de Silva to measure fish orientation for optimal cutting for fish processing [10]. The system was later improved using statistical pattern recognition.

Few publications in the literature describe work related to real-time fish species recognition and monitoring. A color restoration technique was developed for touch screen display and process of underwater fish images in an aquatic environment [4]. An automatic method for color cast removal from underwater images was reported. Fish species recognition was attempted by using a variety of geometric features (area, perimeter, roundness ratio, elongation, orientation), color features (hue, gray levels, color histograms, chrominance values), texture features (entropy, correlation), and motion features [4, 34]. Other researchers used invariant moments and shape descriptors to recognize fish species on board a finishing boat [35, 37, 38].

Prior attempts to construct automated systems to monitor fish migration have had limited success. One deployed system was unable to distinguish between bubbles, debris, and fish in a reliable manner. Another system employs software to review images of passing fish recorded on video tapes, but the software cannot distinguish between species and records only the size of each fish. Due to poor image quality, the size measurements are inaccurate. A third commercially available system utilizes arrays of photo diodes and detectors assembled into a submersed "tunnel" that counts and determines the size of fish passing through the detection array. This technology does not have the capability of classifying fish according to species, and measurements of size are affected by fish movement and speed. Other work has investigated spectral and spatial signatures of fish for purposes of classification [2], but challenges related to lighting and computational overhead were not fully addressed.

This chapter discusses the design and development of an automated fish recognition and monitoring system called FishID. The system automatically acquires high resolution digital images of fish, classifies each according to species and size, tracks the movement of the fish so that each passing fish is recorded just once, and then stores the observation data for later use. In principle, species recognition could be based on geometric features, such as size and shape, or appearance features, such as color and surface texture. For example, size could be used to distinguish between smallmouth bass (very small) and coho or chinook salmon (very large), all of which can be found in the Columbia River basin where the FishID system was tested. Although similar distinctions between certain species can be made reliably based on color or texture, our experience suggests that the *shape* of the fish is often the most reliable general characteristic in determining its species. The FishID system employs a novel shape-based approach to classification that we describe in detail.

The FishID system could be installed at any dam with a viewing window, or at any facility where migrating fish must pass through a narrow passage that allows images to be captured. For example, the FishID system could be used to monitor the inflows and outflows of lakes, as long as fish can be guided to pass through narrow passages.

In Sect. 8.2 we present an overview of the FishID system, including a flowchart of the processing steps involved, a discussion of required equipment, and constraints in the setup of the system. In Sect. 8.3 we discuss shape extraction methods and present a modified curve evolution method for shape representation. In Sect. 8.4 we briefly discuss shape characteristics and the limitations of previously developed methods for matching whole shapes. Section 8.5 presents our novel algorithm for measurements of shape similarity based on *Turn Angle Distribution Analysis* (TADA). Results using the TADA algorithm are presented in Sect. 8.6, and we offer conclusions in Sect. 8.7.

8.2 System

8.2.1 Processing Flowchart

The overall design and operation of the FishID system is illustrated by the processing flowchart shown in Fig. 8.2. All computation and control is performed by software written in C++ on a standard personal computer running Microsoft Windows. Five distinct steps occur in processing.

The first and most important step is image acquisition. It is imperative that images be of high quality, and this in turn depends on lighting, camera selection, and setup. The second step determines whether or not an object is present in the image. If an object is detected, the third step is to extract its contour and classify the species using shape-based analysis. At this step, the size of the fish is also determined. The fourth step is to track the movement of the imaged fish, allowing the identification of the best image frame for recognition and to avoid counting the same fish multiple times. The fifth and final step is storing the observation data (including time-stamped images, if desired) for further analysis by the user. Details of each of these steps are described in the following sections.

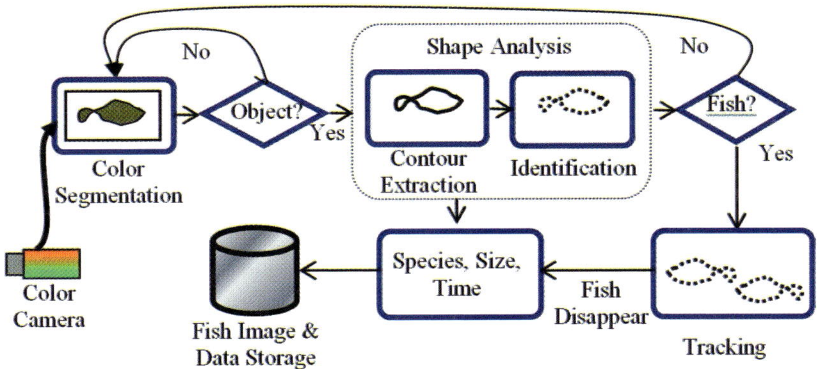

Fig. 8.2. Flowchart of the FishID algorithm

8.2.2 Image Acquisition

The FishID system uses a high resolution color camera. An early prototype of the system – developed for initial software development, data collection, and field testing – used a near-infrared (NIR) camera [26, 27], but the high resolution color images allow the fish to be more easily *segmented*, or distinguished from the background. A blue background plate was used to increase color contrast for this purpose. Identification and tracking software was first developed in the lab, and then the vision system was brought to the field for testing and data collection.

To minimize any possible negative impact on the fish, it is essential that a monitoring system be non-intrusive in its surveillance. Since FishID itself is a passive system, the only area of concern is illumination that may be required to obtain suitable images in a dark fish ladder. An extensive study of the impact of illumination on fish behavior was conducted by the USBR in 2000 [13]. The study concluded that illumination does not have a noticeable effect on fish behavior as they pass through a fish passageway. The study also tested NIR versus visible light and concluded that the fish showed no noticeable change in behavior under these two different types of lighting. Based on these results, we concluded that FishID did not require NIR light and that regular fluorescent light sources will not affect the fish.

With a version of the image acquisition software suitable for high resolution color cameras, we field tested a Sony DFW-SX910 color camera with a resolution of 1380×960 pixels capable of acquiring images at 7.5 frames per second. Testing was performed on the fish ladder at the Ice Harbor Dam in Washington State, which is run by the US Army Corps of Engineers. Our initial analysis showed that resulting color images could be segmented effectively, but that the camera did not provide sufficient light sensitivity. We switched to a CameraLink camera from CIS. This camera is well suited for this application because of its special features, including flexible gain and offset adjustments, a high-speed electronic shutter for acquiring images of fast moving objects, and progressive scanning for full frame resolution to avoid motion-induced interlacing artifacts.

8.2.3 Field Setup

Pictures in Fig. 8.3 show (a) the system's graphical user interface, (b) the fish viewing window at the test site, (c) an image captured with original white background, and (d) an image captured with added blue background. The viewing window is built in a vault approximately 30 feet below the top surface of the dam. A camcorder was in place to record video for off-site manual counting, and we mounted our color camera alongside the camcorder as shown in Fig. 8.3(b). The viewing window was 4 ft x 4 ft, and our camera was mounted approximately 4 ft from the window. To illuminate the recorded video, two 40-Watt fluorescent lights had previously been mounted on top of the window, with another two mounted on the bottom (not visible in Fig. 8.3(b)). We determined that this lighting arrangement was adequate for the collection of our color images. Setup and adjustments in the field were simplified by FishID's user-friendly graphical user interface that provides calibration and parameter adjustment via mouse clicks.

During field testing, we determined that a camera shutter speed of 1/125 second resulted in the best overall image quality. A shorter exposure time

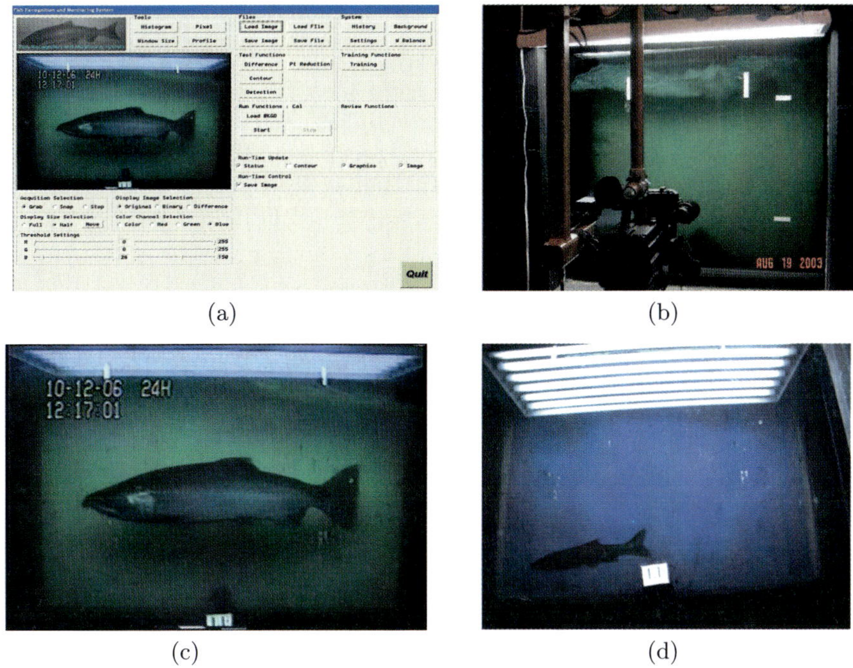

(a)   (b)

(c)   (d)

Fig. 8.3. Functional graphical user interface and pictures from the test site at Prosser Dam

would have produced slightly sharper images of the moving fish, but the lighting was not sufficient to reduce the shutter speed to 1/250 second. Given the constraints on camera placement and field of view, we also determined the optimal lens focal length to be 6 mm using a C-mount lens. Using these settings and optics, we were able to acquire sharp, good quality images of the entire window for analysis.

8.3 Fish Detection, Tracking, and Representation

8.3.1 Color Segmentation and Fish Contour Extraction

For this application, detecting objects in the images is a simple matter of separating the known background from the foreground. Subtraction of images acquired at different times can detect the motion of an object [36]. It is also a simple way to detect the presence of an object assuming a stationary camera position and constant illumination. However, our original design [27] that subtracted a reference image from each acquired image did not work well due to insufficient difference between the fish and the background.

To enhance the color contrast, a piece of plywood was painted blue and fastened to the back of the fish passageway (Fig. 8.4(a)). Given this background, we can make two useful assumptions about acquired images. First, pixels with high blue-channel values (close to 255) and low red-channel values (close to 0) are part of the background. Second, pixels belonging to a fish will have lower blue-channel values and higher red-channel values than the background pixels. These assumptions are reflected in the following relations:

$$B_{blue} = High \tag{8.1}$$

$$B_{red} = Low \tag{8.2}$$

$$F_{blue} = High - \Delta B$$
$$= B_{blue} - \Delta B \tag{8.3}$$

$$F_{red} = Low + \Delta R$$
$$= B_{red} + \Delta R \tag{8.4}$$

where B and F (with subscripts indicating the color channel) indicate background and foreground (fish), respectively, and the terms ΔB and ΔR denote the difference in intensity between typical foreground and background pixels for the blue and red channels respectively.

Background and fish pixels can therefore be determined simply by subtracting the red channel from the blue channel of the image. If $B_{difference}$ and $F_{difference}$ represent, respectively, the resulting values when red is subtracted from blue for typical background and foreground pixels, then the following relationships hold:

$$B_{difference} = High - Low \tag{8.5}$$

$$F_{difference} = High - Low - \Delta B - \Delta R$$
$$= B_{difference} - \Delta B - \Delta R \tag{8.6}$$

From these equations we note that a simple threshold can be applied to distinguish between background and fish pixels. If the difference between red and blue channels for a given pixel is greater than or equal to $B_{difference}$ then the pixel is marked as a background pixel. If the difference is less than $B_{difference}$, the pixel is marked as part of a fish. This information is used to create a binary image, with a '1' denoting each fish pixel and a '0' denoting each background pixel. Note that this segmentation is accomplished without subtracting a background or reference image from the acquired image.

Figure 8.4(b) shows the binary image after fish pixels have been identified. In practice, the binary image sometimes contains small pixel clusters (blobs) from water turbulence or image noise. These small blobs in the binary image are removed with a morphological opening operator before the fish contour is extracted. After small blobs have been removed, the size and location of remaining (large) blobs are used to determine if a fish is actually present and to initiate the edge detection process.

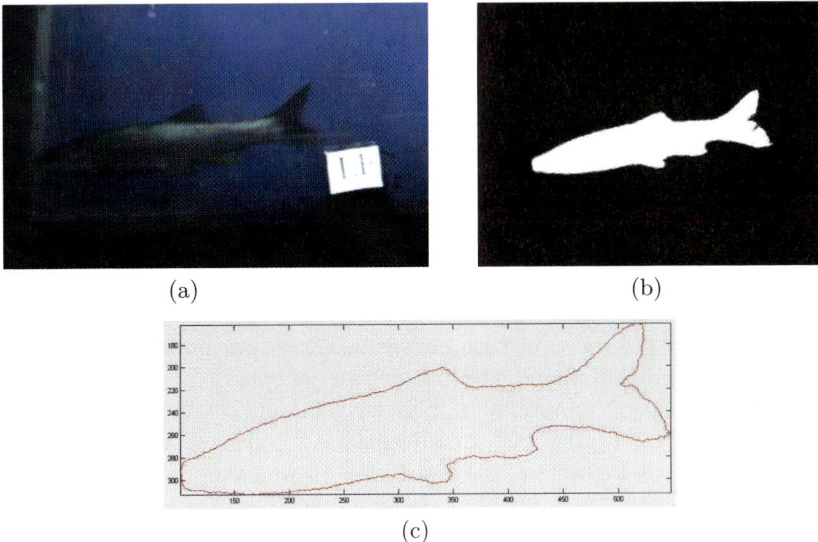

Fig. 8.4. (a) Fish image with blue background, (b) binary image after segmentation, and (c) extracted fish contour.

A fast and simple eight-neighborhood contour trace algorithm was developed to extract the x and y coordinates of the fish contour as shown in Fig. 8.4(c). As will be shown, characterizations of this contour are used to classify the species of each fish.

8.3.2 Fish Movement Tracking

As long as each fish remains in the window, its movement must be tracked in order to select the best image for processing, to ensure that each fish is counted just once, and to maintain an accurate count for any fish moving downstream. The frame-grabber and software are triggered to acquire an image sequence as soon as a foreign object is detected in the viewing window. The recording of images continues at the camera frame rate until no foreign objects are left in the viewing window. In our approach, the recorded image sequence is stored in a large *short-term memory*. Software processes each recorded image sequence, tracking objects from the time they enter the viewing area until they disappear from view. The tracking software also selects the best image for each fish from the recorded sequence, at which point the chosen image is saved to disk and all other images in the short-term memory are deleted, freeing up the memory space for the recording of other image sequences. At present, the object-tracking code takes a relatively simple approach to the tracking of multiple objects; in an image sequence with multiple fish, each will be tracked correctly provided it does not overlap with another object in the image sequence.

8.3.3 Data Reduction

In order to classify fish species using the contour, it is first necessary to reduce the number of data points on the contour to a reasonable number that can be evaluated using shape similarity measurement. Many of the data points obtained in the contour extraction algorithm are redundant. Moreover, it is desirable to filter out data points that contain edge noise. Experimentally, we found that a reduced data set with as few as 30 contour points was sufficient to retain the important shape features for comparison.

One method of data reduction is a curve evolution technique that iteratively compares all the relevance measures of the vertices on the contour [1, 18, 19]. A higher relevance measure means that the vertex makes a larger contribution to the overall shape of the contour, and thus is more important to be retained. For each iteration, the vertex with the lowest relevance is removed and a new polygon is created by connecting the remaining vertices with a straight line. We modified the relevance measure, K, for the curve evolution method used in [1, 18, 19] to remove redundant points while maintaining the significance of the contours. The new relevance measure is:

$$K(s_1, s_2) = \frac{|(\beta(s_1, s_2) - 180)| \, l(s_1)l(s_2)}{l(s_1) + l(s_2)} \tag{8.7}$$

where β is the turn angle on the vertex between adjacent line segments s_1 and s_2, and $l(s_1)$ and $l(s_2)$ are the normalized lengths from the vertex to the two adjacent vertices [20, 21].

This modified curve evolution method reduces short, straight line segments that provide little information about the overall shape of the object. In our implementation, this method preserves a fixed number of data points for each contour, making it easier to measure shape similarity.

A second data reduction method is to sample data points on the contour at a fixed interval. This method is much simpler, but it may lose data points containing significant shape information. Figure 8.5(a) shows the original contour of a fish obtained using the technique described above. Figure 8.5(b) shows the contour resulting from 40 data points obtained using the curve evolution technique and Eq. (8.7). Although there is some distortion, the basic shape and detail of the fish is retained. The data points were selected because they make the most significant contribution to the shape. Figure 8.5(c) shows a contour resulting from 40 data points obtained by fixed interval sampling. The technique preserves a reasonable subset of the shape information, but it contains redundant data points (resulting in consecutive segments that form a straight line) that should be removed. The data set also loses fine detail of the fish contour, especially on the edges corresponding to fins.

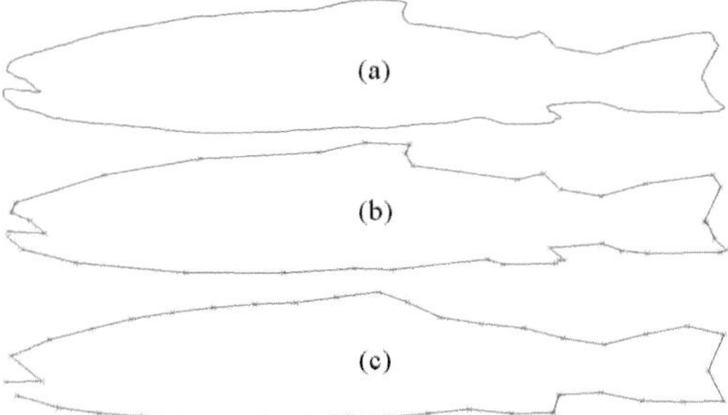

Fig. 8.5. (a) Original data set, (b) reduced to 40 points using curve evolution, and (c) reduced to 40 points using fixed interval sampling.

8.3.4 Tangent Space and Bend Angle

The reduced set of data points can be represented using polygon approximation, expressed either as *turn angle* vs. normalized length (tangent space), or *bend angle* vs. normalized length. The turn angle is calculated for each segment with respect to a fixed horizontal line that defines angle 0. The bend angle is calculated with respect to the previous segment, so that a clockwise turn gives a negative angle and a counter-clockwise turn gives a positive angle. The two alternate representations of these two functions are shown in Fig. 8.6 for representative data sets.

It is important to note that both polygon representation techniques use functions of normalized length so that the results are scaling-invariant. Ideally, shape descriptors and classification algorithms should be invariant to translation, rotation, and scaling because objects can vary in size and be viewed from different angles, locations, and distances. The representation techniques are translation-invariant because the turn angle or bend angle and length do not contain information about the shape location. The object rotation angle is irrelevant when calculating the bend angle; it simply shifts the values of the turn-angle function along the y axis. In other words, both functions stay essentially the same if the object is rotated, translated, or resized. (Depending on the data point selected as the first point in the data sequence, these two functions may shift along the x axis.)

8.4 Shape Features and Whole Shape Matching

In order to perform fish species recognition in real time, we have developed and tested several efficient shape-based recognition methods. Our previous work focused on shape characterization and landmark point analysis [26, 28, 39] and whole shape matching [27]. In this section, we briefly review these methods,

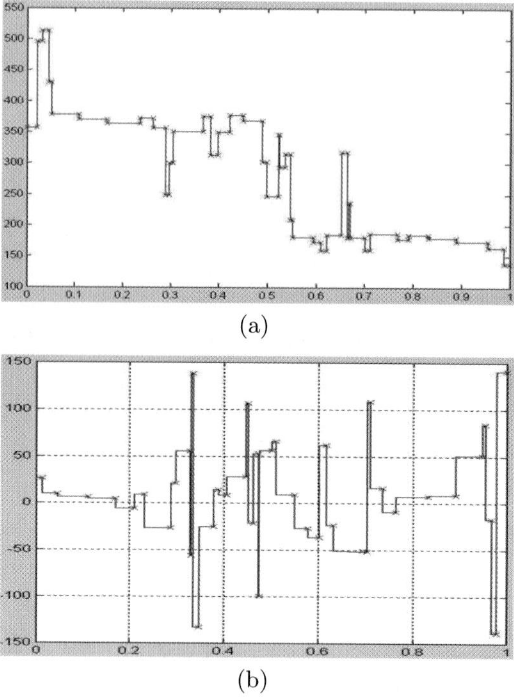

(a)

(b)

Fig. 8.6. (a) Turn angle vs. normalized length, and (b) bend angle vs. normalized length.

compare their performance, and motivate the development of a new method with increased performance.

Preliminary investigations suggested that shape analysis and matching would outperform alternative approaches for fish recognition. Initially, our work focused on finding critical landmark points on the fish contour using curvature function analysis [26, 28, 39]. While it yielded reasonable results, this approach is problematic in that landmark points sometimes cannot be located precisely. To overcome this limitation, methods were investigated that match the whole shape [27]. Several shape descriptors, such as Fourier descriptors, polygon approximation, and line segments were tested. A power cepstrum technique was developed in order to improve the categorization speed using contours represented in tangent space with normalized length. Some whole shape matching methods were shown to perform better than landmark point matching, but their recognition accuracy seldom exceeded 60%. Critical to the success of the FishID system was the development of a more accurate classification technique.

8.4.1 Shape Characterization and Landmark Points

Since the proposed fish species recognition and monitoring system is non-invasive, there is no control of fish movement. The algorithms for fish shape representation and analysis must therefore satisfy requirements of geometrical invariance, or independence of translation, orientation, scaling, and the contour starting point. To meet these requirements, we extract shape features and construct a feature vector for recognition. The shape characteristics included in the feature vector clearly depend on the species to be classified and their species-specific traits.

An initial comparison of key attributes of multiple species resulted in the following list of distinguishing factors:

1. Adipose fin
2. Anal fin
3. Caudal fin
4. Head and body shape
5. Size
6. Length/depth ratio of body.

Figure 8.7(a) illustrates the location of the various fins. Specific characteristics associated with these attributes include the specific shape of the fins, their locations (as a normalized distance from the fish nose), the tail shape (forked or not). While the characteristics included in the feature vector generally allow species to be distinguished, they are not unique for every species. For example, multiple species have forked tails. Thus, multiple shape characteristics must be used to classify many different species.

Since pectoral and pelvic fins are on the sides of the fish and change positions as the fish swims, they are difficult to detect, particularly in an image taken from the side. Similarly, the mouth may be open or closed. In contrast, the dorsal, adipose, and anal fins can be consistently and reliably detected, so the locations of these fins were chosen as classification features for the tested system [26]. A further study of these features and their shape representations resulted in five specific feature measurements:

1. The length between the nose and the front end of the dorsal fin
2. The width of the dorsal fin
3. The distance between the dorsal fin and adipose fin
4. The width of the adipose fin
5. The width of the anal fin

These five significant features are illustrated in Fig. 8.7(b).

To investigate the classification accuracy of the system using these five features, seven fish species with similar shape characters were chosen for the study [26, 28, 39]. Fish selected were chinook salmon (SA), winter coho (WC), brown trout (BT), Bonneville cutthroat (BC), Colorado River cutthroat trout (CRC), Yellowstone cutthroat (YC), and mountain whitefish (WF). A total of

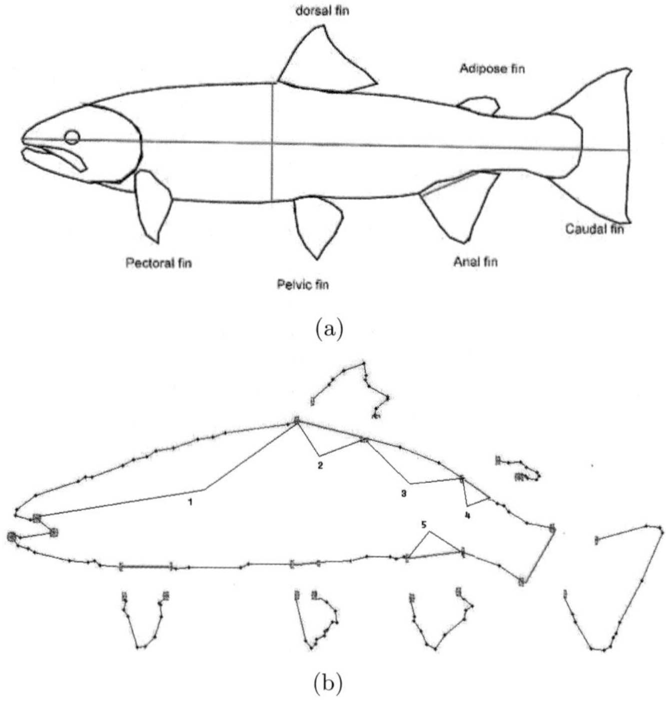

Fig. 8.7. (a) Fish shape characteristics and (b) the five significant features selected for fish recognition.

64 representative images were selected, including 10 chinook salmon, 8 winter coho, 10 brown trout, 7 Bonneville cutthroat, 10 Colorado River cutthroat, 10 Yellowstone cutthroat and 9 mountain whitefish. Shapes were generated for each of the 64 images and then classified using landmark point measurements. Table 8.1 shows the accuracy of recognition over the set of images. The high level of misidentification stems from the principal drawback of this approach: landmark points cannot always be located accurately.

Table 8.1. Fish species recognition results using landmark point measurements

	SA	WC	BT	BC	CRC	YC	WF	Overall
Correct #	7	6	7	5	6	8	9	48
Incorrect #	3	2	3	2	4	2	0	16
Accuracy	70%	75%	70%	71%	60%	80%	100%	75%

8.4.2 Whole Shape Matching

To avoid the problem of locating precise landmark points, several whole shape matching methods were developed or evaluated. Studied methods matched a variety of global shape characteristics, including perimeter, convex perimeter, major axis length and angle, minor axis length and angle, compactness, roughness, invariant moments, matching in tangent space (introduced in Sect. 8.3.4), and Fourier descriptors using bend-angle functions. From prior research [20, 21, 26, 28], we learned that the first two methods do not yield accurate matching results especially for minor shape variations. In [27], we used a turn-angle function to represent shapes in tangent space and computed shape similarity using a 12-norm. We discovered that alignment of two turn-angle functions for similarity measurement is a time consuming process. We also developed a much faster matching algorithm that uses power cepstrum to align two similar signals before the similarity can be measured. We also used a bend-angle function for shape representation and computed Fourier descriptors for performance evaluation.

The power spectrum of the bend-angle function is invariant to the shift in length. Because of this property, Fourier descriptors of a bend-angle function (a function of normalized length) meet all invariant requirements for shape description suitable for shape matching. Power spectrum and phase angle information can be calculated as shape descriptors for recognition [11, 27, 40]. Shape similarity can be measured by calculating the 12-norm of the two Fourier descriptors.

Another method was developed to match turn-angle functions from two contours to measure shape similarity. In most cases, the two turn-angle functions are not identical because of differences in shape. The alignment can be achieved only by minimizing the distance while shifting one turn-angle function. Another approach is to reduce the search to one dimension by calculating the best value of the turn angle [1, 18, 19]. This process is illustrated in Fig. 8.8. A measure of shape similarity is given by the overall distance between two aligned turn-angle functions.

Although matching in tangent space can be reduced to a 1-D search, searching in turn-angle space is still a time consuming process. We developed a new searching technique based on the *power cepstrum*, defined as the power spectrum of the logarithm of the power spectrum of a signal [6, 16, 24]. The cepstrum technique was first described by Bogert and colleagues in 1962 as a means of analyzing data containing echoes [3, 6, 9, 16, 24]. This technique was extended to analyze 2-D signals for image registration and 3-D vision [22–24, 31].

The result of applying the power cepstrum to the summation of two similar turn-angle functions is the power spectrum of the reference turn-angle function plus a train of impulses occurring at integer multiples of the shift between the two functions. By detecting the occurrence of these impulses, a shape indexing shift can be determined. The distance (dissimilarity) between

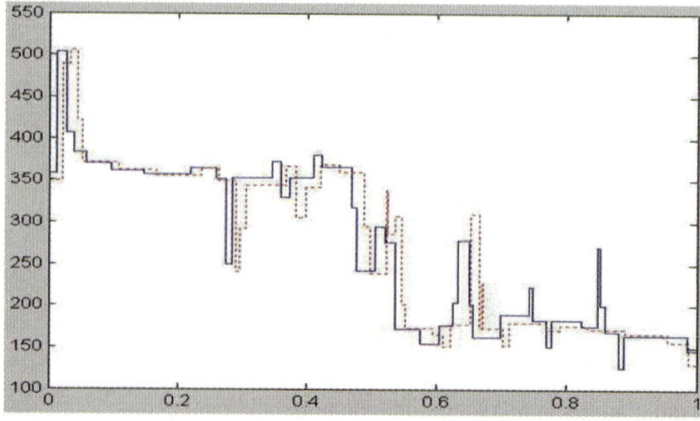

Fig. 8.8. Two aligned normalized turn-angle functions in tangent space.

the two turn-angle functions can then be calculated by aligning the two functions according to the detected shape indexing shift. Figure 8.9(a) shows the turn-angle functions of two similar contours and Fig. 8.9(b) shows the power cepstrum resulting from the two turn-angle functions. Table 8.2 summarizes the accuracy of the most promising of the whole shape matching approaches that were investigated. As can be seen, the best approach is correct for just 64% of the test images.

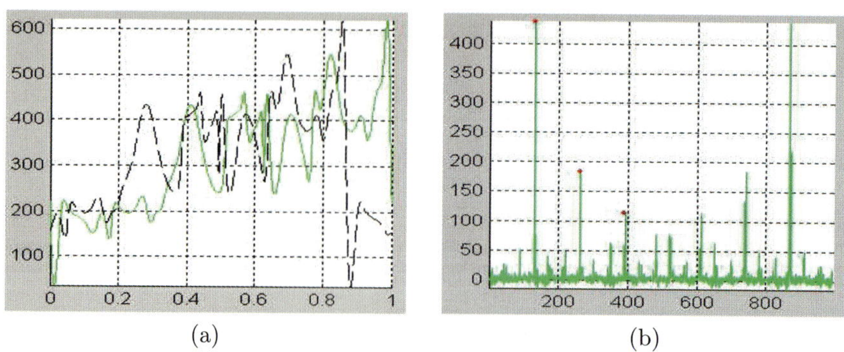

(a)   (b)

Fig. 8.9. (a) Turn-angle functions of two similar contours, and (b) resulting power cepstrum.

8.5 Turn Angle Distribution Analysis

The disappointing accuracy of even the best whole-shape matching approaches motivated the development of a new technique with significantly improved performance. As before, a contour is extracted from the fish image, and a reduced set of data points is generated. In the new approach, the data set is

Table 8.2. Fish species recognition results using whole shape matching

Category	Invariant Moments	Fourier Descriptors	Tangent Space	Power Cepstrum
SA	7/10	10/10	5/10	7/10
WC	5/8	3/8	4/8	4/8
BT	3/10	6/10	4/10	6/10
BC	0/7	3/7	2/7	4/7
CRC	4/10	6/10	8/10	5/10
YC	3/10	6/10	2/10	4/10
WF	8/9	7/9	9/9	9/9
%	47%	64%	53%	61%

then characterized by a fast and accurate shape matching method called *Turn Angle Distribution Analysis* (TADA) that allows the contour for the current image to be matched against species-specific contours in the FishID database. The TADA approach results in significantly improved accuracy, and it is the key to the overall effectiveness of the FishID system. Figure 8.10 illustrates the complete process of generating a fish model using the TADA approach. The individual steps are discussed in the following subsections.

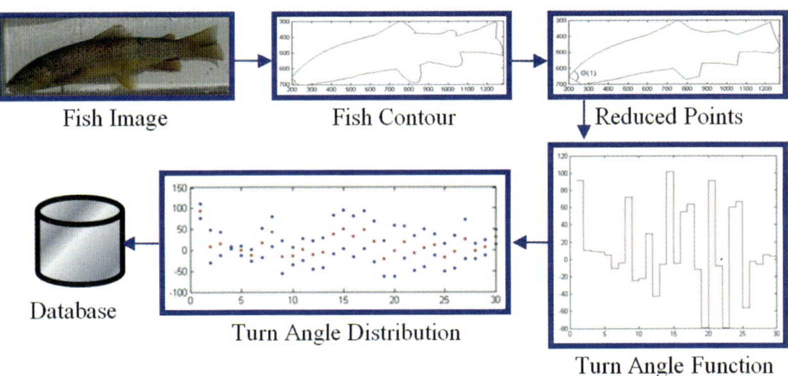

Fig. 8.10. TADA fish angle distribution model generation

8.5.1 Image Selection from Short-term Memory

The first step in species recognition is to choose a single image of the fish for processing from the associated image sequence stored in short-term memory. Two principal factors are considered when choosing the image. The first is the proximity of the fish to the middle of the viewing window. This helps ensure that the fish is entirely visible in the image, and not partially entering or exiting the frame. The second factor is the length of the fish. Fish move their tail fins back and forth to propel themselves through the water, but this distorts the appearance of the tail in profile. An ideal image for classification will show the tail fin extended directly behind the fish, in which case the fish contour is the longest from head to tail. Therefore, the FishID system picks the image that shows the longest fish that is near the center of the image.

8.5.2 Turn-Angle Function Generation

The system converts the chosen image into a binary image using the segmentation approach described in Sec. 8.3.1. From this binary image, the fish outline or contour is determined. The contour is represented by a list of points, with the leftmost fish pixel as the starting point and traced in the clockwise direction. Next, the total number of contour points is reduced to 30 equally spaced points along the outline of the fish. The use of fixed-interval sampling implicitly makes our shape representation scaling-invariant. (As noted in Section 8.3.3, fixed interval sampling preserves less of the original shape than curve evolution techniques; the choice reduced the development time of our prototype system.)

The reduced-point contour of the fish forms a closed polygon. The FishID system calculates the turn angles at each vertex of this polygon, again starting from the leftmost point and working clockwise. This list of turn angles contains no information concerning the position of the fish, and the shape description is therefore translation-invariant. Also, as long as the leftmost point on the fish is the mouth of the fish (which is generally true) then the turn angles are rotation-invariant as well.

8.5.3 Building Turn Angle Distribution Model

In order to perform accurate recognition, a turn-angle distribution model for each possible species must be generated and stored in the database as a similarity measure. Based on available resources (50 images per species), 20 representative images of each species were selected for the generation of turn-angle distribution models, and 30 equally spaced points on each fish contour were used to calculate a turn-angle function for each image. These 20 turn-angle functions (one for each image) were then used to calculate a turn-angle distribution model for each species. Because of slight variations in fish contours due to fish movement, fish size variation, and segmentation error, turn angles at these 30 data points are not identical for fish of the same species.

For each of the seven species, the mean and standard deviation of the 20 turn-angle functions were computed at each of 30 data points. Figure 8.11 shows a plot of the mean and standard deviation of the turn angle versus data point number. Each black '+' symbol corresponds to the mean angle for a single species, while gray dots indicate the standard deviation over the same image set. As can be seen, the species-specific turn-angle distributions are very similar at certain data points (dots are close together) and measurably different at others. Note that observed data values overlap at virtually all data points, so no single measurement of turn angle can serve to distinguish between species.

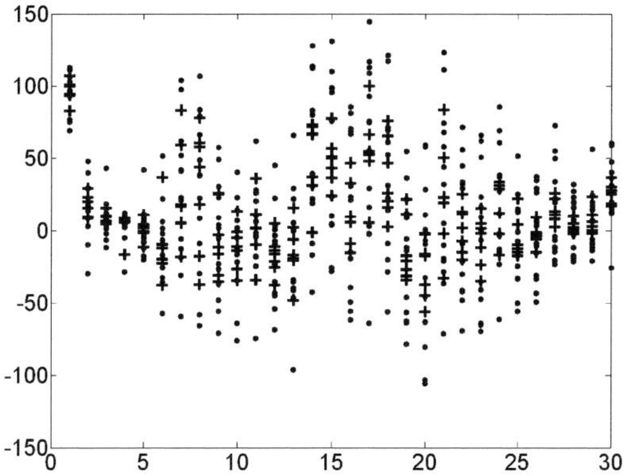

Fig. 8.11. Turn-angle function distribution model for training data set generation.

8.5.4 TADA Training Dataset Generation

From each species-specific model of turn-angle distribution, a set of training data was generated. In order to improve the recognition results, data points that show little difference in turn-angle distribution between species were not used in the calculation of shape similarity. In other words, points with substantial overlap in turn-angle distribution were excluded from the training dataset.

8.5.5 TADA for Fish Species Recognition

Fish species recognition in the FishID systems is done by comparing the turn angles of a test fish with the turn angles of fish stored in a database. A Euclidean distance measure is used to calculate the distance of the turn-angle function of the test fish from the turn-angle functions of species in the database. As noted above, data points with significant overlap between species-specific turn angles are excluded from the distance calculation. If the distance to any of the known species is above a specified threshold, the fish is classified as "unknown." This may occur when a fish passes that is not in the database or when a large piece of debris washes by. Images from "unknown" objects can be saved to disk for later examination by a human operator.

This classification method can also be used to determine which direction a fish is facing. Normally fish are facing upstream, but occasionally a fish will swim down the fish ladder. Distinguishing between the two is accomplished by performing the classification process twice. The first time, turn angle calculation starts with the leftmost point on the fish, as usual. The second time, it begins with the rightmost point on the fish and goes counter-clockwise. If the fish was facing upstream then the first turn angle is somewhere on the tail of the fish, and the resulting set of turn angles will be far different than any in the database. The effect is that the object will be classified as a particular fish in one direction and as "unknown" in the other direction. If the "unknown" classification was obtained starting with the leftmost point, then the fish is headed downstream; otherwise the fish is headed upstream.

8.6 Results

8.6.1 Image Data Collection

The prototype FishID system was tested at two facilities, Ice Harbor Dam on the Snake River and Prosser Dam on the Yakima River, both in Washington State, USA. In both cases, a limited number of species were available for testing, including salmon, steelhead, pikeminnow, and smallmouth bass. To more fully test the robustness of the TADA algorithm, we included some additional species in the database locally available to the (Utah-based) authors. A small fish tank was used to hold individual fish briefly for imaging, after which they were set free. A large number of fish images were collected, and a total of 300 images were selected for testing based on image quality and the location of the fish in the tank. (Images with fish touching the bottom of the tank were excluded.) Fifty fish images for each of six species made up the database. The six species are: brown trout (BT), cottid (C), salmon (S), speckled dace (SD), Utah sucker (US), and whitefish (W). Because the small fish tank used for imaging did not have suitable illumination and background, fish images of the local species had to be segmented manually using Adobe Photoshop.

8.6.2 Performance Analysis

For these results, the contour coordinates were converted into a list of fifty turn-angle functions. Each of the six species in our database is represented by a single representative "mean" fish. The mean for each of the thirty bend angles is calculated across all fifty fish for each species. Datapoints with high species overlap (providing little discrimination) are excluded, and the remaining set of turn angles defines the "mean" fish. Classification is performed by comparing each test fish with the six "mean" fish, rather than comparing it to all 300 fish in the database. Our recognition method was tested using N-fold cross-validation on our database with N having a value of 10. The classification results are shown in Table 8.3.

Table 8.3. Fish species recognition results using TADA

Species	2 Species	3 Species	4 Species	5 Species	6 Species
Speckled Dace	97%	93%	93%	77%	77%
Whitefish	97%	97%	97%	93%	93%
Cottid	-	100%	100%	97%	97%
Utah Sucker	-	-	100%	80%	77%
Salmon	-	-	-	90%	83%
Brown Trout	-	-	-	-	13%
Average %	97.0%	96.7%	97.5%	87.4%	73.3%

The fish species were ranked based on their similarity to each other, and the pair with the greatest dissimilarity was tested first. Then, the species with the greatest distance from those already selected was added, and this was repeated until all six species were tested together. In this way, the species with the most distinctive contour characteristics were considered first.

Overall, the results show that good accuracy is obtained using the relatively simple classification algorithm based on contour matching. As can be seen, the accuracy of the TADA recognition algorithm generally decreases as the number of species in the test set increases. In part, this results from adding increasingly less-distinctive species to the test group. Testing with just two species resulted in accuracies of 97%, while classification with four species was performed with at least 93% accuracy. When salmon was added as the fifth species, the similarity of its contour with that of speckled dace reduced the overall accuracy of classification, particularly for speckled dace. When the final species was added, the similarities between brown trout and salmon further reduced the overall classification accuracy, mostly the result of incorrect classifications of brown trout.

Relative to the common practice of manual viewing of recorded video, the overall accuracy of the prototype system is sufficient to warrant its use in automated observation. We note that the species hardest to distinguish in our test set are rarely present at the same facility at the same time. Moreover, biologists are typically concerned about monitoring a small number of species at each facility.

While the classification accuracy of TADA is also an improvement over previous whole-shape matching algorithms, our approach could be further improved by adding a filter to consider fish size. Species that exhibit very similar shape characteristics often differ markedly in size. Furthermore, the system's capabilities could be augmented by considering color features, since color can often be used to distinguish between similarly-shaped species.

If size and hue were used in the classification process, the image quality would become even more critical. Field tests of the prototype system underscored the importance of the operational setup in producing good images. In particular, lights must illuminate the entire observation window consistently and without shadows. In our case, improved lighting would have allowed the use of a faster shutter speed, resulting in improved image quality. Furthermore, the background color must be selected carefully so that it appears to the camera to have the desired hue when viewed through the water and with the available lighting. In general, site-specific details of each observation point would likely require fine tuning of the lighting, the camera settings, and the processing software.

8.7 Conclusion

This chapter presents the design of an automated fish species recognition and monitoring system. Optics, background color, and lighting considerations were discussed. Several shape-based recognition methods were implemented for testing on the prototype system. We measured shape similarity between the test fish contour and the contours stored in a database to determine the fish species. We developed landmark point measurement, shape property, invariant moments, Fourier descriptors, tangent space matching, and power cepstrum to align two functions in tangent space for comparison. The performance of these whole-shape matching methods did not give satisfactory result.

A new method, called Turn Angle Distribution Analysis (TADA), was developed and tested on a larger database than those used in our previous work. 300 images of 6 species were tested for recognition with impressive results. Although recognition accuracy dropped to 73.3% when brown trout was added to the test, the overall system performance is considered adequate for biological and environment research. The TADA algorithm is easy to set up and does not require extensive training. It uses a small number of parameters that require user input, and it has low computational overhead because Euclidean distance is calculated only for a small number of turn angles. One important feature of the proposed recognition algorithm is that the similarity measure can be used as the recognition confidence measure as well. The confidence measure can be used to determine if an image should be saved for later review by a human operator.

The prototype system has been tested at two facilities to demonstrate its robustness. While the FishID system would require more work to become a viable commercial product, the prototype system and the shape-matching approach it employs show tremendous promise in improving the accuracy of observational data through cost-effective automation. While technology that improves the accuracy of collected data is of practical interest, we note that long-term, sustainable solutions to the acute problems facing the nation's fisheries will require wide-spread systemic changes.

Acknowledgment

The project was supported by the Small Business Innovation Research program of the US Department of Agriculture through grant 2004-33610-14804.

The authors are grateful for the field test support provided by the US Bureau of Reclamation and the Yakima Nation.

References

1. Arkin EM, Chew LP, Huttenlocher DP, Kedem K, Mitchell JSB (1991) An efficient computable metric for comparing polygon shapes. IEEE Trans. on Pattern Analysis and Machine Intelligence 13:209–216
2. Bendall C, Hiebert SD, Mueller G (1999) Experiments in in situ fish recognition systems using fish spectral and spatial signatures. US Department of the Interior, US Geological Survey
3. Bogert GM, Healy MJ, Tukey JW (1963) The quefrency analysis of time series for echoes: cepstrum and saphe cracking. In: Rosenblatt M (ed) Proc. of a Symposium on Time Series Analysis, pp 209–243. John Wiley, New York
4. Chambah M, Semani D, Renouf A, Courtellemont P, Rizzi A (2004) Underwater color constancy: enhancement of automatic live fish recognition. Proceedings of the SPIE 5293:157–168
5. Chan D, Hockaday S, Tillett RD, Ross LG (1999) A trainable n-tuple pattern classifier and its application for monitoring fish underwater. In: Proc. Seventh Int. Conf. on Image Processing And Its Applications, vol 1, pp 255–259
6. Childers DG, Skinner DP, Kemerait RC (1977) The cepstrum: A guide to processing. Proceedings of the IEEE, 65: 1428–1443
7. Cunningham DJ, Anderson WH, Anthony RM (2006) An image-processing program for automated counting. Wildlife Society Bulletin 24:345–346
8. Dauble DD, Mueller RP (2000) Upstream passage monitoring: difficulties in estimating survival for adult Chinook salmon in the Columbia and Snake Rivers. Fisheries 25:24–34
9. Dudgeon D (1977) The computation of two-dimensional cepstra. IEEE Trans. Acoustics, Speech, and Signal Processing 25:276–484
10. Gamage LB, de Silva CW (1990) Use of image processing for the measurement of orientation with application to automated fish processing. In: Proc. 16th Annual Conf. IEEE Industrial Electronics Society, pp 482–487
11. Gonzalez RC, Woods RE (2002) Digital image processing. Prentice-Hall Inc., Upper Saddle River, New Jersey
12. Gregory S, Li H, Li J (2002) The conceptual basis for ecological responses to dam removal. BioScience 52:713–723
13. Hiebert S, Helfrich LA, Weigmann DL, Liston C (2000) Anadromous salmonid passage and video image quality under infrared and visible light at Prosser Dam, Yakima River, Washington. North American Journal of Fisheries Management 20:827–832
14. Huettmann F (1993) Use of a video camera and digitized video pictures in wildlife biology. Proc. of XXI IUGB (Int. Union of Game Biologists) Congress, pp 187–191
15. Huettmann F (1995) Recognizing animal species with Artificial Intelligence (AI) software on digitized video pictures; an application using roe deer and red fox. Proc. of XXII IUGB (Int. Union of Game Biologists) Congress, pp 129–138
16. Kemerait RC, Childers DG (1972) Signal detection and extraction by cepstrum techniques. IEEE Trans. Information Theory 18:745–759

17. Laliberte AS, Ripple WJ (2003) Automated wildlife counts from remotely sensed imagery. Wildlife Society Bulletin 31:362–371
18. Latecki LJ, Lakämper R (2001) Shape description and search for similar objects in image databases. In: State-of-the-Art in Content-Based Image and Video Retrieval, pp 69–95, Kluwer, Deventer, The Netherlands
19. Latecki LJ, Lakämper R (2002) Application of planar shape comparison to object retrieval in image databases. Pattern Recognition 35:15–29
20. Lee DJ, Bates D, Dromey C, Xu X (2003) A vision system performing lip shape analysis for speech pathology research. In: Proc. 29th Annual Conf. IEEE Industrial Electronics Society, pp 1086–1091
21. Lee DJ, Bates D, Dromey C, Xu X, Antani S (2003) An imaging system correlating lip shapes with tongue contact patterns for speech pathology research. In: Proc. 16th IEEE Symposium on Computer-Based Medical Systems, pp 307–313
22. Lee DJ, Krile TF, Mitra S (1988) Power spectrum and cepstrum techniques applied to image registration. Applied Optics 27:1099–1106
23. Lee DJ, Mitra S, Krile TF (1988) Noise tolerance of power cepstra and phase correlation in image registration. Optical Society of America Meeting, Santa Clara, California
24. Lee DJ, Mitra S, Krile TF (1989) Analysis of sequential complex images using feature extraction and 2-D cepstrum techniques. Journal of Optical Society of America 6:863–871
25. Lee DJ, Mitra S, Krile TF (1990) Accuracy of depth information from cepstrum-disparities of a sequence of 2-D projections. Proceedings of the SPIE 1192:778–788
26. Lee DJ, Redd S, Schoenberger R, Xu X, Zhan P (2003) An automated fish species classification and migration monitoring system. In: Proc. 29th Annual Conf. IEEE Industrial Electronics Society, pp 1080–1085
27. Lee DJ, Schoenberger RB, Shiozawa DK, Xu XQ, Zhan P (2004) Contour matching for a fish recognition and migration monitoring system. Proceedings of the SPIE 5606:37–48
28. Lee DJ, Zhan P, Shiozawa DK, Schoenberger R (2004) An automated fish recognition and migration monitoring system for biology research. Annual Meeting of the Western Division of the American Fisheries Society, Salt Lake City, UT, March
29. Lichatowich JA (2001) Salmon without rivers: a history of the pacific salmon crisis. Island Press, Washington D.C.
30. Menard M, Loonis P, Shahin A (1997) A priori minimization in pattern recognition: Application to industrial fish sorting and face recognition by computer vision. In: Proc. Sixth IEEE Int. Conf. on Fuzzy Systems, vol 2, pp 1045–1050
31. Mitra S, Lee DJ, Krile TF (1990) 3-D representation from time-sequenced biomedical images using 2-D cepstrum. In: Proc. IEEE Conference on Visualization in Biomedical Computing, pp 401–408
32. Naiberg A, Little JJ (1994) A unified recognition and stereo vision system for size assessment of fish. In: Proc. Second IEEE Workshop on Applications of Computer Vision, pp 2–9
33. Nogita S, Baba K, Yahagi H, Watanabe S, Mori S (1988) Acute toxicant warning system based on a fish movement analysis by use of AI concept. In: Proc. Int. Workshop on Artificial Intelligence for Industrial Applications, pp 273–276

34. Semani D, Bouwmans T, Frélicot C, Courtellemont P (2002) Automatic fish recognition in interactive live video. In: Proc. Int. Workshop on IVRCIA, The 6th World Multi-Conference on Systemics, Cybernetics and Informatics, pp 14–18
35. Semani D, Saint-Jean C, Frélicot C, Bouwmans T, Courtellemont P (2002) Alive fishes species characterization from video sequences. In: Proc. Joint IAPR Int. Workshop on Structural, and Statistical Pattern Recognition, pp 689–698
36. Sonka M, Hlavac V, Boyle R (1999) Image processing, analysis, and machine vision. PWS Publishing, Pacific Grove, California
37. Strachan NJC (1993) Recognition of fish species by colour and shape. Image and Vision Computing 11:2–10
38. Strachan NJC, Nesvadba P, Allen AR (1990) Fish species recognition by shape analysis of images. Pattern Recognition 23:539–544
39. Strout C, Shiozawa DK, Lee DJ (2004) Computerized fish imaging and population count analysis. Annual Meeting of the Western Division of the American Fisheries Society, Salt Lake City, UT, March
40. Zahn CT, Roskie RZ (1972) Fourier descriptors for plane closed curves. IEEE Trans. on Computers 21:269–281

Using Random Forests to Provide Predicted Species Distribution Maps as a Metric for Ecological Inventory & Monitoring Programs

Dawn R. Magness[1], Falk Huettmann[2], and John M. Morton[3]

[1] University of Alaska, EWHALE Lab, Department of Biology & Wildlife, Fairbanks, Alaska, 99775 and Kenai National Wildlife Refuge, U.S. Fish and Wildlife Service, Soldotna, Alaska 99669, USA
dawn.magness@uaf.edu
[2] University of Alaska, EWHALE Lab, Institute of Arctic Biology, Department of Biology & Wildlife, Fairbanks, Alaska 99775, USA
fffh@uaf.edu
[3] Kenai National Wildlife Refuge, U.S. Fish and Wildlife Service, Soldotna, Alaska 99669, USA
john_m_morton@fws.gov

Summary. Sustainable management efforts are currently hindered by a lack of basic information about the spatial distribution of species on large landscapes. Based on complex ecological databases, computationally advanced species distribution models can provide great progress for solving this ecological problem. However, current lack of knowledge about the ecological relationships that drive species distributions reduces the capacity for classical statistical approaches to produce accurate predictive maps. Advancements in machine learning, like classification and bagging algorithms, provide a powerful tool for quickly building accurate predictive models of species distributions even when little ecological knowledge is readily available. Such approaches are also well known for their robustness when dealing with large data sets that have low quality. Here, we used Random Forests (Salford System's Ltd. and R language), a highly accurate bagging classification algorithm originally developed by L. Breiman and A. Cutler, to build multi-species avian distribution models using data collected as part of the Kenai National Wildlife Refuge Long-term Ecological Monitoring Program (LTEMP). Distribution maps are a useful monitoring metric because they can be used to document range expansions or contractions and can also be linked to population estimates. We utilized variable radius point count data collected in 2004 and 2006 at 255 points arranged in a 4.8 km resolution, systematic grid spanning the 7722 km^2 spatial extent of Alaska's Kenai National Wildlife Refuge. We built distribution models for 40 bird species that are present within 200m of 2–56% of the sampling points resulting in models that represent species which are both rare and common on the landscape. All models were built using a common set of 157 environmental predictor variables representing topographical features, climatic space, vegetation, anthropogenic variables, spatial structure, and 5 randomly generated neutral landscape variables for quality assessment. Models

D.R. Magness et al.: *Using Random Forests to Provide Predicted Species Distribution Maps as a Metric for Ecological Inventory & Monitoring Programs*, Studies in Computational Intelligence (SCI) **122**, 209–229 (2008)
www.springerlink.com

with that many predictors have not been used before in avian modeling, but are commonly used in similar types of applications in commercial disciplines. Random Forests produced strong models (ROC >0.8) for 16 bird species, marginal models (0.7 >ROC <0.8) for 13 species, and weak models (ROC <0.7) for 11 species. The ability of Random Forests to provide accurate predictive models was independent of how common or rare a bird was on the landscape. Random Forests did not rank any of the 5 neutral landscape variables as important for any of the 41 bird species. We argue that for inventory and monitoring programs the interpretive focus and confidence in reliability should be placed in the predictive ability of the map, and not in the assumed ecological meaning of the predictors or their linear relationships to the response variable. Given this focus, computer learning algorithms would provide a very powerful, cost-saving approach for building reliable predictions of species occurrence on the landscape given the current lack of knowledge on the ecological drivers for many species. Land management agencies need reliable predictions of current species distributions in order to detect and understand how climate change and other landscape drivers will affect future biodiversity.

9.1 Introduction

Accurate species distribution maps are desperately needed for sustainable wildlife management and the conservation of biological diversity [1]. Basic information about how plants and animals are distributed across the landscape is often cited as a data gap that must be filled before management issues may be addressed. However, the need for inventory and monitoring continually resurfaces and this type of basic information remains generally unavailable. For example, the National Wildlife Refuge System, the only U.S. land-holding agency with a mission to conserve biological diversity, still lacks basic species inventories and therefore has little information about the spatial distributions of species [2].

Within the past 20 years, remotely-sensed data have become more widely available for ecological analysis due to advancements in data processing algorithms, increased hardware capabilities, and the deployment of comprehensive satellite systems [4] [3]. Linking remotely-sensed data with available information about wildlife species occurrence can provide a powerful solution for the lack of information on how species are distributed across the landscape. Species distribution modeling is an emerging discipline in ecology that has utilized many different available algorithms, including computer learning algorithms like Random Forests. A trade-off exists between optimizing distribution models for prediction within a study area and optimizing for generality across regions under an assumption that predictors are relevant to a species ecology [5]. We argue that for inventory and monitoring programs, the interpretive focus and confidence in reliability should be placed in the predictive ability of the map and not in the ecological meaning of the predictors. Given this focus, computer-learning algorithms provide a powerful, cost-saving approach for building reliable predictions of current species occurrence on the landscape given the current lack of knowledge on the ecological drivers for many species.

In this chapter, we explore the utility of Random Forests, a highly accurate regression and classification algorithm, for building distribution maps for a multi-species ensemble of bird species using a large, common set of predictor variables. We also outline an approach for using distribution maps to monitor changes in species distributions that could be used by agency personnel with little or no experience in working with computer intelligence algorithms.

9.2 Background

9.2.1 Inventory & Monitoring Programs

The National Wildlife Refuge System Improvement Act of 1997 organizes all refuges into a system with the common mission 'to ensure that the biological integrity, diversity, and environmental health of the system are maintained.' As part of the mission, managers are required to inventory and monitor biological diversity on refuge lands. An inventory refers to the systematic determination of ecosystem status for a single point in time and monitoring refers to collecting information across time to determine trends in ecosystem status [6]. Inventory and monitoring programs must be designed based on the goals of the program and the ecological focus (i.e. habitat, population, species). Generally, monitoring programs goals and focus can be categorized as targeted, cause-and-effect, or context [7]. Targeted monitoring is focused on the condition and response of specific species and habitats to management actions. Cause-and-effect monitoring is concerned with understanding the mechanisms that drive ecological response to management conditions. Context monitoring is used to document a broad array of ecosystem components, at multiple scales, without reference to management actions. Context monitoring often generates information that is rich for data mining and therefore appropriate for the application of computer intelligence algorithms.

The Kenai National Wildlife Refuge Long-term Ecological Monitoring Program (LTEMP) utilizes a context monitoring approach to document ecological condition and biodiversity without consideration of specific management actions. LTEMP collects information in conjunction with the U.S. Forest Service's Forest Inventory and Analysis program (FIA) on a 4.8 km resolution systematic sampling grid. Ecotypes are sampled in direct proportion to their prevalence on the landscape. At each sampling location, FIA collects detailed vegetation information but only in forested ecotypes. LTEMP extends the sampling framework to non-forested ecotypes and links other information about terrestrial flora and fauna to the vegetation plots. In the first 3 years after the establishment of LTEMP, all plots were visited once to establish a baseline sample for a variety of metrics, including avian diversity. Although many monitoring metrics are possible with the LTEMP dataset, we focus on species distribution models because range size was identified by Angermeier and Karr [8] as an appropriate metric for ecosystem change. In addition, distribution maps can also be linked to population estimates [9] [10].

9.2.2 Species Distribution Models

Species distribution models describe the relationship between the occurrence (or density, abundance) of species and a set of predictor variables that quantify habitat and other limiting variables. In the classical approach, distribution models build on ecological niche theory under the assumption that predictable relationships can describe the range of environmental conditions where an animal occurs [11]. Species distribution modelers usually define the niche as either the fundamental niche that expects a species to occur in all suitable environmental conditions or as the realized niche that excludes species from some of the fundamental niche due to biotic interactions like competition or predation. Some species distribution models further reduce the fundamental niche by overlaying source-sink dynamics that exclude species from areas that do not have population growth or immigration. In addition, dispersal limitation effects reduce the fundamental niche based on historical and current dispersal barriers [1]. Because there are no assumptions, computer learning algorithms would likely identify patterns that include all constraints on the fundamental niche.

The purpose of a distribution modeling exercise can range from testing hypotheses about the mechanisms driving species distributions to accurate prediction. For mechanistic models, the focus is generality because the processes driving relationships should be transferable in space. However, predictive power may be lower for mechanistic models because the context of the landscape of interest, like unknown historical effects or metapopulation dynamics, will not be captured. Alternatively, predictive models can be highly accurate within the response space, but do not provide information about the underlying ecological mechanisms and may not be useful outside of the landscape in which they were developed [5]. Although active management often requires the former approach to modeling, we argue that specific information about where species occur within the management area is needed for planning and, as we subsequently discuss, for monitoring. Therefore, maximizing predictive ability within the study area is a priority.

9.2.3 Random Forests

Random Forests, a data-mining algorithm developed by Leo Breiman and Adele Cutler, produces accurate predictions without overfitting [12]. Random Forests constructs a classification or regression tree by successively splitting data based on single predictors. Each binary node, or split, forms a branch in the decision tree and trees are grown without pruning. However, Random Forests does not grow only one tree. Instead, Random Forests utilizes bagging, or 'bootstrap aggregation', a technique that builds a large number of trees and averages the output. For bagging, a bootstrap [13] sample of the data set is randomly drawn to build each tree. Data not in the bootstrap sample, termed 'out-of-bag', are used to estimate an unbiased error rate and to rank variable importance. Before each tree is constructed, the data is reordered and a new bootstrap sample is randomly drawn. Resampling the training data for each tree reduces output error caused by the structure of the data set. In order to decrease bias due to correlation among trees, Random Forests also perturbs tree construction by only considering a random subset of all predictors while searching for the best predictor to use at each node. Although Random Forests provides highly accurate prediction, interpretation is very difficult because numerous trees generate a plurality vote to provide the output [14]. Random Forests does provide a ranking of variable importance based on the change in prediction error caused by randomly changing the data values of a given predictor.

Computer intelligence algorithms have generally been touted as a data mining method to be applied secondarily to very large datasets [15]. However, some machine-learning algorithms do not require that prediction variables be restricted based on sample size. For example, Random Forests has been applied to an 81-sample microarray lymphoma data set using 4,882 predictor variables with accurate classification results [14]. In this application, we apply a computer intelligence algorithm to a small monitoring data set (255 records) in order to find patterns with a relatively large number of predictors (157) given the dataset.

9.3 Methods

9.3.1 Study Area

The Kenai National Wildlife Refuge (KENWR) comprises 7722 km^2 on the Kenai Peninsula in Southcentral Alaska 9.1. KENWR lies at the interface between the boreal forest and coastal rainforest ecoregions. The location at an eco-region boundary and the broad elevation range of the refuge (sea level to 2000 m) create a diverse array of habitat types. West of the Kenai Mountains, refuge lands are boreal lowlands characterized by pothole lakes, extensive peatlands, and forests dominated by black spruce (*Picea mariana*) and white spruce (*Picea glauca*). Aspen (*Populus tremuloides*) and birch (*Betula neoalaskana*) stands are interspersed within the spruce forests. Sitka spruce-dominated (*Picea sitchensis*) stands extend from the coastal rainforest located along the southern portions of the Kenai Peninsula. Mountain hemlock (*Tsuga mertensiana*) and sub-alpine shrub habitats turn to lichen-dominated tundra along an elevational gradient in the Kenai Mountains and Caribou Hills. The refuge also includes portions of the Harding Ice Field.

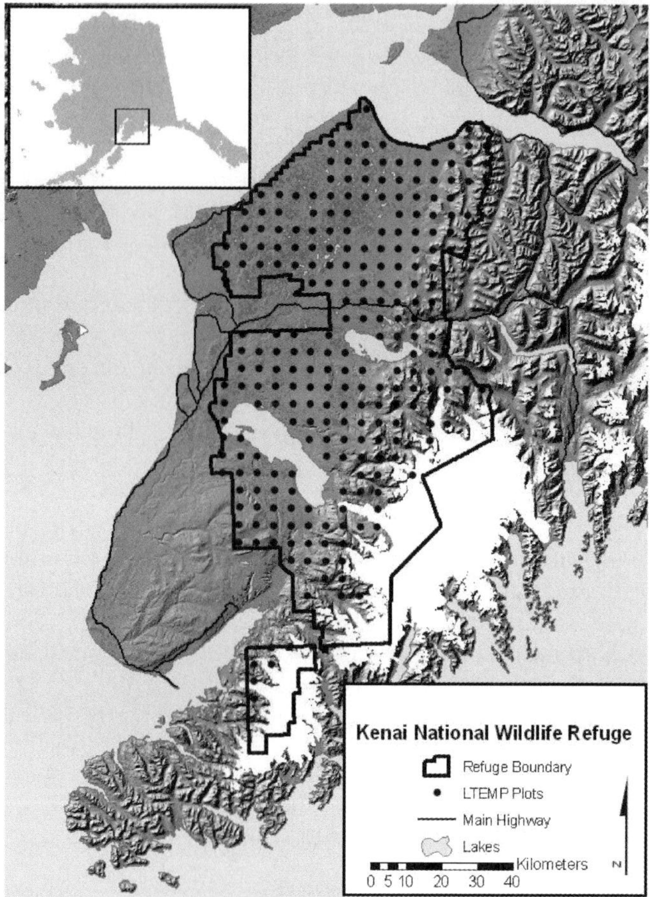

Fig. 9.1. Map of the Kenai National Wildlife Refuge

9.3.2 Target Variables: LTEMP Field Collection Methods

We sampled breeding landbird populations on 255 LTEMP plots 9.1 using variable circular point count methodology [16]; 152 in 2004 and 103 in 2006. Sampled points are regularly spaced in a 4.8 km systematic grid and were accessed via helicopter. Observers walked from the helicopter landing site to the point count sampling locations. During each point count, an observer recorded the species and distance from point center of all birds seen and heard within a 10 minute sampling period. Two skilled observers conducted the point counts with each observer sampling across the range of habitat types. Point counts were conducted during the last 3 weeks of June when weather conditions were clear, with little or no precipitation, and light wind. We excluded birds that were observed flying over the sampling area from the analysis. We defined the occurrence of a species as a bird being located within 200 m of the point count within the 10 minute sampling period. If a species was not encountered within the 12.56 ha (200 m radius) sampling area during the 10 minute sampling period, we assumed the species was absent. However, the probability of detecting a given species within a sampling window differs by species and increases with increasing sampling time [17].

9.3.3 Development of Predictor Variables

We linked the 255 sampled LTEMP plots (dataset rows) with 157 predictor variables (dataset columns). We developed 157 predictor variables in ArcView 9.2 Geographic Information System (GIS) using layers available from the KENWR or freely available on the internet (Alaska Geospatial Data Clearinghouse). Predictor variables are both categorical and continuous data and vary in terms of temporal and spatial resolution 9.1. We generated 152 variables to represent topographic conditions, climatic information, landscape vegetation types, spatial structure, and anthropogenic factors. We created 5 predictor variables randomly to represent neutral landscapes in order to test the ability of the Random Forests algorithm to delineate biologically meaningful variables. In the remainder of this section, we provide a general description of all predictor variables used in this analysis.

Topographic Variables

We used a digital elevation model to estimate elevation. Spatial Analyst surface analysis tool in GIS converted the digital elevation model into a percent slope surface and categorical aspect surface (flat, N, NW, W, SW, S, SE, E, and NE). We also calculated surfaces representing the shortest linear distance to streams and to glaciers.

Climatic Variables

We used PRISM (www.ocs.orst.edu/prism) monthly and yearly average precipitation and average temperature surfaces. The PRISM methodology fits local linear regressions of climate versus elevation, but also includes information about terrain barriers, terrain induced climate transitions, cold air drainage, inversions, and coastal effects [19]. WORLDCLIM (www.worldclim.org) surfaces represent minimum and maximum temperature by month. WORLDCLIM surfaces interpolate weather station data using the thin-plated smoothing spline algorithm with latitude, longitude, and elevation as independent variables [20]. In addition, WORLDCLIM provides 19 bioclimatic surfaces; annual mean temperature, mean diurnal range, isothermality, temperature seasonality (index of SD), maximum temperature in the warmest month, minimum temperature in the coldest month, annual temperature range, mean temperature in the wettest quarter, mean temperature in the driest quarter, mean temperature in the warmest quarter, mean temperature of the coldest quarter, annual precipitation, precipitation in the wettest month, precipitation in the

Table 9.1. Prediction variables used to build species occurrence models

Type	No.	Unit	Time Scale	Spatial Scale	Source
Topographic Variables					
Elevation	1	meters	N/A	50 m^2	KENWR
Slope	1	%	N/A	50 m^2	KENWR
Aspect	1	9 Categories	N/A	50 m^2	KENWR
Dist. to Stream	1	meters	2004	50 m^2	KENWR
Dist. to Glacier	1	meters	1991	50 m^2	AGDC
Climatic Variables					
Ave. Precipitation	13	mm	1996 –2002	2 km^2	KENWR
Ave. Temperature	13	C*10	1996 –2002	2 km^2	KENWR
Minimum Temperature	12	C*10	1950 –2002	1 km^2	WORLD CLIM
Maximum Temperature	12	C*10	1950 –2002	1 km^2	WORLD CLIM
Bioclimatic Variables	19	Varies	1950 –2002	1 km^2	WORLD CLIM
Vegetation Variables					
Duck's Unlimited Vegetation	1	18 Categories	1989	30 m^2	KENWR
Alaska Transect Vegetation	1	21 Categories	1992	1 km^2	AGDC
KENWR Landcover	1	26 Categories	2002	50 m^2	KENWR
Area/Density/Edge Metrics	36	Varies	2002	50 m^2	KENWR
Shape Metrics	6	Varies	2002	50 m^2	KENWR
Isolation/Proximity Metrics	3	Varies	2002	50 m^2	KENWR
Contagion/Interspersion Metrics	4	Varies	2002	50 m^2	KENWR
Connectivity Metrics	1	Index	2002	50 m^2	KENWR
Diversity Metrics	3	Varies	2002	50 m^2	KENWR
Stand Age	1	11 Categories	2000	50 m^2	KENWR
Landsat TM Bands	6	Spectral Index	2002	30 m^2	KENWR
Greenness Maximum	1	(NDVI*100)+100	1991	50 m^2	AGDC
Greeness Mean	1	(NDVI*100)+100	1991	50 m^2	AGDC
Greeness Onset	1	7 Categories	1991	50 m^2	AGDC
Greeness Mean	1	NDVI	2002	30 m^2	KENWR
Spruce Bark Beetle Infestation	1	Binomial	1996 –2002	50 m^2	AGDC
Dist. to Spruce Bark Infestation	1	meters	1996 –2002	50 m^2	AGDC
Anthropogenic Variables					
Road Density	2	meters/km2	2004	50 m^2	KENWR
Linear Feature Density	2	meters/km^2	2004	50 m^2	KENWR
Dist. to Float Plane Lake	1	meters	2004	50 m^2	KENWR
Spatial Structure					
Latitude	1	UTM	N/A	N/A	KENWR
Longitude	1	UTM	N/A	N/A	KENWR
Neutral Landscape Variables					
Neutral Layers	5	N/A	N/A	50 m^2	
Total	**157**				

driest month, precipitation seasonality, precipitation of the wettest quarter, precipitation of the driest quarter, precipitation in the warmest quarter, and the precipitation in the coldest quarter.

Vegetation Variables

We used 3 independent classifications of LandSat imagery from different time periods to provide categorical representations of landcover. We quantified landscape structure across the study area with a 200 m radius (12.56 ha; equal to area sampled for birds) moving window analysis in program FRAGSTATS [18] using the most recent landcover classification. We calculated the percentage of the 200 m radius moving window in each of the 26 landcover types to quantify landscape composition. We added the percentage of the moving window that contained 9 forest types to generate a forest area layer, 4 deciduous forest types for a deciduous area layer, and 5 shrub types for a shrub area layer. In landscape ecology, patches are a basic unit within a landscape and can be defined as a relatively homogeneous area that differs from the surrounding area. The patch concept can be used to quantify landscape structure using patch descriptors like patch area, indices of patch shape, or patch edge length. We used edge density, largest patch index, patch density, 3 patch area distribution metrics (area-weighted mean, mean, and coefficient of variation), and the largest shape index to represent landscape composition and structure with area, density, and edge metrics. We also described patch shape using 3 shape index distribution metrics and 3 contiguity index distribution metrics (area-weighted mean, coefficient of variation, and mean). We quantified the isolation of patches within the 200 m radius moving window with the mean, median, and area-weighted Euclidean nearest neighbor distance between all patches of the same type. We used contagion (an index of overall clumpiness of vegetation categories), an aggregation index, percentage of like adjacencies, and a landscape division index to represent the spatial configuration of the landscape (contagion / interspersion in 9.1). We used a patch cohesion index as a metric of landscape connectivity. Finally with FRAGSTATS, we used patch richness density, Shannon diversity index, and Shannon evenness index within the 200 m radius moving window to provide diversity metric surfaces.

We created a categorical layer of forest stand age based on a reconstruction of fire and spruce bark beetle disturbance events. We used geometrically corrected and radiometrically normalized Landsat TM spectral bands to provide information about landcover which is independent of subjective classification processes. We utilized 4 greenness indices calculated using the normalized vegetation index (NDVI). Maximum greenness represents the maximum photosynthetic activity in the growing season, mean greenness is a surrogate of net primary productivity, and onset is a categorical variable that summarizes vegetation emergence in 2-week intervals. Spruce bark beetle activity is a major disturbance factor on the Kenai Peninsula. We used Alaska Department of

Natural Resources transect surveys to map areas of beetle activity. We summarized all beetle activity over a 10 year timeframe into a binomial surface of areas that had activity versus areas with no activity. Finally, we created a surface of the shortest linear distance to beetle activity.

Anthropogenic Variables

All roads, including highways, secondary roads, oil infrastructure access, logging roads, and in-holder access, contributed to estimates of road density. We calculated 2 road density surfaces; the first used a 200 m search radius and the second used a 1 km search radius. We generated 2 density surfaces for all linear features using the same search radii. Linear features include all roads, snowmachine trails, canoe trails, hiking trails, power-lines, pipelines and seismic lines. We mapped all lakes that allow float plane assess and created a surface representing the shortest linear distance to these lakes.

Neutral Landscape Variables

In order to test the performance of the landscape layers, 5 neutral landscape variables were created. Neutral landscape variables are 'biologically meaningless', but unbiased. Therefore, we generated 2 neutral landscape layers as a density function of randomly generated points in the study area. We also generated 3 neutral landscape layers by randomly placing points, assigning a random number between 0 and 100 to each point, and then kriging the random numbers to create a continuous surface.

9.3.4 Model Development with Random Forests

We used Salford System's commercial version of Random Forests because the software includes a user-friendly interface. Models were run for 40 bird species present at 4 or more sites (minimum of 1.6% of sampled sites) using the 157 predictor variables. In classification, Random Forests has the ability to weight observations based on the proportion of observations in each class. We allowed Random Forests to pick these weights, so the presence values for rare species had a larger weight than the absence values. Without weighting, classes with more observations (i.e. absences of rare species) would have a greater influence on the model than classes with fewer observations. With weighting, observations from classes with larger weights have more influence and this reweighting of influence balances the data. Users may vary the number of predictors randomly selected for consideration at each node and classification error may be influenced by this parameter, termed the 'mtry' parameter. We initially built 20 trees while varying the number of predictors considered. For each species, we conducted 14 runs with different values of the mtry parameter (1, 3, 6, 10, 15, and from 30–100 in intervals of 10). We used output from the 20 trees with different levels of mtry to select the parameter value that had the lowest out-of-bag (see section 2.2) error rate for prediction. The mtry parameter affects model fit, but no rules of thumb are available with regard to the maximum proportion of predictors that should be considered. Therefore, we explored a wide range of mtry values.

After this initial exercise, we built 5000 trees using the selected mtry parameter value. We used the out-of-bag prediction error rates to understand the predictive ability of each model. We also used the Receiver Operating Characteristic (ROC) to provide a metric of predictive ability that is independent of the classification threshold [21]. A ROC value of 0.5 indicates no predictive ability and a value of 1.0 indicated perfect predictive ability [22]. We used a ROC cutoff of 0.8 as criteria to delineate strong models. We considered ROC values of 0.7 >ROC <0.8 to be marginal models and <0.7 to be poor models (ROC <0.5 indicates random performance). We reviewed variable importance to ascertain whether any of the random, neutral landscape variables were ranked as important. In order to explore how tuning the model with the mtry parameter affects model output, we generated models for the second best mtry parameter for Yellow Warbler (*Dendroica petechia*) and Snow Bunting (*Plectophenax nivalis*) based on the 20 tree out-of-bag error rate.

9.3.5 Map Output

We generated a 500 m resolution prediction grid consisting of 32,189 points across the spatial extent of the refuge and linked the predictor variables to the points using Hawth's Analysis Tools (www.spatialecology.com/htools). We scored the prediction grids for all species with strong to moderate models with the Random Forests groves (multiple trees that generate a plurality vote for output). In this chapter, we provide example maps for a common species, a moderately common species, and 2 rare species to obtain a probability of occurrence index for each point. Ruby-crowned Kinglet (*Regulus calendula*) represents the common species and Lincoln's Sparrow (*Melospiza lincolnii*) a moderately common species. Yellow Warbler and Snow Bunting distribution maps represent rare species. We imported the scored points back into GIS and converted the points into a grid with a 500 m^2 pixel size; each pixel value represents the probability of occurrence index for the prediction grid point located at pixel center. Probability of occurrence indices generally range from 0–100%. We binned values into 10% intervals for the map display.

In order to explore how the mtry tuning parameter affects map output, we built maps for Yellow Warbler and Snow Bunting using the second best mtry value, termed the second best model. We generated a binary map of occurrence (threshold of 0.5) for the model with the best out-of-bag prediction error, named out-of-bag best, and the second best model. For each species, we subtracted the predictions of the second best model from the out-of-bag best to generate a spatial representation of model differences.

9.4 Results

We built models for 40 bird species that occurred at 2–56% of the plots sampled for birds. Of these, Random Forests produced strong models (ROC >0.8) for 16 bird species 9.2, marginal models (0.7 >ROC <0.8) for 13 species 9.3, and weak models (ROC <0.7) for 11 species 9.4. The ability of Random Forests to provide accurate predictive models was independent of how common or rare a bird was on the landscape. For the strong models, 78–100% of the presence locations in the out-of-bag dataset were accurately classified (labeled Pres. Pred. in 9.2, 9.3, and 9.4) and 31–89% of the absence locations were correctly classified (labeled Abs. Pred.). In other words, with the out-of-bag test data, Random Forests predicted a species would be present where it was not found more often than Random Forests predicted a species would be absent where it was found. Random Forests did not rank any of the 5 neutral landscape variables as important for any of the 41 bird species. As we increased the number of variables considered at each node, the number of variables tagged as important decreases. Therefore, high mtry values produced models where fewer variables influenced the model output.

Distribution maps for all species with strong to moderate predictive models provided spatial representations of species occurrence that were reasonable based on anecdotal information obtained from refuge biologists. We provide examples of distribution maps for Ruby-crowned Kinglet, Lincoln's Sparrow, Snow Bunting and Yellow Warbler 9.2. Snow Bunting and Yellow Warbler differed in respect to how tuning the mtry parameter affected map outputs. For Snow Bunting, the best out-of-bag error with 20 trees was achieved when 50 variables (out-of-bag error = 2.008) were randomly selected at each node, but the error for mtry of 10 was only slightly higher (out-of-bag error = 2.610). The best mtry parameter for Yellow Warblers was 80 (out-of-bag error = 19.038), but an mtry of 10 produced a similar error (out-of-bag error = 20.461). Snow Bunting distribution was fairly stable with different mtry parameters. Yellow Warbler distribution differed significantly between models 9.3. The out-of-bag best model did not predict that Yellow Warblers would occur in high alpine snow-fields that were misclassified by the second best model. However, unlike the second best model, the out-of-bag best model did predict warblers would occur in a saltwater estuary where they are unlikely. In terms of prediction accuracy, the ROC values were higher for the out-of-bag best model (mtry = 80, ROC = 0.882) than the second best model (mtry = 10, ROC = 0.764). More validation and greater assessment data are needed to understand the range of mtry values that should be chosen from in the model building process.

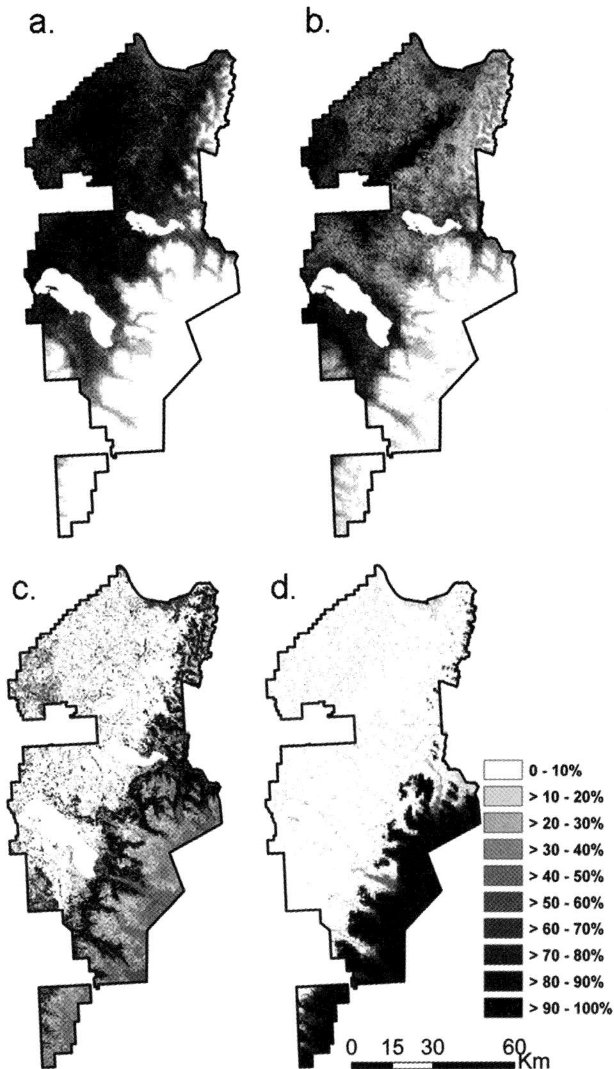

Fig. 9.2. Example distribution maps for (a)Ruby-crowned Kinglet; (b)Lincoln's Sparrow; (c)Yellow Warbler; and (d)Snow Bunting

Table 9.2. Summary of 16 strong models (ROC >0.8); species organized from most to least common

# Plots	% Plots	Species	Abs. Wt.	Pres. Wt.	Mtry	ROC	Abs. Pred. %	Pres. Pred. %
144	56	Ruby-crowned Kinglet (*Regulus calendula*)	1.3	1.0	3	0.807	64	85
131	51	Swainson's Thrush (*Catharus ustulatus*)	1.1	1.0	15	0.867	74	87
128	50	Myrtle Warbler (*Dendroica coronata*)	1.0	1.0	1	0.809	63	94
67	26	Alder Flycatcher (*Empidonax alnorum*)	1.0	2.8	40	0.847	69	81
48	19	White-crowned Sparrow (*Zonotrichia leucophrys*)	1.0	4.3	80	0.852	74	81
47	18	Savannah Sparrow (*Passerculus sandwichensis*)	1.0	4.4	50	0.861	73	87
38	15	Wilson's Warbler (*Wilsonia pusilla*)	1.0	5.7	90	0.919	89	89
38	15	Lincoln's Sparrow (*Melospiza lincolnii*)	1.0	5.7	6	0.844	72	84
36	14	Golden-crowned Sparrow (*Zonotrichia atricapilla*)	1.0	6.1	1	0.894	77	100
34	13	Fox Sparrow (*Passerella iliaca*)	1.0	6.5	10	0.843	76	79
18	7	Townsend's Warbler (*Dendroica townsendi*)	1.0	13.2	40	0.893	82	83
16	6	American Pipit (*Anthus rubescens*)	1.0	14.9	80	0.942	87	100
9	4	Yellow Warbler (*Dendroica petechia*)	1.0	27.3	80	0.882	76	89
9	4	Northern Waterthrush (*Seiurus noveboracensis*)	1.0	27.3	70	0.861	77	78
6	2	Snow Bunting (*Plectrophenax nivalis*)	1.0	41.5	50	0.966	31	100
5	2	Sandhill Crane (*Grus Canadensis*)	1.0	50.0	15	0.873	76	60

9.5 Discussion

Predictive species distribution models can provide value information for biodiversity conservation efforts. However, in the ecological literature, distribution models often focus on the fit or significance of the ecological predictors. In this chapter, we argue that for inventory and monitoring programs the interpretive focus and confidence in reliability of distribution models should be placed in the predictive ability of the map, and not in the ecological meaning

Table 9.3. Summary of 13 marginal models (0.7 >ROC <0.8); species organized from most to least common

# Plots	% Plots	Species	Abs. Wt.	Pres. Wt.	Mtry	ROC	Abs. Pred. %	Pres. Pred. %
143	56	Slate-colored Junco (*Junco hyemalis*)	1.3	1.0	1	0.787	66	92
56	22	Orange-crowned Warbler (*Vermivora celata*)	1.0	3.6	15	0.755	66	75
55	22	Hermit Thrush (*Catharus guttatus*)	1.0	3.6	40	0.767	72	71
36	14	White-winged Crossbill (*Loxia leucoptera*)	1.0	6.1	1	0.734	41	94
34	13	Boreal Chickadee (*Poecile hudsonica*)	1.0	6.5	10	0.749	60	71
32	13	Varied Thrush (*Ixoreus naevius*)	1.0	7.0	3	0.709	59	78
29	11	Wilson's Snipe (*Gallinago delicata*)	1.0	7.8	70	0.722	65	69
20	8	Western Wood-Pewee (*Contopus sordidulus*)	1.0	11.8	6	0.745	64	65
10	4	Lesser Yellowlegs (*Tringa flavipes*)	1.0	24.5	15	0.709	72	70
9	4	Golden-crowned Kinglet (*Regulus satrapa*)	1.0	27.3	90	0.757	72	78
6	2	Common Raven (*Corvus corax*)	1.0	41.5	70	0.751	72	67
6	2	Bohemian Waxwing (*Bombycilla garrulus*)	1.0	41.5	90	0.745	72	83
4	2	Grey-cheeked Thrush (*Catharus minimus*)	1.0	62.8	90	0.729	72	50

of the predictors. Algorithmic models like Random Forests can provide accurate predictive ability without a priori knowledge of the processes influencing species distributions. For monitoring programs focused on biodiversity, biologists often have little information about the underlying processes for many species and these processes are rife with weak signals, multiple interactions, and non-linear relationships.

We believe confidence within the area of interest can be quantified and assessed based on omission and commission error rates. For the 16 bird species with strong models, prediction accuracy for species presence ranged from 78–100%. Accuracy for absence was lower, but we expect higher error for predicting absences based on ecological processes. Distribution models operate under an assumption that all available habitats are saturated, but this may be unjustified. In addition, competition with other species may inhibit species

Table 9.4. Summary of 11 weak models (ROC <0.7); species organized from most to least common

# Plots	% Plots	Species	Abs. Wt.	Pres. Wt.	Mtry	ROC	Abs. Pred. %	Pres. Pred. %
56	22	American Robin (*Turdus migratorius*)	1.0	3.6	1	0.687	50	77
47	18	Gray Jay (*Perisoreus canadensis*)	1.0	4.4	70	0.662	56	64
36	14	Olive-sided Flycatcher (*Contopus cooperi*)	1.0	6.1	90	0.663	60	56
20	8	Black-capped Chickadee (*Poecile atricapilla*)	1.0	11.8	50	0.657	62	50
13	5	Pine Grosbeak (*Pinicola enucleator*)	1.0	18.6	10	0.613	66	54
12	5	Greater Yellowlegs (*Tringa melanoleuca*)	1.0	20.3	40	0.653	68	67
10	4	Common Redpoll (*Carduelis flammea*)	1.0	24.5	6	0.660	71	60
7	3	Rusty Blackbird (*Euphagus carolinus*)	1.0	35.4	60	0.617	34	43
7	3	Blackpoll Warbler (*Dendroica striata*)	1.0	35.4	30	0.578	71	57
5	2	Red-breasted Nuthatch (*Sitta canadensis*)	1.0	20.0	15	0.640	66	40
4	2	Downy Woodpecker (*Picoides pubescens*)	1.0	62.8	60	0.651	77	50

from occupying available habitat and territoriality can cause false negatives if the sampling unit is smaller than the minimum distance between individuals [23]. Finally, failure to detect a species within the sampling frame can be a source of false negatives [24]. In logistic regression, false negatives can bias parameter estimation and lead to erroneous conclusions about the driving variables [24] [25]. False negatives likely cause the probability of occurrence index to be underestimated and this bias may not be consistent across habitat types. However, in terms of simple presence/absence prediction (default threshold of 0.5 probability of occurrence index), it is unclear how much of the detection error is contained in the omission error rate of the model. Future work should compare whether high omission error rates coincide with species that are more difficult to detect and whether predicted distribution size coincides with landscape occurrence that is adjusted for detection error. Error rates can be further explored for bias across vegetation or other subjective, human-constructed categories.

A focus on prediction ability will allow models built with data from different timeframes (e.g., the LTEMP grid resampled in 100 years) to be overlaid

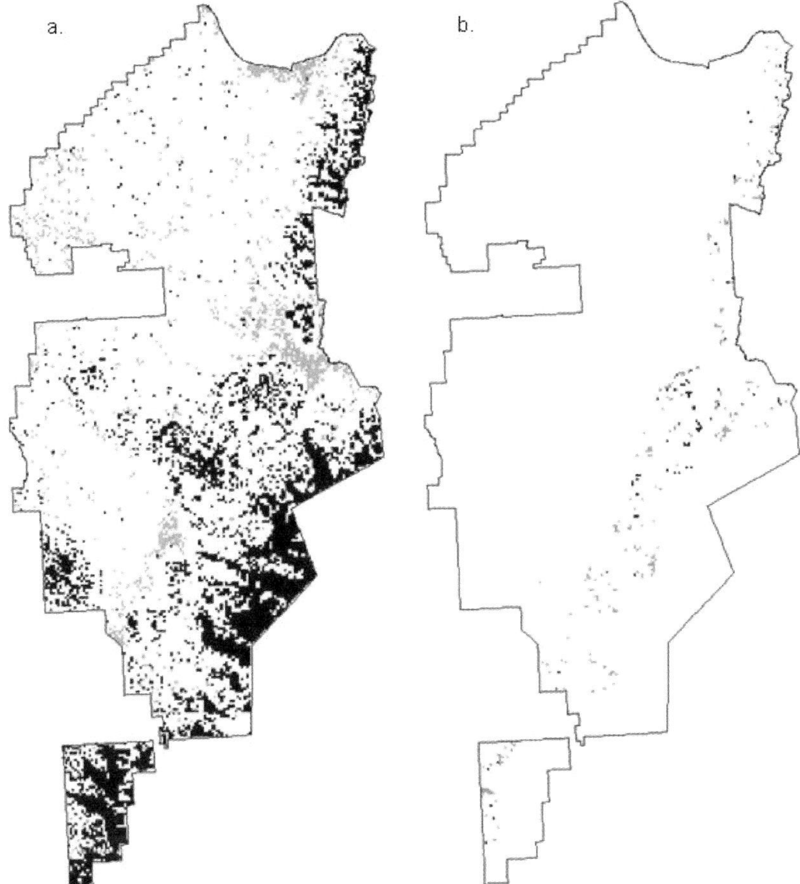

Fig. 9.3. Maps display difference between the out-of-bag best model and the second best models for (a)Yellow Warbler and (b)Snow Bunting. Black represents a cell where the out-of-bag best predicts an absence and the second best predicts a presence. Grey represents a cell where the out-of-bag best predicts a presence and the second best predicts an absence

and compared even though totally different prediction variables might be used to build the models. Prediction error and the quantitative assessment of error rates will put time-series maps into a common framework for comparison. The possibility of overlaying species distribution maps from different timeframes could provide a powerful monitoring metric, especially for species for which there is limited knowledge, although model error in terms of the additive prediction error for each species would need to be transparent and quantified. Pixel by pixel change detection analysis of species distribution maps could provide map output of changes to species distributions. Area in each probability of occurrence category could also provide a monitoring metric (i.e. area with >80% probability of occurrence). In addition, if future sampling rates produce omission or commission error rates that exceed the known model error, this change in predictive ability could be used as a trigger for management concern.

Distribution maps as a monitoring metric are one step in an iterative process and not meant to provide total information about the species of interest (i.e., causal significance of predictors). Changes in the prediction space would need to be met with additional research efforts before effective management could be undertaken. If an expansion or contraction of a distribution were documented, a researcher could easily use the information captured by the maps as an early warning system and to help formulate hypotheses to explain the change. Although the predictors would be transparently available, the researcher would not need to engage the predictors to use this information. For example, the researcher could evaluate the locations of expansion or contraction against areas of little to no change for any number of variables thought to be causal that were excluded in the original distribution models.

A trade-off exists between the focus on predictive ability within a bounded study area and the general applicability of the model to regions outside the study area [5]. For monitoring efforts that utilize grid-based designs, like LTEMP, this limitation is not important because the spatial extent of interest is sampled. In fact, we are more interested in accurate predictions within the study area then the general applicability of the model. Additionally, grid-based designs with adequate resolution provide a representative sample of the landscape because landscape properties like habitat types, distance to roads, and fragmentation levels are surveyed in direct proportion to their prevalence on the landscape. A representative sample ensures that the Random Forests model is optimized for the landscape. Monitoring programs may provide biased models for a given study area due to sample stratification or a non-robust sample frame. Random Forests may also be used to build distribution models with alternative data sources like telemetry approaches (information from collared individuals compared to random locations). However, models from telemetry approaches may not be as robust and will need extra validation efforts if the sample of marked individuals was spatially biased (i.e. roads or other factors that ease trapping costs).

This approach of using predictive models to monitor changes in species distributions would be very applicable to problems raised by global climate change. We know species in the Northern Hemisphere are moving upward in elevation and/or further northward in latitude in response to a generally warming climate [26]. Yet, Crozier [27] argues that current large-scale associations alone are insufficient for predicting the biogeographical consequences of climate change because there are species-specific response times to climate change, issues associated with scaling down climate models to smaller land areas, and a high likelihood of spurious associations between species distributions and climatic conditions. These arguments, plus the fact that empirical distribution data for most species are simply lacking, suggests that mechanistic approaches to modeling species distributions are problematic.

Computer learning algorithms are more flexible, user friendly, and require less time than traditional statistical approaches. For Random Forests, there is no need to limit the number of predictor variables explored (i.e. 1,000 or more are acceptable) and interactions between variables are included which is a benefit for species with little information available. This also allows one set of predictors to be developed for a suite of species which is beneficial for multi-species monitoring programs. Additionally, there are no constraints on predictors like assumptions of normality; correlated predictors are helpful for tree-based methods to obtain better predictions. More conventional general linear models require technical expertise to meet statistical assumptions needed to ensure unbiased output and ecological expertise to capture important variables and interactions within the limiting constraints that comes with caps on the number of predictors that may be analyzed. Programs like Random Forests that utilize computer learning algorithms can be run consistently with default settings requiring very little institutionalization of specific modeling skill sets on a given wildlife refuge or other land management unit [14] [28]. Linking monitoring programs with spatially-explicit modeling provides alternative approaches for detecting and quantifying changes in species distributions.

9.6 Acknowledgments

We would like to thank Salford Systems for providing an evaluation version of Random Forests. The Kenai National Wildlife Refuge, WORLDCLIM, and Alaska Geospatial Data Clearinghouse supported this work by maintaining the numerous GIS layers necessary for this analysis. We also wish to acknowledge the collaborative support of the USDA Forest Service, Pacific Northwest Research Station, Forest Inventory & Analysis program. The University of Alaska's Integrative Graduate Education and Research Traineeship (IGERT) program provided fellowship support to DM. The U.S. Fish & Wildlife Service also provided financial support to DM through an appointment with the Student Career Experience Program (SCEP). The Corporative Institute of Arctic Research and Center for Global Change provided needed computer hardware through the University of Alaska's Global Change Student Research Grant Competition. T. Burke and T. Eskelin from the Kenai National Wildlife Refuge reviewed and provided useful insights about the distribution maps generated for this chapter. Three anonymous reviewers provided helpful comments to improve this chapter. FH acknowledges S. Linke. This is EWHALE Lab publication #41.

References

1. Guisan A, Thuiller W (2005) Ecology Letters 8:993–1009
2. Meretsky V J, Fischman R L, Karr J R, Ashe D M, Scott J M, Noss R F, and Schroeder R L (2006) Bioscience 56:135–143
3. Gottschalk, T K, F Huettmann, and M Ehlers (2005) International Journal of Remote Sensing 26:2631–2656
4. Lunetta, R S, Elvidge C D (eds.) (1998) Remote sensing change detection: environmental monitoring methods and applications. Ann Arbor Press, Chelsea, Michigan
5. Guisan A, Zimmermann N E (2000) Ecological Modelling 135:147–186
6. Busch D E, Trexler J C (eds.) 2003 Monitoring ecosystems: interdisciplinary approaches for evaluating ecoregional initiatives, Island Press, Washington Covelo London
7. Holthausen R, Czaplewski R L, DeLorenzo D, Hayward G, Kessler W B, Manley P, McKelvey K S, Powell D S, Ruggiero L F, Schwartz M K, Van Horne B, Vojta C D (2005) Strategies for monitoring terrestrial animals and habitats. General Technical Report RMRS–GTR–161, U.S. Department of Agriculture, Forest Service, Rocky Mountain Research Station, Fort Collins, Colorado
8. Angermeier P L, Karr J R (1994) Bioscience 44:690–698
9. Boyce M S, McDonald L L (1999) Trends in Ecology and Evolution 14:268–272
10. Yen P P W, Huettmann F, Cooke F (2004) Ecological Modelling 171:395–413
11. Heglund P J (2002) Foundations of species-environment relations. In: Scott J M, Heglund P J, Morrison M L, Haufler J B, Raphael M G, Wall W A, Samson F B (eds.). Predicting species occurrences: issues of accuracy and scale. Island Press, Washington Covelo London
12. Breiman L (2001) Machine Learning 45:5–32
13. Efron B, Tibshirani R J (1993) An Introduction to the Bootstrap, Chapman and Hall, New York
14. Breiman L (2001) Statistical Science 16:199–231
15. Hand D J (1998) The American Statistician: 52:112–118
16. Buckland S T, Anderson D R, Burnham K P, Laake J L, Borchers D L, Thomas L (2001)Introduction to distance sampling: estimating abundance of biological populations. Oxford University Press Inc., New York
17. Dawson D W, Smith D R, Robbins C S (1995) Point count length and detection of forest neotropical migrant birds. In Ralph CJ, Sauer J R, Droege S (eds.). Monitoring bird populations by point counts. General Technical Report RMRS–GTR–149, U.S. Department of Agriculture, Forest Service, Pacific Southwest Research Station, Albany, California
18. McGarigal K, Marks B J (1995) FRAGSTATS: spatial pattern analysis program for quantifying landscape structure. General Technical Report PNW–351, U.S. Department of Agriculture, Forest Service, Pacific Northwest Research Station, Corvallis, Oregon
19. Daly C, (2006) International Journal of Climatology 7:707–721
20. Hijmans R J, Cameron S E, Parra J L, Jones P G, Jarvis A (2005) International Journal of Climatology 25:1965–1978
21. Pearce J, Ferrier S (2000) Ecological Modeling 13:225–245
22. Boyce M S, Vernier P R, Nielsen S E, Schmiegelow F K A (2002) Ecological Modelling 157:218–300

23. Fielding A H, Bell J F (1997) Environmental Conservation 24:38–49
24. MacKenzie D L, Nichols J D, Royle J A, Pollock K H, Bailey L L, Hines J E (2006) Occupancy estimation and modeling:inferring patterns and dynamics of species occurrence. Academic Press, Amsterdam Boston Heidelberg London New York Oxford Paris San Diego San Francisco Singapore Sydney Tokyo
25. Gu W, Swihart R K (2004) Biological Conservation 116:195–203
26. Parmesan C, Yohe G (2003) Nature 421:37–42
27. Crozier L (2002) Climate change and its effect on species range boundaries: a case study of the Sachem Skipper butterfly, *Atalopedes campestris*. In: Schneider S H, Root T L (eds.) Wildlife responses to climate change. Island Press, Washington Covelo London
28. Elith J, Graham C H, Anderson R P, Dudik M, Ferrier S, Guisan A, Hijmans R J, Huettmann F, Leathwick J R, Lehmann A, Li J, Lohmann L G, Loiselle B A, Manion G, Moritz C, Nakamura M, Nakazawa Y, Overton J M, Peterson A T, Phillips S J, Richardson K, Scachetti-Pereira R, Schapire R E, Soberon J, Williams S, Wisz M S, Zimmermann N E (2006) Ecography 29:129–151

Visualization and Interactive Exploration of Large, Multidimensional Data Sets

John T. Langton, Elizabeth A. Gifford, and Timothy J. Hickey

Michtom School of Computer Science, Brandeis University,
Waltham, MA 02254, USA
{psyc,egifford,tim}@cs.brandeis.edu

Summary. As biologists work with more and more data, there is an increasing need for effective tools to analyze it. Visualization has long been used to communicate experimental results. It is now being used for exploratory analysis where users rapidly determine significant trends and features by working with visual projections of data. A basic workflow is to a) record experimental results or simulate some biological system, b) form hypotheses, c) verify hypotheses with interactive visualizations and statistical methods, d) revise hypotheses, and e) confirm computational results with experiments in wet-lab.

In this chapter we describe a number of visualization methods and tools for investigating large, multidimensional data sets. We focus on approaches that have been used to analyze a model neuron simulation database. These methods are best applied to databases resulting from brute force parameter space exploration or uniform sampling of a biological system; however, their wider applicability is currently under investigation. An example analysis is provided using a generalized tool for interactive visualization called NDVis. We include a summary of NDVis, its plugin architecture, and JavaSim which can be used as a plugin for NDVis or as a stand alone tool for investigating model neuron simulations.

10.1 Introduction

Modern biology is becoming an increasingly data intensive enterprise. High throughput data collection devices such as flow cytometers generate megabytes of data for each biological sample. Large-scale simulations of biological systems (often executed in parallel) can generate gigabytes or even terabytes of data. The analysis of these large data sets has stimulated considerable new research in Computer Science such as Basic Local Alignment Search Tool (BLAST) algorithms for analyzing gene sequence alignment, and visualization tools for inspecting their results [1]. The key is to identify relevant methods and how they can be applied to open problems in biology.

Visualization has long been used to communicate experimental results. It is now being used for exploratory analysis where users rapidly determine

J.T. Langton et al.: *Visualization and Interactive Exploration of Large, Multidimensional Data Sets*, Studies in Computational Intelligence (SCI) **122**, 231–255 (2008)
www.springerlink.com   © Springer-Verlag Berlin Heidelberg 2008

significant trends and features by working with visual projections of data. Using interactive visualizations, analysts can form hypothesis, verify them with statistical methods, revise hypothesis, and eventually confirm the computational results with experiments in wet-lab.

In this chapter we describe a number of visualization methods and tools for investigating large, multidimensional data sets. The approaches described here are best applied to data generated from parameter space exploration through brute force grid-based simulations or uniform sampling of biological systems. We first describe the generation of a neuron model simulation database in Section 10.2. We then present a number of common visualization methods in Section 10.3.1. Dimensional stacking and pixelization were used for visual analysis of the neuroscience data and are detailed in Section 10.3.2. Section 10.4.1 presents a number of visualization tools while Section 10.4.2 describes NDVis, a tool that combines dimensional stacking and pixelization with a number of interaction techniques. A case study of using NDVis to motivate hypotheses about the neuroscience data and testing those hypotheses with another simulation and visualization tool called JavaSim is presented in Sections 10.5 and 10.6. We also show how these visualization tools and techniques can be combined with statistical and machine learning algorithms in a way that increases the power of both approaches. We present our conclusions in Section 10.7.

10.2 The Prinz-Billimoria-Marder Neuron Model Simulation Data Set

The neuroscience data set described here provided the impetus for our research and is referenced throughout the rest of this chapter. Prinz, Billimoria, and Marder [2] constructed a neuron simulation database to better understand the role of a neuron's individual membrane currents in shaping the dynamics of membrane potential, the signal neurons use to encode information and communicate. Data from several simulations of a model neuron was classified according to electrical activity and recorded. The model was based on a lobster stomatogastric neuron and its activity was governed by a system of coupled, non-linear, differential equations based on the work of Hodgkin-Huxley. These equations took 8 input parameters which represented maximal conductances of 8 ion channels. The data set was generated by independently varying each of the conductance parameters over 6 equidistant values ranging from 0 to an experimentally estimated maximum for each conductance. The conductance parameters and their possible values are shown in Table 10.1. For each of the $6^8 = 1,679,616$ possible combinations of maximal conductance parameters, the neuron model was simulated until the voltage plots entered a steady state, repeating pattern, or a certain time threshold was exceeded. The maxima and minima of each simulation waveform were then stored along with other information such period length for active neurons and membrane potential for silent neurons. The 1,679,616 neuron model simulations were performed on a 30 processor Beowulf cluster over a 3 month period. Details of the neuron model are described in [2]. An analysis of the data using some of the techniques described in this chapter is presented in [3].

Table 10.1. Model neuron conductance parameters.

Name	Definition	min value mS/cm^2	increment mS/cm^2	max value mS/cm^2
Na	a Na$^+$ current, I$_{\text{Na}}$	0	100	500
CaT	a fast Ca^{2+} current, I$_{\text{CaT}}$	0	2.5	12.5
CaS	a slow Ca^{2+} current, I$_{\text{CaS}}$	0	2	10
A	a transient K$^+$ current, I$_{\text{A}}$	0	10	50
KCa	a Ca^{2+}-dependent K$^+$ current, I$_{\text{KCa}}$	0	5	25
Kd	a delayed rectifier K$^+$ current, I$_{\text{Kd}}$	0	25	125
H	hyperpolar.-activated inward current, I$_{\text{H}}$	0	.01	.05
leak	a leak current, I$_{\text{leak}}$	0	.01	.05

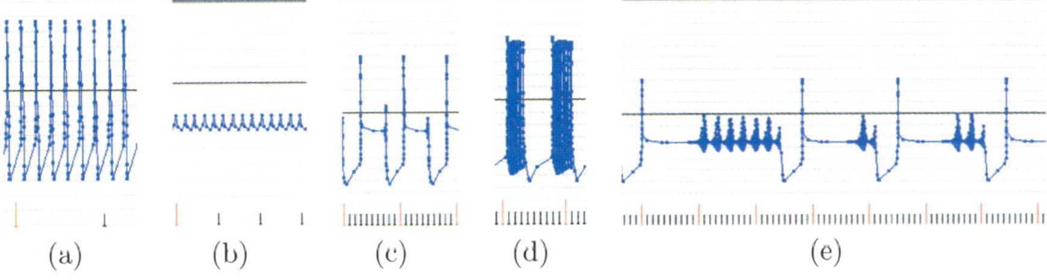

(a)   (b)   (c)   (d)   (e)

Fig. 10.1. Voltage plots generated from JavaSim's neuron simulator. (a) and (b) are tonically spiking, (c) and (d) are bursting, and (e) is classified as irregular. The horizontal gray lines are in 10mV increments where the horizontal black line is the origin. The red tick marks indicate seconds, the black are 1/10 seconds.

Each simulation output waveform was classified as belonging to one of four groups based on activity type: (a) silent (i.e. with a constant membrane potential), (b) tonically spiking, (c) bursting, or (d) unclassifiable (only 8694 simulations were unclassifiable representing 0.52% of the data set). Figure 10.1 shows several types of voltage plots generated by the neuron simulator. Spiking neuron behavior was classified as either fast or slow. The difference between the two their period or duration of time between two spikes. Fast spikers have a period of < 0.09 seconds while slow spikers have a period ≥ 0.09 seconds.

This method of creating a database from simulations over every possible combination of a set of discrete input values is commonly called "parameter space exploration." Similar methods can be used for any computational model of a biological system. Some of the main advantages are that the simulations can be performed in parallel and the results give a view of the entire parameter space rather than isolated regions.

10.3 Visualization Methods

10.3.1 General Methods

In this section we summarize a number of methods for visualizing multidimensional data. The more dimensions there are in a data set, the more sparse the data. This is commonly referred to as "the curse of dimensionality." A number of techniques exist for selecting dimensions of interest and reducing the total number of dimensions to facilitate visualization such as Principle Components Analysis (PCA), Multi-dimensional Scaling (MDS), and Projection Pursuit. Here we focus on visualization methods for displaying the selected dimensions.

Scatter plot matrices are the oldest and probably most common technique for visualizing multidimensional data. In this approach, every pair-wise combination of dimensions is shown in a 2D plot, and each of these 2D plots is assembled into a matrix. This requires N choose 2 plots where N is the number of dimensions (some plots are eliminated such as transforms from flipping the x and y axis of a plot and plotting a dimension against itself). The difficulty with using this approach is that it does not scale well with large numbers of dimensions (the 8 dimensions of our neuroscience data require 8 choose 2 = 28 plots) and it is difficult to determine relationships between more than 3 dimensions.

Some methods deal very well with a large number of dimensions but fewer data points such as parallel coordinates [4] and multidimensional glyphs [5]. In parallel coordinates, the y-axis serves as a normalized value interval, while the rest of the dimensions are plotted on the x-axis. Each data point is represented by a line that runs through each dimension on the x-axis at its value on the y-axis. For glyphs, each data point is associated with a shape where its features are determined by the values of the data point. For instance, with a star glyph, lines for each dimension protrude from a central point and the length of each line is determined by a data point's value for its associated dimension. Thus, every data point appears as a slightly different star. The major draw back of these approaches are that they do not deal well with large data sets, primarily because of obfuscation. For instance, trying to fit 1.7 million stars into a 2D display for the data points of our data set results in glyphs overlapping.

Pixelization is a method specifically designed to deal with large data sets [6]. In this method, each pixel of an image is associated with a single data point. The locations of pixels are arbitrary, and their color is specified by the values of their associated data points. A window is created for each dimension and the user toggles back and forth to determine relations between dimensions. While this method allows users to view much more data in a single image, it has problems similar to that of scatter plot matrices in that more dimensions mean more windows and it can become difficult to determine complex relationships between more than 3 dimensions.

In order to visualize the neuroscience data set we needed to address not only many dimensions but also many data points. As described above, there are no established methods to do both. We therefore combined a layout technique called dimensional stacking that deals with multiple dimensions with pixelization to deal with many data points. The following section describes our approach.

10.3.2 Dimensional Stacking

Dimensional Stacking is a method to project data with 3 or more dimensions into a 2 dimensional image [7]. To describe dimensional stacking we'll start with visualizing two dimensions of the neuron model data and "stack" up to eight dimensions. We make observations along the way to illustrate how these images can be used during analysis. One should note that all images in this section are examples of dimensional stacking but not pixelization. Only the 8D images in the following sections represent a combination of dimensional stacking and pixelization where each pixel is associated with one row of the neuron model simulation database. Further, the order in which dimensions are stacked has a profound impact on the resulting image. This is discussed with examples in Section 10.5.

Selecting a Color Map

To generate a dimensional stacking image one must first select a set of attributes to map to colors (or gray scale) and how this mapping will occur. Color mappings are often called "transfer functions" in the field of visualization. In our approach, independent variables are typically mapped to the axes of an image whereas dependent variables are mapped to attributes of pixels such as color, hue, saturation, and brightness. For the neuroscience database, the simulation input conductance parameters are mapped to the axes of images while the simulation output values such as spike frequency are mapped to pixel color.

Figures 10.2 through 10.4 all share the same color map. Each of their grid squares correspond to a subset of neuron models from the simulation database. The RGB (red, green, blue) color component values of each grid square can be considered to vary between 0 and 1. The amount of red is determined by the proportion of fast spikers, green is determined by the proportion of slow spikers, and blue is determined by the proportion of blue spikers. These images therefore map the concentration of certain neuron activity types into a color gradient. For instance, if almost all neurons in a grid square are spikers, half are fast and half are slow, then the grid square will have the RGB color

(0.5,0.5,0) which is a dull yellow. Silent (or irregular) neurons don't contribute to the color map. If almost all neurons in a grid square are silent or irregular, its color will be mostly black (r = 0.0, g = 0.0, b = 0.0).

It is possible to map any continuous value to the RGB color components, alpha component, or any other attributes of a pixel. The number of attributes that can be mapped is primarily limited by how perceptible those mappings are to the user. In our analysis we have never mapped more than 3 values for coloring an image and always use the RGB color components of pixels. However, we have discussed using transparency and overlays.

Standard Plot with 2 Dimensions and a Color Map

Figure 10.2 shows the average activity types for all neuron models with KCa on the x-axis and CaT on the y-axis. We selected KCa and CaT for the dimensions of this plot because we know from our analysis that they are important conductances in determining fast or slow spiking behavior. Selecting a different pair of parameters might reveal other properties of the data. Because each conductance takes on 6 discrete values as specified in Table 10.1, the image is a 6 X 6 grid and each rectangle in the grid represents $6^6 = 46,656$ neuron models. The amount of red in each grid cell corresponds to the number of fast spikers, the amount of green corresponds to the number of slow spikers, and the amount of blue corresponds to the number of bursters.

Observe that this simple image allows us to immediately start inferring properties of the data set. For example, there is a high concentration of green in every cell in the bottom row where $CaT = 0$ which reveals that very low values of CaT commonly lead to neurons having a slow spiking activity pattern. Similarly, there is a high concentration of red in the first column (where $KCa = 0$) above the bottom row (where $CaT = 0$). This can mean that KCa is an inhibitor of fast spiking activity, meaning that lower levels of KCa often lead to neurons with a fast spiking activity pattern. As the values of both KCa and CaT increase together, there is a high concentration of blue, symbolizing a large number of neurons with a bursting activity pattern.

Dimensionally Stacked Image with 4 Dimensions

Figure 10.3 uses dimensional stacking to display an image where four parameters vary independently. Every value of Na is stacked inside each value of KCa on the x-axis. Every value of Kd is stacked inside each value of CaT on the y-axis. Thus, the image is a 6x6 grid of 6x6 grids, and each one of the small grid cells represents a set of $6^4 = 1296$ neuron models. This can be thought of as a refinement of the image in Figure 10.2 where fewer neuron models are aggregated per grid square. A cursory examination of this image shows the following interesting patterns:

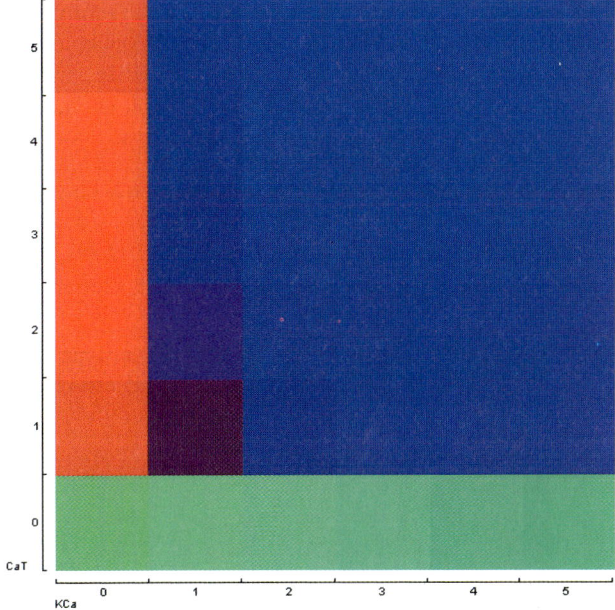

Fig. 10.2. 2D image of the neuron simulation database. Every matrix cell maps to 46656 data points. The concentration of red is the number of fast spikers, green is the number of slow spikers, and blue is the number of bursters.

- The black regions occur where there is no spiking or bursting activity (i.e. no RGB values) and therefore represents mostly silent neurons. From the plot we can see that this silent behavior occurs when $KCa = 0$ (the first column) and either $Na = 0$ (the first sub-column) or $Kd = 0$ (the first sub-row of each row).
- In the first column $(CaT = 0)$ there are five blue squares which were not visible in the 2D plot in Figure 10.2. This clearly indicates that there are sets of neurons where $KCa = 0$ and $CaT >= 3$ that exhibit bursting behavior.
- In the column where $KCa = 1$ and $CaT \geq 1$ we find primarily bursters or fast spikers and moreover the probability a neuron is a fast spiker increases as $Na + Kd$ increases. This behavior is slightly more pronounced for smaller values of CaT.
- There is now greater detail in the bottom row (where $CaT = 0$). For each fixed value of KCa, as Na and Kd increase, so does the concentration of green or slow spiking neurons. This means that these two conductances are also linked to the slow spiking behavior, where higher levels of $Na + Kd$ lead to a higher probability of a neuron having a slow spiking activity pattern.

- In the bottom row where $CaT = 0$ and the left two columns where KCa is 0 or 1, low values of Na or Kd lead to less green and more red - equivalently to fewer slow spikers and more fast spikers. Further, as KCa increases, this trend decreases. For instance, where $KCa \geq 3$, no values of Na or Kd produce a high concentration of red (or fast spikers).

By using only four dimensions with dimensional stacking we can already see correlations that involve three or four dimensions simultaneously. This would require toggling back and forth between 4 windows with traditional pixelization or several different plots with scatter plot matrices.

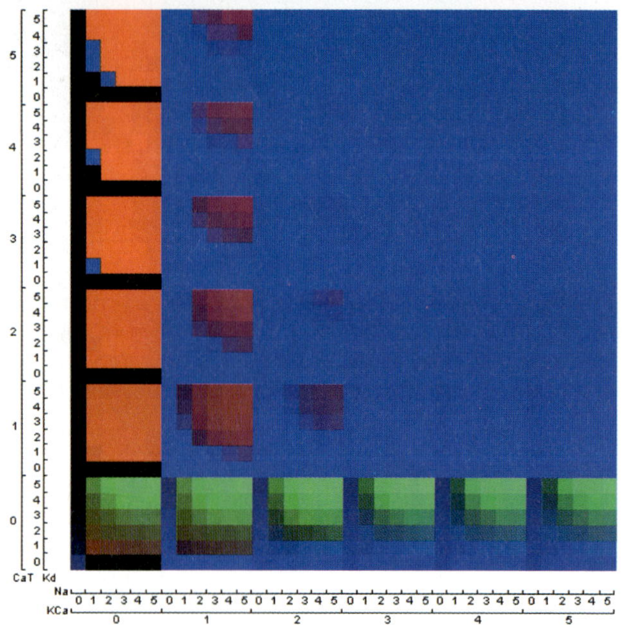

Fig. 10.3. Image of neuron simulation database with 4 dimensions stacked. The smallest cells of the matrix each map to 1296 data points. The projection used here is $x = (KCa, Na)$ and $y = (CaT, Kd)$, thus the top level 6x6 grid is where KCa and CaT vary between 0 and 5 on the x and y axes respectively. The innermost 6x6 grids are where Na and Kd vary along the x and y axes respectively. The RGB color components are assigned so that the concentration of red is the number of fast spikers, green is the number of slow spikers, and blue is the number of bursters.

Dimensionally Stacked Image with 6 Dimensions

We go from 4 dimensions stacked in Figure 10.3 to 6 dimensions stacked in Figure 10.4. Each of the smallest grid cells in the image represent $6^4 = 1296$ neuron simulations. All of the observations made for the 4D image hold for the

6D image. An interesting feature that is revealed in the 6D image but not the 4D image is the dependence of spiking behavior on the innermost dimensions, A and CaS. Observe that in the squares where $CaT = 0$ and $KCa <= 2$, for each fixed value of Kd and Na, the green region appears mostly above the diagonal of these small squares and the red regions appear mostly below. This suggests that the value of $A - CaS$ is a determinant in whether a spiker will be fast or slow when $CaT = 0$ and $KCa < 2$.

We can now make the general hypothesis that a linear hyperplane separates the fast and slow spikers in parameter space. An analyst could test this hypothesis by applying statistical or machine learning tools to search for such a hyperplane. One method would be to identify boundary points between the fast and slow spikers then use linear regression to create a plane that approximates the boundary. To do so we first find pixels or grid-squares in the image (or parameter space) that are adjacent but lie in different regions i.e. one in the fast spikers region and one in the slow spikers region. We then take the midpoint between these pixels which should be close to the boundary between the fast and slow spiker regions (assuming the boundary is a hyperplane). After collecting some number of these points we run a linear regression to obtain coefficients for an equation that tests for whether a point lies above or below the hyperplane boundary between regions. The following formula was obtained in such a manner and is a test for whether a spiking neuron is a slow spiker:

$$0.96 * CaT + 0.17 * CaS + 0.0044 * H$$
$$< 0.12 * A + 0.077 * Kd + 0.073 * KCa + 0.02 * Na + 0.0033 * leak + 0.13$$

This equation correctly predicts 99.75% of the slow spikers and 97.33% of the fast spikers and incorrectly predicts the fast/slow property of only 2.2% of the spikers. Observe also that this hyperplane criteria validates our observations made above. As Na+Kd increases, the probability that a neuron model is a fast spiker decreases and similarly for A-CaS.

Dimensionally Stacked Image with 8 Dimensions

An 8D dimensional stacking image shows the entire model neuron simulation database in one display, where all 8 conductance parameters are assigned to one of the axes. At the resolution of this printed page, the 8D dimensionally stacked image would appear essentially identical to the 6D image in Figure 10.4 so we do not include it here. However, the differences are noticeable on a standard computer screen with a resolution at or above 1296X1296. This is because the 4 dimensions on each axis assign each of the 1,679,616 model neurons to one pixel resulting in a 1296 X 1296 image. Again, all of the previous observations are preserved however greater detail can be seen especially when zooming and panning.

The amount of information aggregated in each pixel decreases as you add dimensions. When all 8 dimensions are represented, no aggregations occurs,

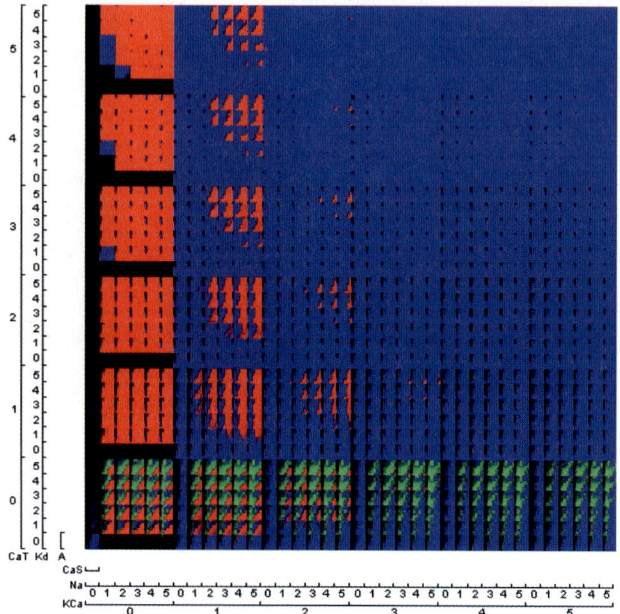

Fig. 10.4. Image of neuron simulation database with 6 dimensions stacked. The smallest cells of the matrix each map to 36 data points. The projection for this image is $x = (KCa, Na, CaS)$ and $y = (CaT, Kd, A)$. So the top level 6x6 grid corresponds to the values of KCa and CaT, while the innermost 6x6 grids correspond to A x CaS. The RGB color components are assigned so that the concentration of red is the number of fast spikers, green is the number of slow spikers, and blue is the number of bursters.

and each data point is mapped to exactly one pixel. To further illustrate this point, Figure 10.5 shows the lower left corner of the 6D and 8D images. Section 10.5 includes a number of 8D images albeit at a low resolution.

10.4 Visualization Tools

10.4.1 General Tools

There are a number of popular tools for the analysis and visualization of data. Weka [8] provides a number of data mining algorithms and standard visualizations such as scatter plot matrices. R [9] and Octave [10] (and their commercial counterparts S-Plus and Matlab) contain many statistical analysis packages and also generate standard visualizations such as scatter plot matrices. GGobi is a software tool that integrates with R and combines its algorithms with interactive visualizations to facilitate a more intuitive data exploration [11]. IBM's DataVisualizer provided similar features to these other

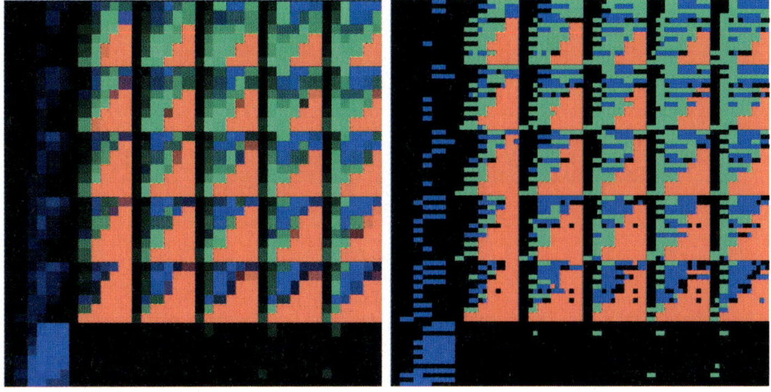

Fig. 10.5. Closeup of the lower left corner of the 6D (on the left) and 8D (on the right) dimensionally stacked images of the model neuron simulation database. Each grid square in the 6D image is mapped to 1296 model neurons where the amount of red corresponds to fast spikers, green corresponds to slow spikers, and blue corresponds to bursters. The 8D image has no color gradient as each pixel is mapped to exactly one data point and thus its colors are discrete.

packages however is not as popular though it now has an open source version called OpenDX. There also exist visualization APIs that can be customized for various purposes such as Prefuse, however, at the time of our research, Prefuse concentrated on network visualization (which ironically was not appropriate for our purposes).

Many of the tools listed above are quite useful for analysis of medium to small sized data sets. Weka and GGobi were unable to load the entire neuron model simulation database. Matlab could load the database (we did not attempt to load it into Octave), however specific functions in R (SPSS's open source engine) could not operate on the entire database. It was desired for a tool to interface with a database such as MySQL such that powerful queries could be specified for rapid analysis. None of the tools have such support although there are interfaces that could be coded for Matlab, Octave, and R. The visualizations of these tools were standard and suffered from the drawbacks described in Section 10.3.1 for analyzing the neuroscience data. We therefore implemented dimensional stacking and pixelization in our own tool called NDVis.

10.4.2 NDVis: Visualization as Interface

NDVis is a software tool (available for download at http:// neuron.cs.brandeis.edu) that combines dimensional stacking and pixelization with a number of interaction techniques. Its visualization methods are described in [12], while its interactive interface is described in [13] (NDVis was called NeuroVis at the time of publication - it is now used for other data sets thus the name change). NDVis was built to facilitate the investigation of the neuron model simulation database described in Section 10.2. Its features are briefly summarized here.

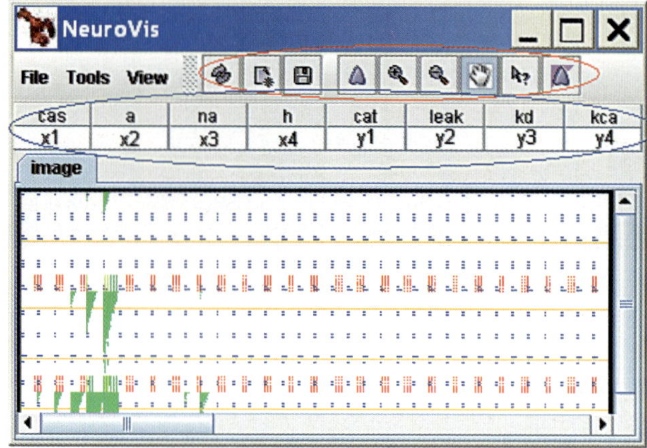

Fig. 10.6. NDVis. The buttons circled in red are for basic functions like panning and zooming. The table row circled in blue shows the dimension values associated with each pixel as the mouse passes over. The order of the parameters can be changed by dragging the column headers, which has an affect on the appearance of the image. The user can click on pixels to execute tools on their associated data points (e.g. rerunning a simulation).

The NDVis interface is shown in Figure 10.6. The buttons circled in red are for basic functionality such as panning and zooming images, while the projection order (the first four parameters are on the x-axis and the second four are on the y-axis) for the image is shown in the single row table circled in blue. This table also reports the parameter values for each pixel as the mouse passes over.

Query-based Color Maps

The primary NDVis feature used for this investigation was the Color Mapper shown in Figure 10.8. This allows users to specify a color map with SQL queries. For each query entered, the user can map its result set to a discrete color or color gradient. When a discrete color is specified, the pixels associated with any data points (or database rows) in the query's result set are assigned that color. This type of color map is used in Section 10.5. When a gradient is specified, NDVis looks for one or more continuous values in the query's result set and maps each data point's values into a color gradient for its associated pixel using the RGB color components. This type of color map was used in Section 10.3.2 (although grid squares represented several data points instead of pixels representing 1). The gradient is determined by the interval between a starting color and ending color that the user specifies (e.g. red to green) then each data point's values are normalized within that interval.

Interactive Dimension Reordering

The order of dimensions has a profound effect on the appearance of a dimensionally stacked image. Yang has provided methods for dimension reordering to reduce clutter and determine trends [14]. NDVis allows users to interactively reorder dimensions and also provides automatic methods for finding "interesting" orders. This is discussed in detail in [13]. The effect of dimension reordering is also shown in Section 10.5.

Click-through Data Access

NDVis allows users to execute tools on data points by clicking on the pixels associated with them. Further, it has a plug-in architecture that allows users to custom build these tools for their data sets. For the neuroscience data, one plug-in tool is the neuron model simulator. Users can click on a pixel in an 8D image to run the simulator for a particular neuron model (or more specifically for a particular set of conductance parameter values). This is useful for "drilling down" around boundaries between neuron activity type regions to better understand the surrounding dynamics. We are currently expanding the available plug-ins to allow users to interact with data mining routines through multidimensional brushing (or selecting multiple pixels) and clicking on pixels to select parameters for algorithms.

10.5 Case Study Part 1: Investigating Neuron Activity

We now illustrate the use of visualization to investigate a fundamental question about the Prinz-Billimoria-Marder neuron model: what are the conditions on the maximal conductance parameters that determine whether a neuron model is active or silent? To answer this question we find a collection of simple regions $(S_1, ..., S_n)$ in conductance grid space such that their union captures most of the silent neurons and relatively few of the active neurons. Our eventual goal is to find regions in the continuous parameter space in addition to the grid parameter space which predict the behavior of the neuron model. In this section we focus on the grid parameter space for making initial hypothesis. The following section refines these hypothesis for the continuous parameter space by inspecting 2D slices of parameter space and re-running simulations at a finer resolution.

The images in this section reflect all 8 dimensions and data points of the neuron model simulation database. As such, we do not label their axes as they would simply be unreadable. Instead, we describe them in text.

10.5.1 Distribution of Silent Neurons and Color Map Specification

There are 286,400 neuron models that are silent or show no activity. This is about 17.05% of all neuron models in the database. Silent neurons have a constant value for the membrane potential after steady state is reached. Figure 10.7 shows the distribution of membrane potential values for the silent neurons. It is clear from this plot that the resting potential partitions the silent neurons into seven groups as shown in Table 10.2, although the four largest groups (S3,S4,S5,S7) account for over 95% of the silent neurons. The data in this table translates directly into a color map where each of the seven regions S1-S7 is defined by an SQL query and associated with a discrete color. This color map is shown in a screen shot of the NDVis ColorMapper tool in Figure 10.8. We will use this color map for all images in this section except 10.12 which shows an error map of silent classification.

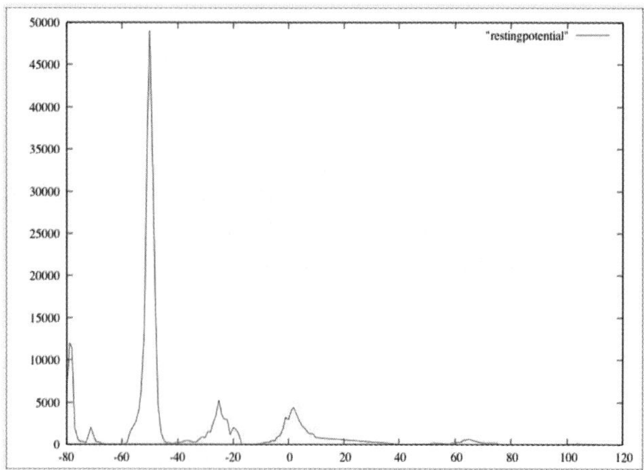

Fig. 10.7. Distribution of resting potential (from -80mV to 120mV) for silent neurons.

Table 10.2. Regions of silent neuron models.

Regions	min	max	count	percent
S1	80mV	120mV	1116	0.39%
S2	40mV	80mV	6308	2.20%
S3	-18mV	40V	35533	12.41%
S4	-41mV	-18mV	35253	12.31%
S5	-61mV	-41mV	170477	59.52%
S6	-72mV	-61mV	4660	1.63%
S7	-80mV	-72mV	33052	11.54%

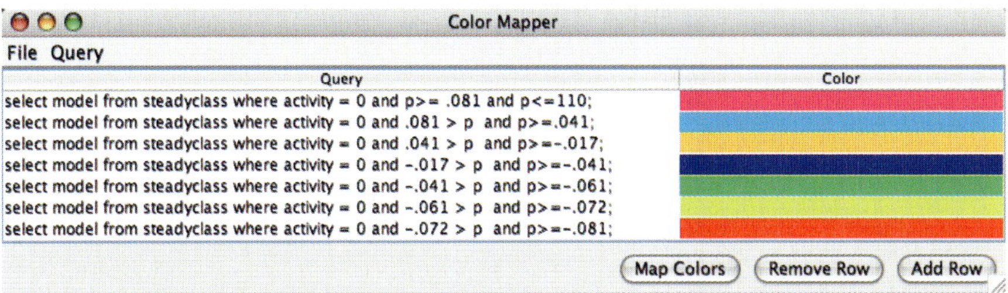

Fig. 10.8. ColorMapper for NDVis defining the domains in Table 10.2 as SQL queries and assigning them colors. "activity $= 0$" indicates silent behavior and "p" refers to resting potential which is specified in terms of min and max in Table. This color map is used in the remaining figures of this section. 10.2.

10.5.2 Exploring Different Projections of Conductance Parameter Space

As noted earlier, the order of dimensions in a dimensional stacking image profoundly influences the resulting image. Once the color map is selected, the analyst can select different dimension orders to draw out different patterns. Although there are 20160 different projections of our 8 dimensions (8! divided by 2 to eliminate simple transforms from flipping the x and y axis of our square images), one can move from any permutation to any other in seven swaps. We have a simple method for finding interesting projects and are working on other methods, however, there is no "best" projection for any color map. Different projections highlight different properties in different parts of the image. Nevertheless, more research on automatic projection finding is clearly warranted. In our case, it only takes a few minutes of exploring to find several interesting projections as shown in the figures below. Each one highlights a different silent region (S1-S7) and allows us to make a hypothesis about the constraints on that part of the silent domain.

The default projection when one starts NDVis is $x = (Na, CaT, CaS, A)$ and $y = (KCa, Kd, H, Leak)$ and generates the image show in Figure 10.9 for our color map. Thus, from the outermost grid of 6x6 boxes one can read off the Na parameter as the x value and the KCa parameter as the y value. This initial projection fortuitously provides an information rich view of the (orange) S3 and (blue) S4 regions in parameter space. We can approximate them with the following domain definitions:

$$D3 \equiv (KCa = Kd = 0) \wedge not(Na = 0 \wedge CaT = 0)$$
$$D4 \equiv (Na = KCa = 0 \wedge Kd \geq 1) \vee (KCa = 0 \wedge Na = Kd = 1 \wedge CaT \geq 4)$$

10.5.3 The S1, S2, S3 Regions

Figure 10.10 shows the projection $x = (KCa, CaS, Na, H)$ and $y = (Kd, Leak, A, CaT)$. Thus, the outermost 6x6 grid provides the values for

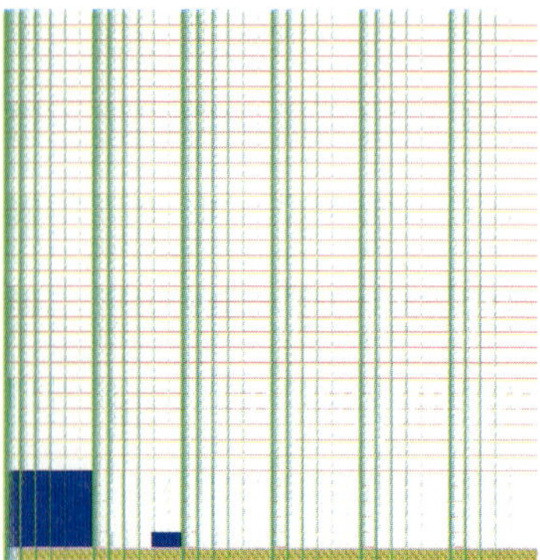

Fig. 10.9. Silent Neuron Visualization for the Default Projection. $x = (Na, CaT, CaS, A)$ and $y = (KCa, Kd, H, Leak)$.

KCa along the x-axis and Kd along the y-axis. We can see immediately from this projection that S1, S2, S3 are well approximated by three domains D1, D2, D3 defined as follows:

$$D3 \equiv (KCa = 0 \wedge Kd = 0)$$
$$D2 \equiv (KCa = 0 \wedge Kd = 0 \wedge Leak = 0)$$
$$D1 \equiv (KCa = 0 \wedge Kd = 0 \wedge Leak = 0 \wedge A = 0)$$

S1 corresponds to D1, S2 to D2-D1 and S3 to D3-D2. This description is not quite accurate though as we see that the region of D1 where $Leak = 0$ and $A > 0$ is actually S2. Also, the green S5 region intersects D1,D2, and D3

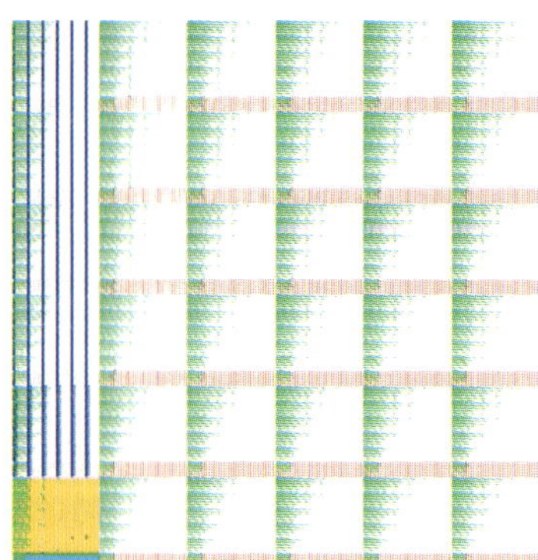

Fig. 10.10. The projection $x = (KCa, CaS, Na, H)$ and $y = (Kd, Leak, A, CaT)$ showing the S1,S2, and S3 regions in orange, aqua, and purple respectively.

10.5.4 The S7 Region

The red S7 region is shown by the projection in Figure 10.11 in which the outer most grid corresponds to H on the x-axis and $Leak$ on the y-axis. It is clear from the image that the yellow S6 region is simply the intersection of the green S5 region and the red S7 region. In other words, there appear to be regions $D5$ and $D7$ in conductance space such that

$$S5 = D5 - (D5 \cap D7)$$

$$S6 = D5 \cap D7$$

$$S7 = D7 - (D5 \cap D7)$$

This observation is immediate from the visualizations, but might be more difficult to infer using purely statistical techniques. Moreover, the region D7 is fairly easy to describe from the two projections in Figure 10.11:

$$D7 \equiv (H = 0 \wedge Leak = 0)$$

The image on the left shows that the region in $D7$ where $KCa = 0$ (i.e. the bottom left of the image) has relatively few S7 (red) neurons while the image on the right shows clearly that the yellow S6 region is the intersection of $D5$ and $D7$.

We will see below that region S7 is actually a "phantom region" that corresponds to silent neurons only if H and Leak are exactly equal to zero. If they deviate from zero even slightly the neuron models are not silent.

The underlying assumption that one makes when using a grid model is that the properties of the underlying biological system are well captured by the grid and that if a particular grid point has a certain property then that property also holds in a continuous region around the grid point. That assumption is shown to fail for this particular region and hence these assumptions must be checked using other means before any firm conclusions are drawn.

10.5.5 The S5 Region

The largest region of silents is S5. Each of the projections we've seen so far and the left side of Figure 10.12 suggest that S5 is defined by a linear condition. Based on these visualizations we are able to provide an estimate for this equation. We can verify this estimate using standard statistical techniques to determine that the S5 region is closely described by the domain D5 defined by following linear equation:

$$D5 \equiv$$

$$CaS + 0.55CaT + 0.21 * Na + 0.03H <$$

$$0.32A + 0.24Leak + 0.036Kd + 0.0083KCa + 0.031$$

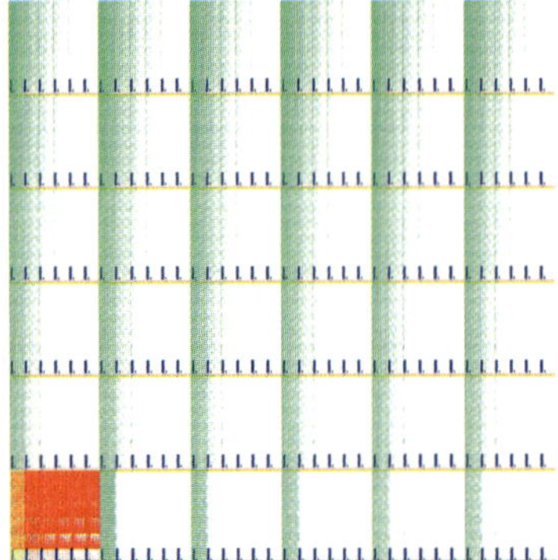

Fig. 10.11. Two projections showing the S7 region in red and S6 in yellow. The projection on the left is given by $x = (H, CaS, Na, CaT)$ and $y = (Leak, KCa, Kd, A)$, while the projection on the right is $x = (H, CaS, A, Na)$ and $y = (Leak, CaT, Kd, KCa)$.

The right side of Figure 10.12 shows the error between D5 and S5. The light green region represents the correctly predicted neurons $D5 \cap S5$. The red region denotes the false positives $D5 - (D5 \cap S5)$. These are the non-silent neurons that satisfy the linear equation D5. The blue denotes the false negatives $S5 - (D5 \cap S5)$. These are the silent neurons that do not satisfy the linear equation. Statistically, the linear equation correctly predicts 88.2% of the S5 neurons, so 11.8% of the S5 neurons are false negatives. The D5 region itself contains mostly silent neurons. Indeed, 92.9% of the D5 neurons are silent and so 7.1% of the D5 neurons are false positives.

If we look at the error visualization there are clearly parts of the S5 (green) region that would not fit into any linearly perturbed version of D5. This suggests a more precise constraint may require a non-linear (e.g. quadratic) formula, or that S5 might be best described as a union of linearly bounded domains. Indeed, the error image suggests that we look at the neuron models where Na=0 and CaT=0. It turns out that this region contains many non-silent neurons whose voltage never rises over -10mV and in many cases oscillates between -25mv and -15mv. Thus, visual analysis of the error plot provides interesting information that would motivate us to refine our definition of a silent neuron or to expand our classification to include non-silent but non-spiking neurons.

10.5.6 Summary

In this section we have shown the use of dimensional stacking and pixelization with NDVis to infer a set of criteria for when a model neuron will be silent. We've used different dimension orderings to investigate different patterns and form testable hypotheses. In particular we have shown how to map the silent neurons to a region in conductance space that is a union of four regions with

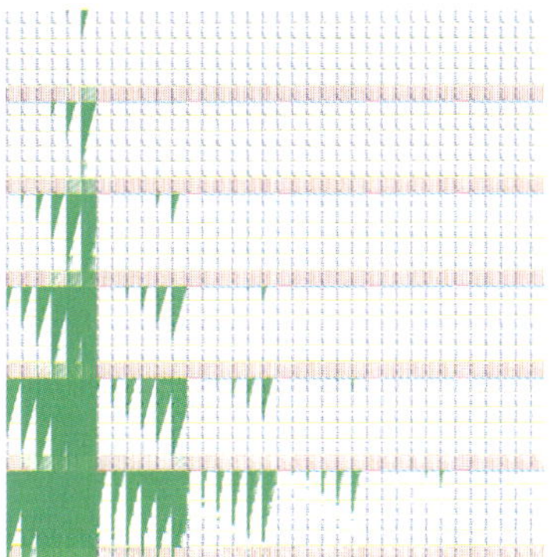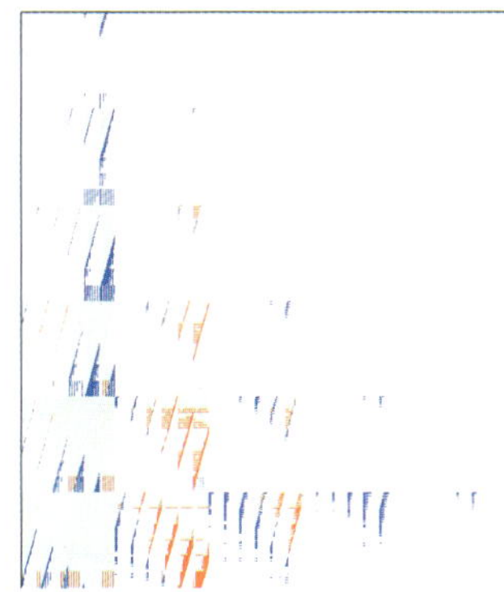

Fig. 10.12. On the left is the projection $x = (CaS, A, Na, H)$ and $y = (CaT, Leak, Kd, KCa)$ showing the S5 region in green. On the right is an error map which compares the region $S5$ with the domain $D5$ defined by the linear equation 10.1. The correct predictions are in light green, false positives in red, and false negatives in blue.

linear boundaries. The results of that analysis are as follows. The silent region is well approximated by the union D of the domains $D3, D4, D5, D7$:

$$D = D3 \cup D4 \cup D5 \cup D7$$
$$D3 \equiv (KCa = 0 \wedge Kd = 0)$$
$$D4 \equiv (KCa = 0 \wedge Na = 0)$$
$$\vee (KCa = 0 \wedge Na = 1 \wedge Kd = 1 \wedge CaT \in [4,5])$$
$$D5 \equiv (CaS + 0.55 CaT + 0.21 * Na + 0.03 H)$$
$$< (0.32 A + 0.24 * Leak + 0.036 Kd + 0.0083 KCa + 0.031)$$
$$D7 \equiv (H = 0 \wedge Leak = 0)$$

The regions S1,S2,S3 all reside in domain D3, and the region S6 is in the intersection of D5 and D7.

10.6 Case Study Part 2: Refining Predictions of Neuron Activity

One problem with parameter space exploration through brute force grid-based simulation is that the grid may be too sparse to adequately sample and capture the continuous parameter space. In other words, interesting parts of the parameter space may slip between the grid points or fixed parameter values that are used for the simulations. For this reason it is worthwhile to test any hypothesis made with the grid-based analysis by inspecting the model with a higher resolution of parameter value inputs (and possibly other experiments).

In this section, we show how one can use simulations along 2D slices of the parameter space to refine the hypothesis generated in the previous Section. Consider the S7 region which is well approximated in the simulation database by the domain $D7 = (H = 0 \wedge Leak = 0)$ (although this domain also contains S6). How do we convert these constraints to the underlying biological model? A natural approximation would be to replace $H = 0$ by $H < 0.5$ or more generally $H < \alpha$ for some constant α between 0 and 1. Thus, the region could be defined by an intersection of linear constraints

$$H < \alpha \wedge Leak < \beta$$

But, the same grid constraint could arise from a single linear constraint

$$\alpha * H + \beta * Leak < \gamma$$

where $\alpha < \gamma$ and $\beta < \gamma$. or it could really be a single point constraint

$$H = 0 \wedge Leak = 0$$

We demonstrate below that all three of these cases occur for the various silent regions S3-S7, but first we describe a tool that was developed to help with this type of analysis.

10.6.1 The JavaSim tool

To refine our hypotheses based on the 8D images, we take 2D slices through conductance space that pass through a single point, conduct simulations at a greater resolution, and generate a 50x50 2D grid which is color mapped to reveal the properties of interest. First we run simulations on 2500 points for a 2D section of the model. We then explore that smaller data set with a visualization tool we call JavaSim. (This tool is available from the authors).

Figure 10.13 shows 2D sections in which Kd and KCa vary and all other parameters are fixed. Clicking on one of the colored boxes brings up the voltage plot for its associated model simulation in the panel at the top of the screen. The colors red, green, and blue are determined by the number of spikes per period, the period length, and the number of spikes per second. This results with fast spikers in dark blue, slow spikers in light blue (none appear in this image), silent neurons in aqua, bursters as reddish, single spike bursters in purple, and bursters with hundreds of spikes per period in dark red. This figure shows the clear boundaries between spikers and bursters and also between single spike bursters and multi-spike bursters. The lines are approximately linear. There is also some noise which arises from activity type classification issues based on the problem of determining the period when the wave form consists of a sequence of 2-10 slightly different burst patterns which itself repeats.

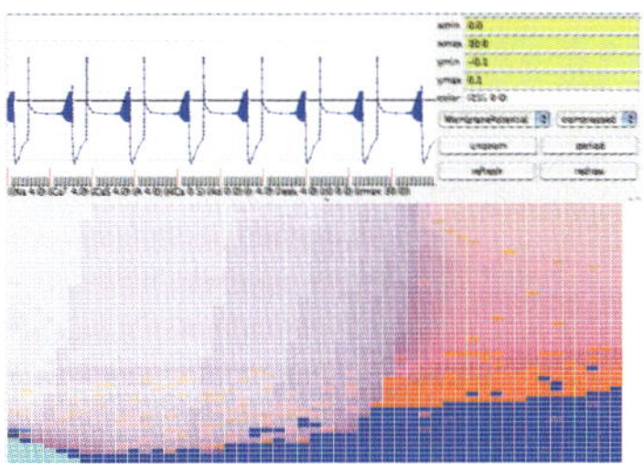 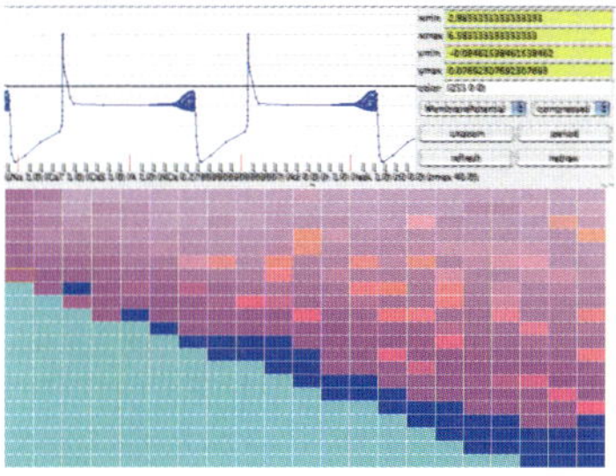

Fig. 10.13. This image on the left shows the JavaSim screen for the 2D section through the point where all conductance parameters have value 4 and the two conductances KCa and Kd are allowed to vary between 0 and 5 in 0.1 step increments. This yields 51 points on both axes. The x-axis is Kd and the y-axis is KCa. Blue rectangles are spikers, red are bursters (darker having more spikes per burst), and aqua is silent. The image on the right is the same except that it represents a slice through the point (1,1,1,1,1,1,1,1) and KCa and Kd vary from 0 to 1.0 in 0.05 step increments. Its voltage plot is the purple neuron on the boundary with Kd=0 and KCa=0.20

10.6.2 Linear Boundaries: Refining the Domain for the S3 Region

Notice that the silent regions in the Kd x KCa sections shown in Figure 10.13 are clearly linear. The slice on the right goes through the point (4,4,4,4,4,4,4,4) and the (aqua) silent region in its lower left corner is bounded by the hyperplane:

$$KCa + 0.8 * Kd < 0.4$$

This suggests that the condition defining the $S3$ region would be a linear hyperplane. The 2D slice in the right half of Figure 10.13 goes through a different point (1,1,1,1,1,1,1,1) and shows that the silent region in this 2D section is also linear, this time with a slightly different formula:

$$KCa + 0.76Kd < 0.26$$

This suggests that the constant term might depend on some of the other parameters. At this point, the hypothesis that the boundary of the S3 region is a hyperplane can be tested using standard statistical techniques. The more interesting observation is that the boundary of the S3 region seems to consist of a thin region of fast spikers (dark blue) for large Kd and slow bursters for small Kd.

10.6.3 Polyhedral Domains: Refining the Domain for the S4 Region

The images in Figure 10.14 show the 2D slices where Na and KCa vary. The aqua section in the lower right is the S4 region and in both cases it is well approximated by a box with the constraints:

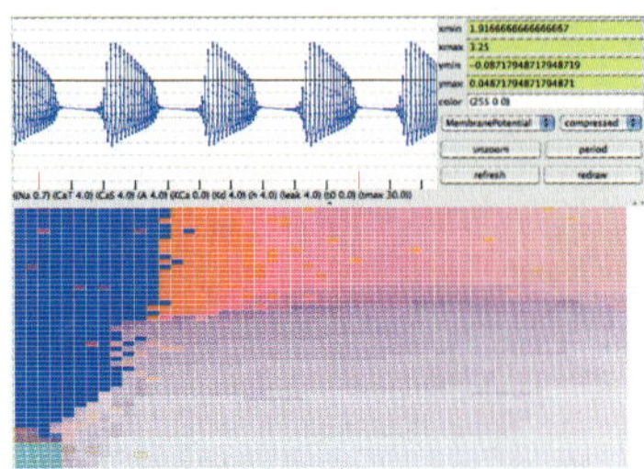 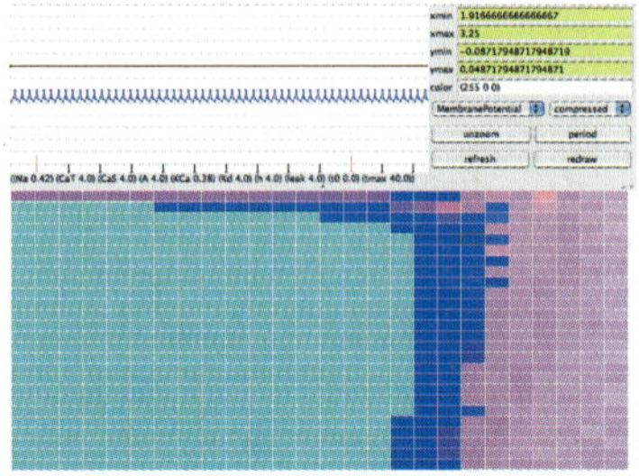

Fig. 10.14. The plot on the left shows the 2D section through the point where all conductance parameters have value 4 and the two conductances KCa and Na are allowed to vary between 0 and 5 in 0.1 step increments. The plot on the right is the same except that x and y vary between 0 and 0.5 in 0.02 step increments. The x-axis is KCa and the y-axis is Na. The voltage plot for the left image is for a neuron in the purple section above the S4 region with KCa=0.0 and Na=0.7. The voltage plot on the right is for a neuron in the blue section near the upper left corner of the silent boundary at KCa=0.38 and Na=0.42.

$$D4' \equiv (Na < 0.48 \wedge KCa < 0.32)$$

Recall that the S4 region corresponds to two disjoint regions in the grid-based analysis. It is likely that these two regions are both part of a single domain. This 2D section indicates that the boundary of the domain is probably the intersection of two or more hyperplanes.

It is interesting to note that the blue region to the right of the silent region consists of "tonic spikers" but these spikers are really just short perturbations around the -20mV line. It might be more appropriate to classify such neurons along with the silent neurons since they're voltage plots are so dampened. In such a case, the size of the D4 region would change, but not its shape.

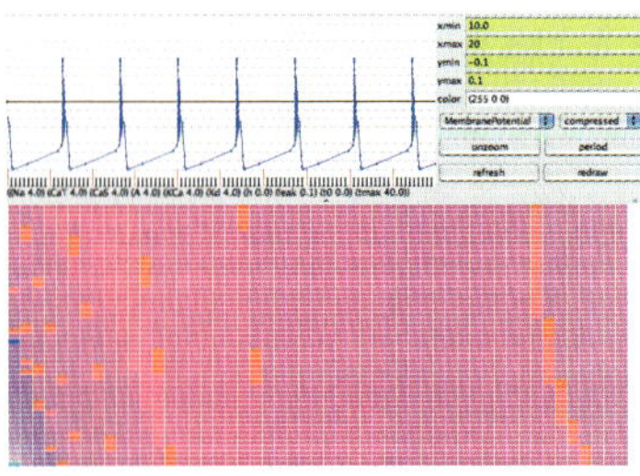

Fig. 10.15. The 2D section through the point where all conductance parameters have value 4 and the two conductances Leak and H are allowed to vary along the x and y axes (respectively) between 0 and 5 in 0.1 step increments. The voltage plot is for the neuron nearest (0,0) on the y-axis (Leak=0, H=0.1).

10.6.4 Point Domains: Refining the Domain for the S7 Region

Next, we repeat the 2D slice process to examine the boundary of the S7 region. The image in Figure 10.15 shows the 2D section going through the point (4,4,4,4,4,4,4,4) where Leak and H vary along the x and y axes respectively between 0 and 5 in steps of 0.1 For this section, the silent S6 region consists of a single point which is barely visible in the lower left corner of the 2D slice. In this case, the continuous constraint for the S7 domain is that both parameters H and Leak must be equal to zero. If either varies slightly from zero, the neuron model is no longer silent. The image in Figure 10.15 also suggests that the spikes per second decreases as one nears the point (0,0).

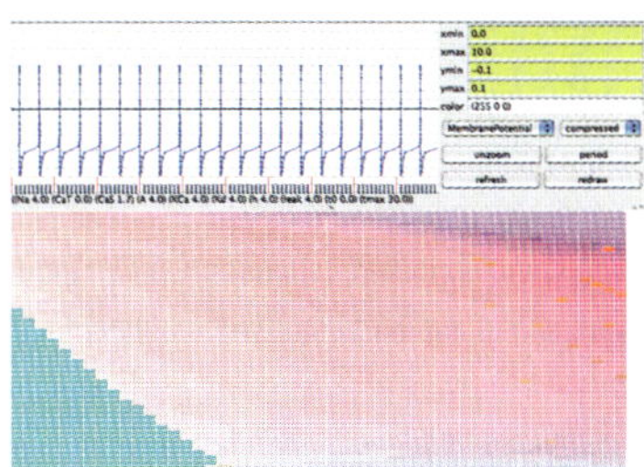

Fig. 10.16. The 2D section through the point where all conductance parameters have value 4 and the two conductances CaS and CaT are allowed to vary along the x and y axes (respectively) between 0 and 5 in 0.1 step increments. The voltage plot is for the first non-silent neuron on the bottom row CaT=0, CaS=1.7

10.6.5 Linear Boundaries: Refining the Domain for the S5 Region

The plot in Figure 10.16 shows the CaS x CaT section through the point (4,4,4,4,4,4,4,4) and the S5 region appears here as a plane

$$CaS + 0.54CaT < 1.7$$

Using the formula for D5 from the previous section, we would expect to get the formula

$$CaS + 0.55CaT < 1.5$$

which indicates that the formula computed from the grid data gives a reasonably good prediction for the 2D slices.

10.6.6 Completing the Analysis

At this stage we have developed a fairly interesting set of criteria for determining which neuron models will be silent and which will be active. The next step is to refine these formula by repeating the process (generating another database but with values concentrated near the boundaries of the regions we are interested in, e.g. S3, S4, and S5) or by generating large samples of random points near the approximated boundaries and using statistical techniques to generate separating hyperplanes and hypersurfaces.

10.7 Conclusion

This chapter presented a number of visualization and analysis techniques for investigating very large multidimensional data sets. We specifically described the combination of dimensional stacking and pixelization with interaction techniques in the software tool NDVis. Users can view an entire data set in one image with this tool to identify global patterns and form hypotheses. These hypotheses can then be tested with more traditional visualizations such as the 2D plots in JavaSim or with statistical and data mining algorithms. These methods are best applied to data generated from brute force grid-based simulations or uniform sampling of biological systems. Further, we have shown how to use these techniques by analyzing a model neuron simulation database.

10.7.1 Open Problems and Future Work

We leave you with a brief description of open problems and future work in data mining and visualization for the analysis of biological databases.

Generalizability, Data Binning and Aggregation: In the description of dimensional stacking in Section 10.3.2, the 2D, 4D, and 6D images have more than one data point mapped to only one visual element (i.e. a square in the matrix). For these situations, visualizations can translate the density of data points within a visual element to translucence, color saturation, intensity, etc.. However, we must also account for how to assign data points to each visual element. This task is analogous to creating a histogram and is particularly difficult for data sets that have a highly non-uniform distribution. Some binning and aggregation techniques must be derived so that the distribution of data to visual elements communicates the data landscape to analysts in an intuitive yet accurate manner.

Integrating with Data Mining and Statistical Algorithms: A key issue to future research is identifying data mining algorithms that can be applied to open biological problems and engineering visualizations and tools for analysts to use them. There are several supervised data mining and statistical algorithms that can benefit from user input. One such example is K-Means clustering. Users can select the initial seeds of the algorithm to help it converge more quickly on a solution. With the visualization methods presented here, users could also watch the progress of optimization techniques such as evolutionary algorithms as they traverse the data landscape. With an interactive tool they could manipulate the evolutionary algorithm, spawn new individuals, and act as fitness functions. By identifying supervised algorithms with points of interaction and how to use a visualization to interface with those points, we can allow users to participate in mixed initiative data mining. This will lead to faster more effective analysis of complicated data spaces.

References

1. Rasko DA, Myers GSA, Ravel J (2005) Visualization of comparative genomic analyses by blast score ratio., BMC Bioinformatics 6:2
2. Prinz AA, Billimoria CP, Marder E (2003) An alternative to hand-tuning conductance-based models: Construction and analysis of data bases of model neurons, Journal of Neurophysiology 90:3998–4015
3. Taylor AL, Hickey TJ, Prinz AA, Marder E (2006) Structure and visualization of high-dimensional conductance spaces, Journal of Neurophysiology 96:891–905
4. Inselberg A, Dimsdale B (1990) Parallel coordinates: a tool for visualizing multidimensional geometry, In: VIS '90: Proceedings of the 1st conference on Visualization '90, 361–378. IEEE Computer Society Press, Los Alamitos, CA, USA
5. Chernoff H (1973) The use of faces to represent points in k-dimensional space graphically, Journal of the American Statistical Association 68:361–368
6. Keim DA (2001) Visual exploration of large data sets, Commun ACM 44(8):38–44
7. Ward MO (1994) XmdvTool: integrating multiple methods for visualizing multivariate data, In: VIS '94: Proceedings of the conference on Visualization '94, 326–333. IEEE Computer Society Press, Los Alamitos, CA, USA
8. Witten IH, Frank E (2005) Practical machine learning tools and techniques. Morgan Kaufmann, San Francisco, 2nd edn.
9. R Development Core Team (2007) R: A Language and Environment for Statistical Computing. R Foundation for Statistical Computing, Vienna, Austria
10. Eaton JW (2002) GNU Octave Manual. Network Theory Limited
11. Cook D, Swayne DF (2007) Interactive and Dynamic Graphics for Data Analysis With R and GGobi. Springer
12. Langton JT, Prinz AA, Hickey TJ (2006) Combining Pixelization and Dimensional Stacking, In: Proceedings of the 2nd International Symposium on Visual Computing (ISVC 2006), vol. 4292 of *Lecture Notes in Computer Science*, 617–626. Springer
13. Langton JT, Prinz AA, Hickey TJ (2007) NeuroVis: combining dimensional stacking and pixelization to visually explore, analyze, and mine multidimensional multivariate data, In: Proceedings of SPIE: Visualization and Data Analysis (VDA 2007), vol. 6495, 64950H-1–64950H-12. SPIE
14. Yang J, Peng W, Ward MO, Rundensteiner EA (2003) Interactive Hierarchical Dimension Ordering, Spacing and Filtering for Exploration of High Dimensional Datasets, In: IEEE Symposium on Information Visualization, 14

Part III

Open Problems

Phylogenomics, Protein Family Evolution, and the Tree of Life: An Integrated Approach between Molecular Evolution and Computational Intelligence

Laila A. Nahum[1] and Sergio L. Pereira[2]

[1] Bay Paul Center, Marine Biological Laboratory, Woods Hole, MA 02543, USA
laila@nahum.com.br
[2] Department of Natural History, Royal Ontario Museum, Toronto, Ontario, M5S
2C6, Canada
sergio.pereira@utoronto.ca

Summary. Information generated by genomic technologies has opened new frontiers in science by bridging a broad range of disciplines. Many tools and methods have been developed over the past several years to allow the analysis of molecular sequences. Nevertheless, the interpretation of genomic data to determine gene function and phylogenetic relationships of organisms remains challenging. Here, we focus on the application of phylogenomics (*phylo*genetics and *genomics*) to improve functional prediction of genes and gene products, to understand the evolution of protein families, and to resolve phylogenetic relationships of organisms. We point out areas that require further development, such as computational tools and methods to manipulate large and diverse data sets. The application of integrated computational and biological approaches may help to achieve a better system-based understanding of biological processes in different environments. This will help to fully access valuable information regarding the evolution of genes and genomes in the wide diversity of organisms.

11.1 Introduction to Phylogenomics

The massive amount of genomic information generated by sequencing projects has opened new frontiers in science by bridging a broad spectrum of disciplines, including computational biology, molecular biology, molecular evolution, evolutionary biology, and ecology. Many bioinformatics tools and methods have been developed over the past several years to allow the analysis of large data sets of whole genome sequences or highly diverse gene families [1, 10, 25, 33, 35, 43]. Nevertheless, interpretation of genomic data to determine gene function and phylogenetic relationships remains challenging. Some major questions in evolutionary biology and genomics remain open.

L.A. Nahum and S.L. Pereira: *Phylogenomics, Protein Family Evolution, and the Tree of Life: An Integrated Approach between Molecular Evolution and Computational Intelligence*, Studies in Computational Intelligence (SCI) **122**, 259–279 (2008)
www.springerlink.com   © Springer-Verlag Berlin Heidelberg 2008

What are the processes driving the evolution of genes, genomes, and organisms? How do new biological functions emerge within and across species over evolutionary time? How can we predict functions and phenotypes from genomic data? What are the relationships among all living organisms?

Phylogenomics, the intersection between *phylogenetics and genomics*, was originally defined as an evolutionary framework designed to improve functional prediction of uncharacterized genes and gene products [17, 19]. Since then, phylogenomic approaches have been applied to an increasing number of multidisciplinary studies at the interface of computational and biological sciences [5, 11, 17, 40]. The major components in phylogenomic analysis include the selection of genes or gene products of interest (targets), the identification of potential homologs[1] through sequence similarity search-based methods, creation of multiple sequence alignments, reconstruction of evolutionary trees using different methods of phylogenetic inference, and mapping of available information onto the trees. Collectively these components serve as a framework for evolutionary hypothesis testing to support a variety of analyses like searching of conserved motifs, patterns, ancestral sequence reconstruction, and a broad range of other comparative analyses.

Here, we focus on the application of phylogenomics to improve functional prediction of experimentally uncharacterized genes and gene products [5, 17]. We also discuss the importance of phylogenomics as a framework to understand the processes that shape the evolution of protein families over time and across diverse species [7, 13, 66]. We then link to other issues that may benefit from the application of a phylogenomic approach such as inferring the phylogenetic relationships of organisms and reconstructing the Tree of Life [11, 18]. We take the still-unresolved phylogeny of modern birds to illustrate areas that need methodological improvement for data manipulation and phylogenetic inference [2, 24, 48]. The importance of rare genomic events such as organization of gene order (synteny), insertions, deletions, duplications, and inversions at the genomic level is also discussed in a phylogenetic context. Finally, we point out areas that require further development, such as creation of tools to filter growing databases and build up large data sets, and development of algorithms that take into account structural data to perform more efficient and robust analyses.

11.2 Methods of Tree Inference

Evolutionary trees constitute the basis of phylogenomic analysis. They summarize the phylogenetic relationships of molecules (single genes, gene families, protein domains, etc) and organisms (populations, species, and higher taxonomic levels), which are inferred from sets of aligned nucleotide or amino acid sequences and a model of molecular evolution. Methodological details

---

[1] Molecular and non-molecular traits that descend from a common ancestor.

of algorithms used to create multiple sequence alignments, methods of tree inference and models of molecular evolution are vastly described in the literature [22, 45, 69]. We summarize very briefly the methods of tree inference and models of evolution applied in phylogenetics and phylogenomics.

Phylogenetic methods can be classified in two major groups: distance-based and character-based methods. Distance-based methods like UPGMA (unweighted pair-group method using arithmetic averages; [59] and neighbor joining [56] compare the differences for all pairs of aligned sequences, and convert them into a matrix of distances. The matrix represents the number or proportion of differences observed, or the expected number of substitutions per site under the assumptions of a model of molecular evolution. Based on the matrix of distances and a tree search algorithm, distance methods find a tree that best fit the estimated pairwise distances. Distance methods quickly infer evolutionary trees for either a large number of sequences, or long sequences, or both. Unfortunately, the drawbacks of distance methods include loss of information when discrete characters are converted into a distance matrix. Additionally, only a single tree is produced and any suboptimal tree that could be as good as or a better representation of the true phylogenetic relationships is discarded. Distance methods can also perform poorly when sequence divergence is high, causing the estimation of pairwise distances to be a difficult problem.

The second group of methods is based on discrete characters. The most commonly used are maximum parsimony [15], maximum likelihood [21], and Bayesian methods [36, 41, 54]. Maximum parsimony searches for the tree that minimizes the number of steps or changes (in terms of mutations or substitutions) needed to explain the variation observed in the alignment. In other words, it assumes that evolution is parsimonious and hence the number of changes to convert one sequence into another is minimal. Only those nucleotide or amino acid sites with at least two changes present in at least two sequences are informative under the methodological framework of parsimony. Multiple equally parsimonious trees can be recovered by the method, especially if sequences are very similar. Regrettably, phylogenetic inference using maximum parsimony can be misleading because the method does not account for variation in substitution rates among sites, and directional biases towards some types of substitutions over others (e.g., transitions versus transversions). Directional biases can be dealt with by assigning weight to different types of substitutions. In this case, more common substitution types will have a smaller weight, and hence a smaller contribution to the total number of steps for a given tree. However, this practice can be contentious because weight assignment is arbitrary. Moreover, fast-evolving sites could be down-weighted if distantly-related sequences are being compared, but phylogenetic signal could be lost if the data set also include closely-related sequences for which fast evolving sites are more informative than slowly-evolving sites. Rate variation and directional biases can exacerbated the problem of tree inference using maximum parsimony when two or more unrelated sequences evolve at a much faster rate, leading to the phenomenon of long-branch attraction [20]. In this case, homologous sites converge to the same character state by chance, and the fast-evolving sequences will group together, reducing the amount of changes along the tree.

The principle of the maximum likelihood method is to estimate the tree that maximizes the log likelihood (L) of having generated the data D (e.g. a set of aligned sequences), given a model of molecular evolution θ. The mathematical notation of the likelihood function is

$$L(\theta, D) = f(D|\theta). \tag{11.1}$$

During the tree searching process, the topology, branch lengths, and parameters of the model of evolution are estimated interactively. The maximum likelihood method allows for complex models of evolution to be incorporated in the analysis, and multiple trees can be evaluated in a statistical framework. Nevertheless, maximum likelihood is computationally slow, and may also be affected by the issue of long-branch attraction if simplistic models of evolutions are used.

Bayesian methods are related to maximum likelihood, differing only in the use of $f(\theta)$, the prior distributions of model parameters (including the tree topology, parameters of the model of evolution, etc), to estimate the posterior probability $f(\theta|D)$ of the inferred tree. Hence, the posterior distribution is given as

$$f(\theta|D) = f(\theta) \times f(D|\theta), \tag{11.2}$$

and is interpreted as the probability that the tree is true. The method became popular in recent years due to advances in Markov Chain Monte Carlo (MCMC), a stochastic algorithm for drawing samples from a posterior distribution of tree. A detailed description of the algorithm can be found elsewhere [30]. The Bayesian method is also subject to controversies and limitations. The controversies mostly focus on the attribution of priors. For example, if there is some expectation that a given topology is supported by previous analysis of molecular or morphological data, this topology can be given a higher prior probability compared to alternative topologies. On the other hand, all topologies could have *a priori* the same probability of being the true tree. In practice, the choice of priors for all model parameters is a difficult task, and the use of flat priors (i.e., a range of possible values large enough to cover the posterior distribution) is usually chosen. Running time is one limiting factor in Bayesian inference of phylogenetic trees. Large and complex data sets require longer chains to achieve convergence, and it may be hard to determine objectively that the chain has run long enough. Moreover, the choice of priors can have a large influence on how samples are drawn from the posterior distribution, leading the method to perform a small portion of the parameter space.

The models of evolution used in tree inference describe how a DNA or protein sequence will change over time. For example, in the case of DNA sequences, only four states are possible: A, C, G and T, corresponding, respectively, to the bases found in the DNA Adenine, Cytosine, Guanine, and Tymine. The simplest model assumes that each nucleotide in a set of DNA sequences occurs in equal frequency, and that the probability of one state change to any other one is the same. The most complex model, on the other hand, assumes that nucleotide frequency is unequal, and the change of state i to j is not the same as from j to i, for every $i \neq j$. More detailed discussion on models of evolution for DNA and amino acid sequences can be found elsewhere [22, 47, 69].

11.3 Functional Prediction of Genes and Gene Products

11.3.1 Predicting function from sequence and structural data

Genomic technologies have generated an enormous amount of data, much of which has yet to be fully analyzed and interpreted. The vast majority of sequenced genes and gene products is either experimentally uncharacterized and has a predicted function or is annotated as *unknown* (no known function). Even the best-studied organisms such as *E. coli*, yeast, *C. elegans*, among others, contain a large number of genes, whose functions have not yet been experimentally verified.

Predicting function from sequence data using computational tools is the first step in processing such information. A great deal of functional prediction has relied on the transfer of information from the most similar sequence(s) or structure(s) found in the databases to the gene or protein of interest. This common practice has generated systematic errors in the functional prediction of genes, the extent of which is not completely known [4, 27]. Structural analysis improve functional prediction by providing clues for evolutionary studies and by guiding experimental verification. With the significant increase of available data obtained by the structural genomics efforts, the scientific community can now address more complex questions regarding the evolution of sequence, structure, and function of molecules. The Protein Structure Initiative for example, is a broad enterprise including several research centers, which aims to provide full coverage of the protein structure space by selecting specific targets for structural and experimental characterization [46].

The major source of errors in functional prediction is perhaps a result of a common misconception of terms like similarity, divergence, and homology [4, 27]. Although homology and similarity have been used interchangeably by the genomics community, *homology* actually implies common ancestry by descent while *similarity* indicates the degree of proximity between molecular sequences. Therefore, there is no such concept as degree of homology. For example, two homologous genes can be 90% similar or, in other words, 10% divergent from each other, but cannot be referred to as 90% homologous. Similar sequences might have evolved from a common ancestor (i.e., by gene duplication followed by divergence) or might be the result of convergent evolution from unrelated ancestors. Furthermore, sequences diverge and become dissimilar at different rates across different organisms depending on varying mutation and selective pressures, sometimes tightly linked or not to functional shifts.

Other sources of error in functional prediction include the lack of recognition that genes and proteins with similar sequences may have different functions, and that fused genes and modular proteins carrying distinct functions in different parts of the molecules should be detected and processed properly [57]. Propagation of existing errors by transferring incorrect information to uncharacterized genes and gene products also exacerbates the issues discussed here. Despite the uncertainty of sequence- and structure-based approaches, several tools and methods have been developed recently to improve functional prediction of previously uncharacterized genes and gene products, each having particular strengths and weaknesses [5, 27, 55, 64].

11.3.2 Annotating evolutionary trees

The process of integrating phylogenetic trees with information regarding sequence, structure, and function of genes and gene products, besides the taxonomic information of source organisms is called *tree annotation*. It is extremely important to discriminate between genes and gene products whose function is supported by experimental work from those that have received a predicted function or remain unknown. Sequences with a clear label or tag of *experimental evidence* and *predicted function* deposited on curated databases or literature information could facilitate this distinction. Biochemical properties such as reaction mechanism, cofactor usage, and substrate specificity of a particular enzyme as well as its subunit composition, cellular localization, and regulation are particularly helpful to the process of tree annotation. It is important to keep in mind the broad concept of biological function, which goes beyond biochemical properties of molecules, and to bring as much information as possible to fully annotate the tree in a phylogenomic analysis.

Other relevant information may include the genomic context (gene neighborhood), gene organization (exons, introns, etc), protein architecture (sequence and structural domains), and conserved motifs. Prediction of structural folds[2] and mapping of critical residues involved in the function along with their distribution in an evolutionary tree may also be very helpful information for the phylogenomic inferences. Finally, the taxonomy of the source organisms including the lineage information (i.e., Class, Order, Family, Genus, etc.) enriches the tree annotation, as do biological and ecological information about inter- and intra-species variation, ecotypes, and habitats.

Detailed and accurate tree annotation helps tremendously with evolutionary and functional inferences of gene and protein families as well as the

---

[2] The spatial conformation of a protein dictated by its amino acid sequence.

relationships among organisms (Section 11.5). For example, tree annotation facilitates discrimination between orthologs[3] and paralogs[4], which is a big challenge in computational approaches, as well as it helps to identify subfamilies or subgroups in any given family based on the presence of well-supported clusters of related proteins (i.e., structural and/or functional variants). Examples of phylogenomic analyses and assignment of functions at the subfamily level are described elsewhere [5, 66]. Here, we illustrate a hypothetical case of a protein family[5] that contains distinct subfamilies of enzymes (Fig. 11.1). The tree topology combined to functional information suggests that enzymes are grouped based on sequence similarity and their biochemical properties (subfamilies 1, 2, 4, and 5), for instance, sharing the same substrate specificity and/or cofactor use. The tree information also highlight instances that would benefit from experimental testing (e.g. unknown), perhaps by gathering insights from this evolutionary framework. In addition to subfamilies of proteins related by sequence, structure, and/or function, one may detect expansions in the gene and protein family that are organism-specific. This may provide insights into the functional divergence of enzymes in this family and potentially highlight cases of adaptation to the environment in which this organism lives. For example, the phylogenomic approach would show an increase in the number of homologs for one or a few organisms in the same subfamily or subgroup as corroborated by the tree topology. Such additional genes or gene products may indicate potential cases of divergence and adaptation to different environments. In particular, we might expect a significant increase in the application of phylogenomic approaches to the interpretation of data derived from environmental samples (e.g. deep sea, human gut, etc) making a significant contribution to the field of metagenomics[6] [12, 58, 63].

11.3.3 Improving functional prediction through phylogenomics

Phylogenomics improves over sequence similarity methods by providing a robust comparative framework in the evolutionary context. The ultimate goal is to connect genomic information with phenotypic diversity and assess higher order processes such as molecular networks and biological adaptations to diverse environments.

One example that illustrates the application of high-throughput approaches to connect sequence elements in a genome with biological function is the work in the ENCyclopedia Of DNA Elements (ENCODE) Project [8].

---

[3] Homologous sequences that diverged after a speciation event.

[4] Homologous sequences that diverged after a duplication event.

[5] A group of genes or proteins that have evolved from the same (divergent evolution) or different ancestors (convergent evolution), and share similarity at the sequence, structural, and/or functional levels.

[6] Sequencing and analysis of DNA extracted from environmental samples, without the need to isolate and cultivate species in the laboratory. Also referred to as ecogenomics or environmental genomics.

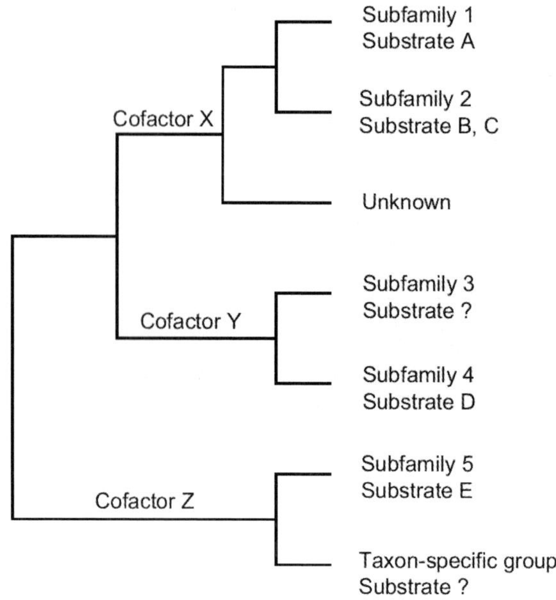

Fig. 11.1. Bridging functional prediction and protein family evolution. The tree information reveals the presence of subfamilies of enzymes sharing cofactor usage and substrate specificity and serves as a framework to study the evolutionary processes that shape the evolution of enzyme families (e.g. gene duplication).

This cross-institutional project aims to create a comprehensive catalog of the structural and functional components in the human genome in order to understand the biology of human health and disease. On the other hand, the Berkeley Phylogenomics Group has created a series of web servers for phylogenomic analysis including PhyloFacts, a structural phylogenomic encyclopedia for functional and structural predictions in an evolutionary framework [25, 35]. PhyloFacts contains thousands of the so-called *books* for protein families and domains with pre-calculated analyses. This is an example of a high-throughput phylogenomic classification.

Together, these considerations highlight the importance of phylogenomics as a framework to investigate sequence relationships as well as structural and functional features of genes and gene products. Furthermore, this approach can make significant contributions to the functional and structural genomics consortia as well as to a broad range of molecular evolution studies. Some tools and resources developed for phylogenomic analysis are listed in Table 11.1.

Table 11.1. Tools and resources for phylogenomic analysis

Resources[7]	URL
BOLI	http://www.dnabarcodes.org
BPG	http://phylogenomics.berkeley.edu
EGenBio	http://egenbio.lsu.edu
ENCODE	http://genome.ucsc.edu/ENCODE
FlowerPower	http://phylogenomics.berkeley.edu/flowerpower
FunShift	http://FunShift.cgb.ki.se
GeneTrees	http://genetrees.vbi.vt.edu
PhIGs	http://phigs.org
Phylemon	http://phylemon.bioinfo.cipf.es
PhyloFacts	http://phylogenomics.berkeley.edu/phylofacts
PSI	http://www.structuralgenomics.org
RIO	http://rio.janelia.org/
SCaFoS	http://megasun.bch.umontreal.ca/Software/scafos
ToL	http://www.tolweb.org

11.4 Evolution of Gene and Protein families

11.4.1 Diversity of gene and protein families

Genes or their products that are related by sequence, structure, function or
a combination of these features form families and superfamilies with various
degrees of similarity. Members of these families are believed to share common ancestry. At the molecular level, each family has a unique evolutionary
history that is connected to the *environment* in which it resides that may
result in differences between the gene tree and the taxon-tree topologies (see
below). Families vary in size and composition within an organism and across
diverse taxa as a result of distinct rates and modes of evolution. Processes
such as gene duplication, lineage sorting[8], and lateral gene transfer[9] followed
by divergence at the sequence, structural, and functional levels, increase the

---

[7] BOLI: Barcode of Life Initiative. BPG: Berkeley Phylogenomics Group.
EGenBio: Data Management System for Evolutionary Genomics and Biodiversity.
ENCODE: ENCyclopedia Of DNA Elements. FlowerPower: protein homology
clustering algorithm. FunShift: Functional Shift (divergence) Analysis. GeneTrees:
A Phylogenomics Resource. PhIGs: Phylogenetically Inferred Groups. Phylemon:
Web tools for molecular evolution, phylogenetics and phylogenomics. PhyloFacts:
Phylogenomic encyclopedias across the Tree of Life. PSI: Protein Structure Initia-
tive. RIO: Resampled Inference of Orthologs. SCaFoS: Selection, Concatenation
and Fusion of Sequences for phylogenomics. ToL: The Tree of Life Web Project.

[8] The process in which alternative copies (alleles) of a gene are sorted out into
different lineages, sometimes causing the gene tree to conflict with the species
tree.

[9] Transfer of genetic material between unrelated organisms, also called *horizontal
transfer*.

diversity of families across different organisms [44]. Studying how members of gene and protein families evolve may help understand the biology of organisms, adaptations to different environments, and history of life on Earth from the molecular perspective.

11.4.2 Major mechanisms of molecular evolution

The predominant mechanism of molecular evolution in all branches of the Tree of Life is gene duplication followed by divergence. This mechanism is sometimes responsible to generate a high proportion of functional genes in a genome [70]. However, most duplicated genes will be lost due to its redundancy in the genome, but those that are kept can diverge from the ancestral gene and acquire a new function at the molecular or cellular levels [44]. Successive events of gene duplication and divergence may take place before and after speciation events, generating further diversification of functions across organisms. In this context, each genome becomes a specific *environment* with its own selection pressures. Furthermore, sequences may diverge so much that similarity cannot be detected, and homology may not be recognized. This has an impact on both functional prediction and molecular evolution studies. Inferring the history of gene and protein families through evolutionary trees involves, for instance, the estimate of how many events of gene duplication took place in any given family and when in the evolutionary time they occurred.

Other mechanisms of molecular evolution exist such as gene gain (i.e. lateral gene transfer), gene loss, gene fusion, domain shuffling, among others, which are described in detailed in molecular evolution textbooks [37, 47]. Together these molecular mechanisms are responsible for a significant increase in the divergence of genes and gene products across organisms over evolutionary time. However, they may also interfere with the reconstruction of the Tree of Life. The presence of multimodular proteins in an organism constitutes one of the biggest issues in the functional prediction from sequence data and in molecular evolution studies if they are not identified and processed properly. Multimodular proteins are encoded by fused genes and carry different functions in distinct parts of the molecule. In *E. coli*, for example, fused genes encode 2.5% of the proteins [57]. To improve functional prediction and/or to identify family membership accurately, it is necessary to detect instances of multimodularity and to separate them into their independent functional units (modules). Protein length and alignment information provide clues to detect these cases, and functional annotation and phylogenetic transfer must be specific to the aligned module or domain.

11.4.3 Phylogenomics applied to protein family evolution studies

Phylogenomics has been applied to a variety of studies addressing the evolution of functions in gene and protein families over time and across diverse species. Several independent studies using a broad range of targets, from slowly to fast evolving molecules, have shown how different processes take place in shaping the diversity of function we see in contemporary organisms. Here, we summarize three examples to illustrate the application of phylogenomics to protein family evolution studies.

Polyketide synthases are multifunctional enzymes involved in the biosynthesis of a broad range of secondary metabolites. Polyketides play a role in pathogen defense and symbiotic interactions encompassing a wide spectrum of biological properties such as antibiotic, antitumor, antifungal, and immunosuppressive. Fatty acid synthases are enzymes evolutionarily related to polyketide synthases and are encoded by a diverse gene family in Eubacteria and Eukaryotes. Recently, the relationships among these enzymes were estimated through phylogenomic analysis of amino acid sequences of eubacterial and eukaryotic type I polyketide synthases and fatty acid synthases using different methods [7]. The analysis revealed that, in addition to possessing fatty acid synthase, organisms such as sea urchin, bird, and fish have a group of polyketide synthase genes that were previously unidentified. Interestingly, polyketide synthase genes in sea urchin, fish and the chicken, fish are more closely related to the homologs in the slime mold *Dictyostelium* and Eubacteria than any other homologs in Fungi and animals. Whether these genes result from events of lateral gene transfer or from other patterns coupled with massive gene loss in several animal lineages remains unclear.

Kinesins are proteins that work as molecular motors in several key cellular processes and are ubiquitous to virtually all eukaryotes identified so far. A holistic analysis of members of the kinesin superfamily has been performed with 486 amino acid sequences from 19 divergent eukaryotes including human, drosophila, yeast, kinetoplastids, red algae, plant, and others [66]. The kinesin motor sequence domain is conserved in all proteins in this superfamily and was used in the analysis. The study revealed the presence of three new kinesin families and identified two new phylum-specific clusters of proteins [66]. This scenario suggests multiple events of gene loss in individual lineages with no single family being ubiquitous to all organisms. Furthermore, this study also contributed to the functional prediction of previously uncharacterized proteins. The distribution of homologs in this superfamily correlates with the presence of cilia or flagella in the source organisms, corroborating the prediction of a flagellar function for members of the two new kinesin families. In this analysis, the authors also provided a set of Hidden Markov Models[10] that can reliably discriminate and place most new sequences into distinct families, including proteins from an organism at a great evolutionary distance from those in the analysis [66].

---

[10] A statistical model where an unknown parameter is estimated based on observable parameters.

A comprehensive analysis of sequence-structure-function relationships in the GIY-YIG[11] superfamily was obtained through phylogenomics [13]. This superfamily includes several nucleases[12] containing a conserved domain with two short motifs (GIY and YIG) in the N-terminal of the protein. The GIY-YIG domain was also identified in mobile genetic elements, including restriction enzymes and non-LTR retrotransposons, and in enzymes involved in DNA repair and recombination. Database searches were performed to identify all members of known GIY-YIG nuclease families. Sequence alignments, secondary structures prediction, and functional information retrieval were combined to infer the relationships among superfamily members. As a result, a comprehensive evolutionary classification of the GIY-YIG superfamily was presented along with the structural annotation of all families and subfamilies. The analysis allowed for extending the GIY-YIG superfamily to include a number of proteins and to predict their function as nucleases potentially involved in DNA recombination and/or DNA repair. The study offers a great potential to guide experimental verification and to facilitate the functional prediction of the newly identified members of this superfamily [13].

The aforementioned examples illustrate the predicting power of phylogenomics when this approach is applied to the analysis of protein families that goes far beyond the use of standard sequence similarity-based methods. Some of the relationships among family members would not be detected without a phylogenomic analysis.

11.5 Reconstructing the Tree of Life

11.5.1 Building phylogenies

The advent of the polymerase chain reaction[13] (PCR) and DNA sequencing techniques along with the development of improved methods of phylogenetic inference using sequence data have significantly contributed to increase our knowledge on how living (and in some case extinct) organisms are related to each other. There is a large and continuing scientific investment dedicated to establish phylogenetic relationships at diverse taxonomic levels, from populations to kingdoms. Initiatives such as the Tree of Life Web Project aim to compile the information on and to stimulate the inference of the evolutionary relationships of all domains of life [39]. Phylogenomics is the ultimate genetic approach to unveil the Tree of Life using complete genome sequences or considerably large genomic regions from a broad spectrum of organisms. It is

---

[11] The one-letter symbol for the amino acids glycine (G), isoleucine (I), and tyrosine (Y).

[12] An enzyme capable of breaking down specific bonds between nucleotides in a DNA or RNA molecule.

[13] A biochemical technique that exponentially replicates a specific DNA segment *in vitro*.

expected that increasing the number or size of sequence data will improve tree resolution [49, 51]. Inferring phylogenies from genomic sequence data involves two major steps. First, it is necessary to determine orthology [10, 14], i.e., to recognize and compare sequences sharing a common ancestor by speciation. Second, based on recognized sets of aligned orthologous sequences, trees must be inferred using traditional phylogenetic methods (see Section 2) [11].

Databases such as those maintained by the European Molecular Biology Laboratory (EMBL; http://www.ebi.ac.uk/embl/), and the National Center for Biotechnology Information (NCBI; http://www.ncbi.nlm.nih.gov/) are valuable sources of sequence data for phylogenomic and phylogenetic analysis. Sequencing efforts are, however, largely biased towards a few genes and taxa within any given taxonomic rank. Taking Aves (taxonomic group that includes all birds) as an example, there were over 1,080,000 mitochondrial and nuclear DNA sequences for 36 orders deposited on NCBI as of May 2007. Among these, 88% are for the Order Galliformes (chicken, turkey, and allies), followed by Passeriformes (song birds) with only 9.1% of the sequences deposited. The remaining avian orders have less than 1% each. Within Galliformes, 97.6% of the approximately 946,000 sequences are for the genus *Gallus*, which contains the domestic chicken, the only bird to have the nuclear genome fully sequenced so far. Sequences are also biased with respect to genes. For example, about 24% of all mitochondrial DNA sequences for birds deposited on NCBI in May 2007 are for the cytochrome *b* gene, followed by 6% for cytochrome oxidase subunit I. Only 63 complete mitochondrial genomes have been sequenced for the 9,000-10,000 known living bird species.

Until gaps in data collection are filled in for genes and taxa, three approaches can be used to building phylogenomic data sets. The first approach is based on gathering sequences for one or a few genes for hundreds or thousands of taxa [31]. Nonetheless, it is unlikely that the sequenced fragments of larger genes (>1.0 kb) will overlap across all taxa. Efforts such as the DNA Barcode Initiative [28, 65], or the Archosaur Tree of Life [2, 6, 49], which aims at sequencing the same fragment of a particular gene for specific taxonomic groups, may help to close this sampling gap in the near future. Evidently, there is no guarantee that the recovered gene tree reflects the taxon tree, especially in cases where the true phylogeny has many short internal branches and few informative characters to support their relationships [24]. The inconsistency observed between gene trees and taxon trees has been a topic of extensive debate [49]. It is important to take into account that sometime the apparent inconsistencies may reflect distinct evolutionary histories of the molecules of interest (e.g. genes or gene products) in different environments (e.g. genomes and organisms).

A second approach for building phylogenies addresses the gene tree versus taxon tree problem based on the hope that multiple genes might recover the correct tree even if some individual gene trees differ from the correct tree [49, 60]. Concatenating genes into a single sequence can reduce the sampling error associated with shorter sequences for a single gene. This approach requires that the same gene fragments be sampled for the same taxa. Simulation and empirical studies have shown that, under some conditions, a considerable amount of missing data can be tolerated, and the same tree topology can be recovered with reasonable support compared to that derived from a more complete data set [9, 67]. A third approach is known as *supertrees* [53]. In this approach, the original data is ignored, and a new data set is created based on the presence and absence of clades in a set of tree topologies inferred from different studies. The topologies may come from single or multiple genes, and even non-molecular characters (e.g., morphological traits), and the input trees have in common some, but not all taxa. A tree is then built based on the combined *pseudo* data, and the raw data used to obtain the input trees are not considered at all. Although this approach has had some success due to its simplicity and computational speed, it is hard to justify such an approach and its general appropriateness is problematic. A preferable approach to supertrees is known as *supermatrix*, where all data available is compiled in a single matrix and the data unavailable for some taxa is coded as missing; a tree is then inferred using traditional phylogenetic methods that may allow for varying amounts of information arising from different data types [23].

11.5.2 Sampling biases, model adequacy, and phylogenetic inference

The above approaches have been applied in studies to establish the phylogenetic relationships of many taxonomic groups. Here, we illustrate the aforementioned issues with studies performed in birds. The mitochondrial cytochrome *b* gene can be easily amplified by PCR and sequenced via standard methods using universal oligonucleotide primers[14]. To resolve issues regarding the relationships of birds, one study compared two cytochrome *b* data matrices, one with 916 species with sequences varying in size from 200 to 1100 base pairs, and another with 713 species with 600-bp overlapping among all sequences [31]. The analyses were performed using maximum parsimony and neighbor joining using corrected and uncorrected genetic distances. Despite the method used, both data sets failed to recover a tree with strongly supported nodes among different groups of birds, regardless of evidence that extensive taxon sampling improves phylogenetic inference [51]. The recovered trees placed the Order Passeriformes (songbirds and allies) as a sister clade to all other birds, contrary to classical views of avian relationships showing tinamous and the flightless ratites (ostrich, emu, kiwi and allies) as a sister group to a clade containing all the other birds (Fig. 2). At the time when the study was performed, Bayesian analysis was still uncommon for phylogenetic inference of large data sets and maximum likelihood analysis would be computationally impractical to perform due to the large amount of sequences. Hence, an appropriate model of DNA evolution was not applied at that time.

---

[14] Small synthetic DNA sequences of about 15-30 nucleotides long used to replicate DNA.

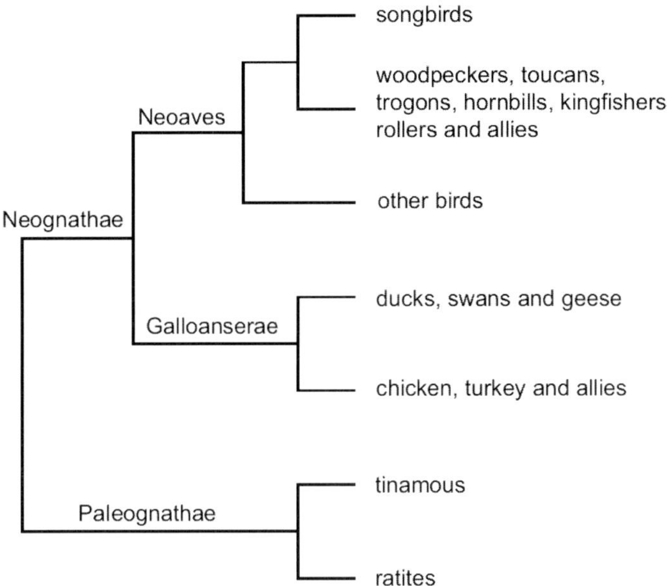

Fig. 11.2. Phylogenetic relationships among major clades of birds. Names above branches are recognized clade names. The depicted relationships are well-accepted based on anatomical and molecular data. There is still a lack of compelling evidence to fully solve the relationships within Neoaves. Simplistic models of DNA evolution places songbirds as a sister lineage to all other Neognathae and Paleognathae.

With the constant improvement in computational power over the last few years, it was possible to analyze a set of complete mitochondrial genome sequences for 40 avian genomes using Bayesian and maximum likelihood methods [24]. The analyses of mitogenomic data[15] improved tree resolution at many basal nodes in the avian Tree of Life, and nested Passeriformes within the avian tree, as expected [38]. We now know that early studies using maximum parsimony or neighbor joining failed to account for higher levels of homoplasy (i.e., same character state randomly acquired in unrelated taxa) present in fast-evolving mitochondrial genes such as cytochrome *b* of Passeriformes [24, 26]. Indeed, the rates of evolution in birds and many other organisms are highly heterogeneous across lineages and genes through time [48]. This makes the inference of phylogenetic relationships non-trivial. The supertree or supermatrix approaches have not yet been applied to solve the deep phylogenetic relationships of birds, but the former approach has been applied to infer phylogenetic relationships of shorebirds [62] and songbirds [32]. In general, the supertree method is able to recover many phylogenetic relationships found in previous published molecular and non-molecular phylogenies,

---

[15] Complete DNA sequences of the mitochondrial genome.

but sometimes it may also recover relationships not previously supported by either molecular or non-molecular data [62]. Hence, the results of a supertree approach have to be carefully scrutinized.

11.5.3 Whole-genome features and rare changes

In recent years, new methods of phylogenomic inference have been developed based on different types of data: whole-genome features and rare genomic changes. Whole-genome features are divided in three main groups: gene order, gene content, and genome signatures. The gene order-based approach compares orthologous regions of the genome among taxa of interest and infers an evolutionary tree that minimizes the number of breakpoints to change one genome organization into another [3]. Alternatively, trees can be inferred based on distances derived from the presence or absence of conserved pairs of orthologous genes across taxa [33]. On the other hand, the gene content approach ignores the genomic organization, and uses a matrix of presence or absence of orthologous genes that can be analyzed by maximum parsimony [68], or using distances that represent the proportion of shared orthologous genes [33]. Finally, a third strategy to extract phylogenomic information from genomes is known as genome signatures or DNA strings [16, 52]. This approach does not require sequences to be aligned, and hence avoids problems associated with homology assessment. In brief, the algorithm calculates the frequency of short nucleotide strings, usually starting from di-nucleotides and up to a specific size, for example, eight-nucleotide long. The frequencies are graphically represented in the form of a colored image, where colors represent string frequency [16]. Analyses of DNA string images can reveal phylogenetic signal otherwise hindered by the definition of genomic homology across deeper taxonomic levels, and appear to complement traditional sequence-based approaches [16, 52].

Rare genomic changes include single or multiple nucleotide or amino acid insertions and or deletions (indels), intron position, gene fissions and fusions, and integration of mobile genetic elements . It is assumed that character state changes are so rare for these characters, that they are virtually homoplasy-free [34, 42]. For example, it is very unlikely that a transposable element will be inserted exactly at the same homologous position in the genome of unrelated taxa; or, it is unlikely that an intron will change its position within a gene without disrupting gene functionality. Hence, rare genomic changes have become increasingly popular to establish deeper phylogenetic relationships such as the basal relationships of placental mammals [34, 42] and rooting the eukaryotic tree [50, 61], in which sequence data have reduced phylogenetic signal due to a larger number of homoplastic characters.

Both whole-genome features and rare genomic changes bring the promise that homoplasies would be non-existent or extremely rare, and the presence of a feature in any two given genomes is likely to have been inherited from a common ancestor. It is necessary to interpret data cautiously, however, since some features may be lost and the phylogenetic signal erased, especially when comparing deeper divergences [11, 29]. These approaches are also more prone to stochastic errors since the numbers of whole-genome characters being compared are smaller than the potential number of sites in nucleotide or amino acid sequences [11].

The aforementioned examples illustrate that despite the improvement achieved in modeling the patterns of DNA and amino acid substitutions, extraction on whole-genome feature for comparative purposes, and advancement in bioinformatics and handling of large data sets, there are still problems when inferring phylogenies from large heterogeneous databases. Some problems are exacerbated at deeper divergences in the Tree of Life because the phylogenetic signal becomes erased by noise. Algorithms and methods have yet to be sufficiently improved to account for particular properties of molecular data such as bias in nucleotide or amino acid composition, differential rates of evolution, secondary structure, and compensatory mutations. The ultimate method will be rapid, yet able to integrate various types of data (e.g., morphological, behavioral, osteological, nucleotide or amino acid sequences, whole-genome features, and rare genomic changes) into a complex mixture model of evolutionary change.

11.6 Conclusions and Perspectives

Phylogenomics has been applied to some major problems such as the functional prediction of uncharacterized genes, studies of protein family evolution, and the reconstructions of the Tree of Life. Questions in evolutionary biology that have been resolved with phylogenomics were illustrated here to clarify the phylogenomics framework. Phylogenomics has great potential to improve the interpretation of genomic and metagenomics data, to connect genotype and phenotype, and to help understand the processes that shape the evolution of genes, genomes, organisms, and communities living in diverse environments.

Despite the increasing number of tools and methods to analyze genomic and metagenomic data, some areas require further development. For example, tools to filter genomic databases and create specific data sets for analysis, such as the selection of specific positions in the DNA or protein sequence, are critically important in this field [43]. On the other hand, algorithms that reduce the computational time required for the analysis of large data sets (length and number of sequences) and for the assessment of statistical support need to be improved. Also, algorithms that take into account structural data are needed in order to obtain more robust evolutionary hypotheses. In summary, the application of an integrated computational and biological approach to access the diversity and evolution of molecules and organisms may help achieve a better system-based understanding of biological processes.

11.7 Acknowledgements

We would like to thank the editors for the opportunity to contribute to this book as part of the series *Studies in Computational Intelligence*. We are grateful to David Pollock (University of Colorado, USA), João Meidanis (Scylla Bioinformatics, University of Campinas - UNICAMP, Brazil), Patrícia Pilisson Cogo (UNICAMP), and Pedro Cipriano Feijão (UNICAMP) for their valuable comments and suggestions to this chapter. We also thank the constructive comments from the three anonymous reviewers. Paulo Nuin (Ontario Cancer Institute - University Health Network, Canada) helped with some of the LaTeX formatting.

References

1. Abhiman S, Sonnhammer EL (2005) FunShift: a database of function shift analysis on protein subfamilies. *Nucleic Acids Res* 33: D197-200
2. Barker FK, Cibois A, Schikler P, Feinstein J, Cracraft J (2004) Phylogeny and diversification of the largest avian radiation. *Proc Natl Acad Sci U S A* 101: 11040-11045
3. Blanchette M, Kunisawa T, Sankoff D (1999) Gene order breakpoint evidence in animal mitochondrial phylogeny. *J Mol Evol* 49: 193-203
4. Brenner SE (1999) Errors in genome annotation. *Trends Genet* 15: 132-133
5. Brown D, Sjolander K (2006) Functional classification using phylogenomic inference. *PLoS Comput Biol* 2: e77
6. Camargo MM, Nahum LA (2005) Adapting to a changing world: RAG genomics and evolution. *Hum Genomics* 2: 132-137
7. Castoe TA, Stephens T, Noonan BP, Calestani C (2007) A novel group of type I polyketide synthases (PKS) in animals and the complex phylogenomics of PKSs. *Gene* 392: 47-58
8. Consortium EP (2004) The ENCODE (ENCyclopedia Of DNA Elements) Project. *Science* 306: 636-640
9. Crowe TM, Bowie RC, Bloomer P, Mandiwana TG, Hedderson TAJ, Randi E, Pereira SL, Wakeling J (2006) Phylogenetics, biogeography and classification of, and character evolution in, gamebirds (Aves: Galliformes): effects of character exclusion, data partitioning and missing data. *Cladistics* 22: 495-532
10. Dehal PS, Boore JL (2006) A phylogenomic gene cluster resource: the Phylogenetically Inferred Groups (PhIGs) database. *BMC Bioinformatics* 7: 201
11. Delsuc F, Brinkmann H, Philippe H (2005) Phylogenomics and the reconstruction of the tree of life. *Nat Rev Genet* 6: 361-375
12. Deutschbauer AM, Chivian D, Arkin AP (2006) Genomics for environmental microbiology. *Curr Opin Biotechnol* 17: 229-235
13. Dunin-Horkawicz S, Feder M, Bujnicki JM (2006) Phylogenomic analysis of the GIY-YIG nuclease superfamily. *BMC Genomics* 7: 98
14. Dutilh BE, van Noort V, van der Heijden RT, Boekhout T, Snel B, Huynen MA (2007) Assessment of phylogenomic and orthology approaches for phylogenetic inference. *Bioinformatics* 23: 815-824

15. Edwards AW, Cavalli-Sforza LL (1963) The reconstruction of evolution. *Ann Hum Genet* 27: 105-106
16. Edwards SV, Fertil B, Giron A, Deschavanne PJ (2002) A genomic schism in birds revealed by phylogenetic analysis of DNA strings. *Syst Biol* 51: 599-613
17. Eisen JA (1998) Phylogenomics: improving functional predictions for uncharacterized genes by evolutionary analysis. *Genome Res* 8: 163-167
18. Eisen JA, Fraser CM (2003) Phylogenomics: intersection of evolution and genomics. *Science* 300: 1706-1707
19. Eisen JA, Wu M (2002) Phylogenetic analysis and gene functional predictions: phylogenomics in action. *Theor Popul Biol* 61: 481-487
20. Felsenstein J (1978) Cases in which parsimony or compatibility methods will be positively misleading. *Syst Zool* 27: 401-410
21. Felsenstein J (1981) Evolutionary trees from DNA sequences: a maximum likelihood approach. *J Mol Evol* 17: 368-376
22. Felsenstein J (2004) *Inferring phylogenies.* Sinauer Associates, Sunderland, Mass.
23. Gatesy J, Matthee C, DeSalle R, Hayashi C (2002) Resolution of a supertree/supermatrix paradox. *Syst Biol* 51: 652-664
24. Gibb GC, Kardailsky O, Kimball RT, Braun EL, Penny D (2007) Mitochondrial genomes and avian phylogeny: complex characters and resolvability without explosive radiations. *Mol Biol Evol* 24: 269-280
25. Glanville JG, Kirshner D, Krishnamurthy N, Sjolander K (2007) Berkeley Phylogenomics Group web servers: resources for structural phylogenomic analysis. *Nucleic Acids Res* 35: W27-W32
26. Groth JG, Barrowclough GF (1999) Basal divergences in birds and the phylogenetic utility of the nuclear RAG-1 gene. *Mol Phylogenet Evol* 12: 115-123
27. Hawkins T, Kihara D (2007) Function prediction of uncharacterized proteins. *J Bioinform Comput Biol* 5: 1-30
28. Hebert PD, Stoeckle MY, Zemlak TS, Francis CM (2004) Identification of birds through DNA Barcodes. *PLoS Biol* 2: e312
29. Hillis DM (1999) SINEs of the perfect character. *Proc Natl Acad Sci U S A* 96: 9979-9981
30. Huelsenbeck JP, Larget B, Miller RE, Ronquist F (2002) Potential applications and pitfalls of Bayesian inference of phylogeny. *Syst Biol* 51: 673-688
31. Johnson KP (2001) Taxon sampling and the phylogenetic position of Passeriformes: evidence from 916 avian cytochrome b sequences. *Syst Biol* 50: 128-136
32. Jnsson KA, Fjelds J (2006) A phylogenetic supertree of Oscine passerine birds (Aves: Passeri). *Zool Scr* 35: 149-186
33. Korbel JO, Snel B, Huynen MA, Bork P (2002) SHOT: a web server for the construction of genome phylogenies. *Trends Genet* 18: 158-162
34. Kriegs JO, Churakov G, Kiefmann M, Jordan U, Brosius J, Schmitz J (2006) Retroposed elements as archives for the evolutionary history of placental mammals. *PLoS Biol* 4: e91
35. Krishnamurthy N, Brown DP, Kirshner D, Sjolander K (2006) PhyloFacts: an online structural phylogenomic encyclopedia for protein functional and structural classification. *Genome Biol* 7: R83
36. Li S, Pearl DK, Doss H (2000) Phylogenetic tree reconstruction using Markov Chain Monte Carlo. *J Am Stat Assoc* 95: 493-508
37. Li W-H (1997) *Molecular evolution.* Sinauer Associates, Sunderland, Mass.

38. Livezey BC, Zusi RL (2007) High-order phylogeny of modern birds (Theropoda, Aves: Neornithes) based on comparative anatomy. II. Analysis and discussion. *Zool J Linn Soc* 149: 1-95

39. Maddison DR, Schulz K-S (2004) *The Tree of Life Web Project.* http://tolweb.org (last accessed in October 2007).

40. Malik HS, Henikoff S (2003) Phylogenomics of the nucleosome. *Nat Struct Biol* 10: 882-891

41. Mau B, Newton MA, Larget B (1999) Bayesian phylogenetic inference via Markov chain Monte Carlo methods. *Biometrics* 55: 1-12

42. Murphy WJ, Pringle TH, Crider TA, Springer MS, Miller W (2007) Using genomic data to unravel the root of the placental mammal phylogeny. *Genome Res* 17: 413-421

43. Nahum LA, Reynolds MT, Wang ZO, Faith JJ, Jonna R, Jiang ZJ, Meyer TJ, Pollock DD (2006) EGenBio: A Data Management System for Evolutionary Genomics and Biodiversity. *BMC Bioinformatics* 7 Suppl 2: S7

44. Nahum LA, Riley M (2001) Divergence of function in sequence-related groups of *Escherichia coli* proteins. *Genome Res* 11: 1375-1381

45. Nei M, Kumar S (2000) *Molecular Evolution and Phylogenetics.* Oxford University Press, Oxford; New York

46. Norvell JC, Machalek AZ (2000) Structural genomics programs at the US National Institute of General Medical Sciences. *Nat Struct Biol* 7 Suppl: 931

47. Page RDM, Holmes EC (1998) *Molecular evolution: a phylogenetic approach.* Blackwell Science, Oxford ; Malden, MA

48. Pereira SL, Baker AJ (2006) A mitogenomics timescale for birds detects variable phylogenetic rates of molecular evolution and refutes the standard molecular clock. *Mol Biol Evol* 23: 1731-1740

49. Pereira SL, Baker AJ, Wajntal A (2002) Combined nuclear and mitochondrial DNA sequences resolve generic relationships within the Cracidae (Galliformes, Aves). *Syst Biol* 51: 946-958

50. Philippe H, Lopez P, Brinkmann H, Budin K, Germot A, Laurent J, Moreira D, Muller M, Le Guyader H (2000) Early-branching or fast-evolving eukaryotes? An answer based on slowly evolving positions. *Proc Biol Sci* 267: 1213-1221

51. Pollock DD (2002) Genomic biodiversity, phylogenetics and coevolution in proteins. *Appl Bioinformatics* 1: 81-92

52. Qi J, Wang B, Hao BI (2004) Whole proteome prokaryote phylogeny without sequence alignment: a K-string composition approach. *J Mol Evol* 58: 1-11

53. Ragan MA (1992) Phylogenetic inference based on matrix representation of trees. *Mol Phylogenet Evol* 1: 53-58

54. Rannala B, Yang Z (1996) Probability distribution of molecular evolutionary trees: a new method of phylogenetic inference. *J Mol Evol* 43: 304-311

55. Reed JL, Patel TR, Chen KH, Joyce AR, Applebee MK, Herring CD, Bui OT, Knight EM, Fong SS, Palsson BO (2006) Systems approach to refining genome annotation. *Proc Natl Acad Sci U S A* 103: 17480-17484

56. Saitou N, Nei M (1987) The neighbor-joining method: a new method for reconstructing phylogenetic trees. *Mol Biol Evol* 4: 406-425

57. Serres MH, Riley M (2005) Gene fusions and gene duplications: relevance to genomic annotation and functional analysis. *BMC Genomics* 6: 33

58. Sogin ML, Morrison HG, Huber JA, Welch DM, Huse SM, Neal PR, Arrieta JM, Herndl GJ (2006) Microbial diversity in the deep sea and the underexplored rare biosphere. *Proc Natl Acad Sci U S A* 103: 12115-12120

59. Sokal RR, Sneath PHA (1963) *Numerical Taxonomy*. W. H. Freeman, San Francisco
60. Soltis DE, Soltis PS, Zanis MJ (2002) Phylogeny of seed plants based on evidence from eight genes. *Am. J. Bot.* 89: 1670-1681
61. Stechmann A, Cavalier-Smith T (2002) Rooting the eukaryote tree by using a derived gene fusion. *Science* 297: 89-91
62. Thomas GH, Wills MA, Szkely T (2004) A supertree approach to shorebird phylogeny. *BMC Evol Biol* 4: 28
63. Venter JC, Remington K, Heidelberg JF, Halpern AL, Rusch D, Eisen JA, Wu D, Paulsen I, Nelson KE, Nelson W, Fouts DE, Levy S, Knap AH, Lomas MW, Nealson K, White O, Peterson J, Hoffman J, Parsons R, Baden-Tillson H, Pfannkoch C, Rogers YH, Smith HO (2004) Environmental genome shotgun sequencing of the Sargasso Sea. *Science* 304: 66-74
64. Watson JD, Sanderson S, Ezersky A, Savchenko A, Edwards A, Orengo C, Joachimiak A, Laskowski RA, Thornton JM (2007) Towards fully automated structure-based function prediction in structural genomics: a case study. *J Mol Biol* 367: 1511-1522
65. Waugh J (2007) DNA barcoding in animal species: progress, potential and pitfalls. *Bioessays* 29: 188-197
66. Wickstead B, Gull K (2006) A holistic kinesin phylogeny reveals new kinesin families and predicts protein functions. *Mol Biol Cell* 17: 1734-1743
67. Wiens JJ (2003) Missing data, incomplete taxa, and phylogenetic accuracy. *Syst Biol* 52: 528-538
68. Wolf YI, Rogozin IB, Grishin NV, Tatusov RL, Koonin EV (2001) Genome trees constructed using five different approaches suggest new major bacterial clades. *BMC Evol Biol* 1: 8
69. Yang Z (2006) *Computational Molecular Evolution*. Oxford University Press, Oxford
70. Zhang J (2003) Evolution by gene duplication: an update. *Trends Ecol Evol* 18: 292-298

Computational Aspects of Aggregation in Biological Systems

Vladik Kreinovich and Max Shpak

University of Texas at El Paso, El Paso, TX 79968, USA
`vladik@utep.edu, mshpak@utep.edu`

Summary. Many biologically relevant dynamical systems are aggregable, in the sense that one can divide their microvariables $x_1, \ldots, x_n$ into several (k) non-intersecting groups and find functions $y_1, \ldots, y_k$ ($k < n$) from these groups (macrovariables) whose dynamics only depend on the initial state of the macrovariable. For example, the state of a population genetic system can be described by listing the frequencies x_i of different genotypes, so that the corresponding dynamical system describe the effects of mutation, recombination, and natural selection. The goal of aggregation approaches in population genetics is to find macrovariables $y_a, \ldots, y_k$ to which aggregated mutation, recombination, and selection functions could be applied. Population genetic models are formally equivalent to genetic algorithms, and are therefore of wide interest in the computational sciences.

Another example of a multi-variable biological system of interest occurs in ecology. Ecosystems contain many interacting species, and because of the complexity of multi-variable nonlinear systems, it would be of value to derive a formal description that reduces the number of variables to some macrostates that are weighted sums of the densities of several species.

In this chapter, we explore different computational aspects of aggregability for linear and non-linear systems. Specifically, we investigate the problem of conditional aggregability (i.e., aggregability restricted to modular states) and aggregation of variables in biologically relevant quadratic dynamical systems.

12.1 Introduction

12.1.1 Dynamical Systems: A Brief Reminder

Many systems in nature can be described as *dynamical systems*, in which the state of a system at each moment of time is characterized by the values of (finitely many) (micro)variables $x_1, \ldots, x_n$, and the change of the state over time is uniquely determined by the initial state.

Definition 1. *Let n be an integer. This integer will be called the* number *of microvariables (or* variables, *for short). These variables will be denoted*

V. Kreinovich and M. Shpak: *Computational Aspects of Aggregation in Biological Systems*,
Studies in Computational Intelligence (SCI) **122**, 281–305 (2008)
`www.springerlink.com`

by $x_1, \ldots, x_n$. *By a* microstate *(or* state*), we mean an n-dimensional vector* $x = (x_1, \ldots, x_n)$.

Definition 2.

- *By a* discrete-time trajectory, *we mean a function which maps natural numbers t into states* $x(t)$.
- *By a* continuous-time trajectory, *we mean a function which maps non-negative real numbers t into states* $x(t)$.

For each trajectory and for each moment of time t, the state $x(t)$ *is called* a state at moment t.

Comment. In our description, we assume that we have a starting point $t = 0$.

Definition 3. *For a given n, by a* dynamical system, *we mean a tuple* $(n, f_1, \ldots, f_n)$, *where* $n \geq 1$ *is an integer, and* $f_1, \ldots, f_n : \mathbb{R}^n \to \mathbb{R}$ *are functions of n variables.*

- *We say that a discrete-time trajectory* $x(t)$ *is* consistent *with the dynamical system* $(n, f_1, \ldots, f_n)$ *if for every t, we have*

$$x_i(t+1) - x_i(t) = f_i(x_1(t), \ldots, x_n(t)). \tag{12.1}$$

- *We say that a continuous-time trajectory* $x(t)$ *is* consistent *with the dynamical system* $(n, f_1, \ldots, f_n)$ *if for every t, we have*

$$\frac{dx_i(t)}{dt} = f_i(x_1(t), \ldots, x_n(t)). \tag{12.2}$$

For example, the state of a biological population can be described by listing the amounts or relative frequencies x_i of different genotypes i; in this example, the corresponding functions $f_i(x_1, \ldots, x_n)$ describe the effects of mutation, recombination, and natural selection.

Equilibria. In general, when we start in some state $x(t)$ at the beginning moment of time t, the above dynamics leads to a different state $x(t+1)$ at the next moment of time. In many natural systems, these changes eventually subside, and we end up with a state which does not change in time, i.e., with an *equilibrium* state. In the equilibrium state x, we have $x_i'(t) = \dfrac{dx_i}{dt} = 0$ or $x_i'(t) = x_i(t+1) - x_i(t) = 0$, i.e., in general, $x_i'(t) = f_i(x_1, \ldots, x_n) = 0$.

12.1.2 Aggregability

For natural systems, the number of variables is often very large. For example, for a system with g loci on a chromosome in which each of these genes can have two possible allelic states, there are $n = 2^g$ possible genotypes. For large g, due to the large number of state variables, the corresponding dynamics are extremely difficult to analyze.

This complexity of this analysis can often be reduced if we take into consideration that in practice, quantities corresponding to different variables x_i can be aggregated into natural clusters. This happens, for example, when interactions within each cluster are much stronger than interactions across different clusters. In mathematical terms, this means that we subdivide the variables $x_1, \ldots, x_n$ into non-overlapping blocks $I_1 = \{i(1,1), \ldots, i(a, n_1)\}, \ldots, I_k = \{i(k,1), \ldots, i(k, n_k)\}$ $(k \ll n)$.

To describe each cluster I_a, it is often not necessary to know the value of each of its "microvariables" $x_{i(a,1)}, \ldots, x_{i(a,n_a)}$. Dynamical systems are sometimes *decomposably aggregable* in the following sense: it is sufficient to characterize the state of each cluster by a single "macrovariable" $y_a = c_a(x_{i(a,1)}, \ldots, x_{i(a,n_a)})$ so that the dynamics of these macrovariables are determined only by their previous values.

Definition 4. *Let us fix an index $i_0 \leq n$.*

- *By a* partition, *we mean a tuple $(i_0, I_1, \ldots, I_k)$ $(k < n)$ where $I_1 \subseteq \{1, \ldots, n\}, \ldots, I_k \subseteq \{1, 2, \ldots, n\}$ are non-empty sets such that $i_0 \in I_1$, $I_1 \cup \ldots \cup I_k = \{1, \ldots, n\}$, and $I_i \cap I_j = \emptyset$ for all $i \neq j$.*
- *For each partition, the number of elements in the set I_a will be denoted by n_a, and these elements will be denoted by $i(a, 1), \ldots, i(a, n_a)$.*
- *We say that a function $c : \mathbb{R}^m \to \mathbb{R}$ actually depends on the variable x_{i_0} if there exist real numbers $x_1, \ldots, x_{i_0-1}, x_{i_0}, x_{i_0+1}, \ldots, x_m$ and a real number $x'_{i_0} \neq x_{i_0}$ for which*

$$c(x_1, \ldots, x_{i_0-1}, x_{i_0}, x_{i_0+1}, \ldots, x_m) \neq c(x_1, \ldots, x_{i_0-1}, x'_{i_0}, x_{i_0+1}, \ldots, x_m).$$

The reason why we select an index i_0 is that we want to avoid a degenerate case $c_a = 0$, and make sure that at least one of the macrovariables depends on some microvariable x_{i_0}. In a partition, this microvariable can belong to one of the blocks. Without loss of generality, we can assume that it belongs to the first block I_1 (if it belong to another block, we can simply rename the blocks).

Definition 5.

- *By a* decomposable aggregation, *we mean a tuple $(i_0, I_1, \ldots, I_k, c_1, \ldots, c_k)$, where $(i_0, I_1, \ldots, I_k)$ is a partition, and for each a from 1 to k, $c_a : \mathbb{R}^{n_a} \to \mathbb{R}$ is a function of n_a variables such that the function c_1 actually depends on x_{i_0}.*
- *For every microstate $x = (x_1, \ldots, x_n)$, by the corresponding* macrostate *we mean a tuple $y = (y_1, \ldots, y_k)$, where $y_a = c_a(x_{i(a,1)}, \ldots, x_{i(a,n_a)})$.*
- *We say that two microstates x and $\widetilde{x}$ are* macroequivalent *if they lead to the same macrostate $y = \widetilde{y}$.*
- *We say that a decomposable aggregation $(I_1, \ldots, I_k, c_1, \ldots, c_k)$ is* consistent *with the dynamical system $(n, f_1, \ldots, f_n)$ if for every two trajectories x and $\widetilde{x}$ for which at some moment of time t, two microstates $x(t)$ and $\widetilde{x}(t)$ are macroequivalent, they remain macroequivalent for all following moments of time $t' > t$.*

- *We say that a dynamical system is* decomposably k-aggregable *if it is consistent with some decomposable aggregation* $(i_0, I_1, \ldots, I_k, c_1, \ldots, c_k)$.

A dynamical system is said to be *decomposably $\leq k$-aggregable* if it is decomposably ℓ-aggregable for some $\ell \leq k$, and *decomposably aggregable* if it is decomposably k-aggregable for some integer k.

Many biological systems (and many systems from other fields such as economics [24] and queuing theory [3] etc.) are decomposably aggregable. In such systems, equations (12.2) or (12.1) lead to simpler equations

$$\frac{dy_a}{dt} = h_a(y_1(t), \ldots, y_k(t)) \tag{12.3}$$

or, correspondingly,

$$y_a(t+1) - y_a(t) = h_i(y_1(t), \ldots, y_k(t)) \tag{12.4}$$

for appropriate functions $h_1, \ldots, h_k$. The aggregability property has been actively studied; see, e.g., [3–5, 13, 18, 19, 22–24].

12.1.3 Discussion

We can have intersecting blocks. Some systems have similar aggregability properties, but with overlapping blocks I_a. In the general case, we have macrovariables $y_a = c_a(x_1, \ldots, x_n)$ each of which may depend on all the microvariables $x_1, \ldots, x_n$. We are still interested in the situation when the dynamics of the macrovariables is determined only by their previous values.

In some cases, such overlapping decomposabilities are not in general useful. For example, for every continuous-time dynamical system, we can define a macrovariable $y_1(x_1, \ldots, x_n)$ as the time t after (or before) which the trajectory starting at a state $(x_1, \ldots, x_n)$ reaches the plane $x_1 = c$ for some constant v (this y_1 is defined at least for values $x_1 \approx a$). The dynamics of the new macrovariable is simple: if in a state x, we reach $x_1 = v$ after time $t = y_1(x)$, then for a state x' which is t_0 seconds later on the same trajectory, the time to reaching $x_1 = v$ is $t - t_0$. In other words, the value of y_1 decreases with time t as $y_1(t_0) = y_1(0) - t_0$, or, in terms of the corresponding differential equation, $\dfrac{dy_1}{dt} = -1$.

From the purely mathematical viewpoint, we have an (overlapping) aggregation. However, the main objective of aggregation is to *simplify* solving the system of equations. In the above example, to find $y_1(x)$ for a given x, we, in effect, first need to solve the system – which defeats the purpose of aggregation.

In view of this observation, and taking into account that most aggregable systems are *decomposable* (i.e., the blocks do not intersect), in this chapter, we will concentrate on decomposable aggregations. Unless otherwise indicated, we will simply refer to decomposable and aggregable systems as *aggregable*.

We can have strong interactions between clusters. In our motivations, we assumed that the interaction within each cluster is much stronger than the interaction among clusters. While this is indeed a useful example, the aggregability property sometimes occurs even when the interaction between clusters is strong – as long it can be appropriately "decomposed". In view of this fact, in the following precise definitions, we do not make any assumptions about the relative strengths of different interactions.

Approximate aggregability. It is worth mentioning that perfect aggregability usually occurs only in idealized mathematical models. In many practical situations, we only have *approximate* aggregability, so that the aggregate dynamics (12.3) or (12.4) differs only marginally from the actual microdynamics of the macrovariables variables $y_a = h_a(x_{i(a,1)}, \ldots, x_{i(a,n_a)})$.

Note that many dynamical systems are only approximately aggregable during certain time intervals in their evolution, or over certain subspaces of their state space [5, 24].

12.1.4 Linear Systems

Linear systems: a brief introduction. In principle, the functions $f_i(x_1, \ldots, x_n)$ can be arbitrarily complex. In practice, we can often simplify the resulting expressions if we expand each function $f_i(x_1, \ldots, x_n)$ in Taylor series in x_i and keep only terms up to a fixed order in this expansion. In particular, when the interactions are weak, we can often use a linear approximation

$$x_i'(t) = a_i(t) + \sum_{j=1}^{n} F_{i,j} \cdot x_j(t). \tag{12.5}$$

In many cases, the i-th variable describes the absolute amount of the i-th entity (such as the i-th genotype). In this case, if we do not have any entities at some moment t, i.e., if we have $x_i(t) = 0$ for all i, then none will appear. So, we will have $x_i'(t) = 0$, and thus, $a_i(t) = 0$. In such cases, the above linear system takes an even simpler form

$$x_i'(t) = \sum_{j=1}^{n} F_{i,j} \cdot x_j(t). \tag{12.6}$$

Let us describe how the general definitions of dynamical systems look in the linear case.

Definition 6. *We say that a dynamical system* $(n, f_1, \ldots, f_n)$ *is* linear *if all the functions f_i are linear, i.e., if* $f_i = \sum_{j=1}^{n} F_{i,j} \cdot x_j$ *for some rational values $F_{i,j}$.*

Comment. In reality, the coefficients $F_{i,j}$ can be arbitrary real numbers. However, our main objective is to analyze the corresponding algorithms. So, instead of the actual (unknown) value of each coefficient, we can only consider the (approximate) value represented in the computer, which are usually rational numbers. In view of this fact, in the computational analysis of problems related to linear dynamical systems, we will always assume that all the values $F_{i,j}$ are rational numbers.

Equilibria. In particular, for linear systems, equilibrium states $x = (x_1, \ldots, x_n)$ are states which satisfy the corresponding (homogeneous) system of linear equations

$$\sum_{j=1}^{n} F_{i,j} \cdot x_j = 0. \tag{12.7}$$

Of course, the state $x = (0, \ldots, 0)$ is always an equilibrium for such systems. In some physical systems, this trivial 0 state is the only equilibrium. However, in biology, there usually exist non-zero equilibrium states. In such cases, the matrix $F_{i,j}$ is singular.

In general, the set of all possible solutions of a homogeneous linear system is a linear space – in the sense that a linear combination of arbitrary solutions is also a solution. In every linear space, we can select a *basis*, i.e., a set of linearly independent vectors such that every other solution is a linear combination of solutions from this basis. The number of these independent vectors is called a *dimension* of the linear space. In principle, we can have matrices for which this linear space has an arbitrary dimension $\leq n$. However, for almost all singular matrices, this dimension is equal to 1.

In view of this fact, it is reasonable to consider only singular matrices in our analysis of biological systems. For such systems, all equilibria states $x = (x_1, \ldots, x_n)$ are proportional to some fixed state $\beta = (\beta_1, \ldots, \beta_n)$, i.e., they can be all characterized by an expression $x_i = y \cdot \beta_i$ for some parameter y.

Linear aggregation: definitions. For linear dynamical systems, we restrict ourselves to *linear* aggregations, i.e., to macrovariables y_a which linearly depend on the the corresponding microvariables x_i, i.e., for which $y_a = \sum_{i \in I_a} \alpha_i \cdot x_i$ for some coefficients ("weights") α_i. As a result, we arrive at the following definition:

Definition 7. *A decomposable aggregation*

$$(i_0, I_1, \ldots, I_k, c_1, \ldots, c_k)$$

is called linear *if all the functions c_a are linear, i.e., have the form $c_a(x_{i(a,1)}, \ldots, x_{i(a,n_a)}) = \sum_{i \in I_a} \alpha_i \cdot x_i$ for some coefficients $\alpha_1, \ldots, \alpha_n$.*

This definition can be reformulated as follows: By a *linear (decomposable) aggregation*, we mean a tuple $(i_0, I_1, \ldots, I_k, \alpha)$, where $(i_0, I_1, \ldots, I_k)$ is a partition, and $\alpha = (\alpha_1, \ldots, \alpha_n)$ is a tuple of real numbers for which $\alpha_{i_0} \neq 0$. For every microstate $x = (x_1, \ldots, x_n)$, by the corresponding *macrostate* we mean a tuple $y = (y_1, \ldots, y_k)$, where $y_a = \sum\limits_{i \in I_a} \alpha_i \cdot x_i$. We say that a dynamical system is *linearly k-aggregable* if it is consistent with some linear decomposable aggregation $(i_0, I_1, \ldots, I_k, \alpha)$.

Similarly, we can define linear $\leq k$-aggregability and linear aggregability.

Formulation of the problem. For every integer $k > 0$, we arrive at the following *linear k-aggregability* problem:

- *given* a linear dynamical system;
- *check* whether the given system is linearly k-aggregable.

As in the previous examples, we also want to compute the partition $I_1, \ldots, I_k$ and the weights α_i for this aggregation.

Analysis of the problem. In matrix terms, a linear dynamic equation has the form $x' = Fx$. Once the partition $I_1, \ldots, I_k$ is fixed, we can represent each n-dimensional state vector x as a combination of vectors $x^{(a)}$ formed by the components x_i, $i \in I_a$. In these terms, the equation $x' = Fx$ can be represented as $x'^{(a)} = \sum\limits_{b} F^{(a),(b)} x^{(b)}$, where $F^{(a),(b)}$ denotes the corresponding block of the matrix F (formed by elements $F_{i,j}$ with $i \in I_a$ and $j \in I_b$).

For the corresponding linear combinations $y_a = \alpha^{(a)\,T} x^{(a)}$, the dynamics takes the form $y'_a = \sum\limits_{b} \alpha^{(a)\,T} F^{(a),(b)} x^{(b)}$. The only possibility for this expression to only depend on the combinations $y_b = \alpha^{(b)\,T} x^{(b)}$ is when for each b, the coefficients of the dependence of y'_a on x_i, $i \in I_b$, are proportional to the corresponding weights α_i, i.e., when for every a and b, we have $\alpha^{(a)\,T} F^{(a),(b)} = \lambda_{a,b} \alpha^{(b)\,T}$ for some number $\lambda_{a,b}$. By transposing this relation, we conclude that

$$F^{(a),(b)\,T} \alpha^{(a)} = \lambda_{a,b} \alpha^{(b)}. \tag{12.8}$$

First known result: the problem is, in general, computationally difficult (NP-hard). The first known result is that in general, the linear aggregability problem is NP-hard even for $k = 2$ [6, 7]. This means that even for linear systems (unless P=NP), there is no hope of finding a *general* feasible method for detecting decomposable aggregability.

Second known result: once we know the partition, finding the weights α_i is possible. The above mentioned result is that in general, finding the partition under which the system is aggregate is computationally difficult (NP-hard).

As we have mentioned, in some practical situations, the partition comes from the natural clustering of the variables and is therefore, known. In the case when the partition is found, it is possible to feasibly find the weights α_i of the corresponding linear macrocombinations y_a [6, 7].

The main idea behind the corresponding algorithm is as follows. From the above equation (12.8), for $a = b$, we conclude that $\alpha^{(a)}$ is an eigenvector of the matrix $F^{(a),(a)}{}^T$. Since the weight vectors $\alpha^{(a)}$ are defined modulo a scalar factor, we can thus select one of the (easily computed) eigenvectors of $F^{(a),(a)}{}^T$ as $\alpha^{(a)}$.

Once we know $\alpha^{(a)}$ for one a, we can determine all other weight vectors $\alpha^{(b)}$ from the condition (12.8), i.e., as $\alpha^{(b)} = F^{(a),(b)}{}^T \alpha^{(a)}$.

12.2 Conditional Aggregation

12.2.1 What is Conditional Aggregability: General Case

Aggregation: reminder. As we have mentioned, in practice, quantities corresponding to different variables x_i can be usually grouped into clusters $I_1, \ldots, I_k$ in such a way that interactions within each cluster are much stronger than interactions across different clusters. In the above text, we considered systems which are (decomposably) aggregable in the sense that in each block I_a, we can find an appropriate combination of variables $y_a = c_a(x_{i(a,1)}, \ldots, x_{i(a,n_a)})$ in such a way that for *all* possible states $x = (x_1, \ldots, x_n)$, the change in the new variables is only determined by the values of these new variables. In other words, we have a simpler system $\dfrac{dy_a}{dt} = h_a(y_1, \ldots, y_k)$. This reduction to a simpler system drastically simplifies computations related to the dynamical behavior of the original system.

In practice, we can restrict ourselves to "modular" states. Systems which are, in this sense, "unconditionally" aggregable, i.e., aggregable for all possible states $x = (x_1, \ldots, x_n)$, are rather rare. However, in practice, we rarely encounter the need to consider arbitrary states x. Specifically, we know that the interaction within each cluster in much stronger than interactions across different clusters.

In the ideal case when a cluster does not interact with other clusters at all, the interaction within the cluster will lead to an equilibrium state of this cluster. The values of the corresponding microvariables variables $x_{i(a,1)}, \ldots, x_{i(a,n_u)}$ will stop changing with time and reach an equilibrium state: $x'_{i(a,k)}(t) = f_{i(a,k)}(x_{i(a,1)}(t), \ldots, x_{i(a,n_a)}(t)) = 0$. Since interactions across clusters are much weaker, it is reasonable to assume that in spite of this interaction, the state within each cluster is very close to an equilibrium state. To a first approximation, we can therefore assume that within each cluster, we have equilibrium.

Towards an exact description of conditional aggregability. As we explained above, a typical biologically relevant dynamical system has a 1-dimensional family of equilibrium states, i.e., a family which is determined by a single parameter y. The values of all other variables x_i are uniquely determined by this value y.

Thus, to describe the combination of equilibrium states corresponding to k different clusters, we must describe the values of the corresponding k variables

y_a, $1 \leq a \leq k$ and the dependence $x_i = F_i(y_a)$ of each microvariable x_i on the "macrovariable" y_a of the corresponding cluster. In these terms, conditional (decomposable) aggregability means that there exist functions $h_a(y_1, \ldots, y_k)$ such that in the equilibrium state, the evolution of the macrovariables is determined by the system $\dfrac{dy_a}{dt} = h_a(y_1, \ldots, y_k)$. In the new state, every cluster a remains in the equilibrium state determined by the new value $y_a(t+1)$ of the corresponding macrovariable.

Formal definition of conditional aggregability. The above analysis leads to the following definitions.

Definition 8.

- *By a* conditional aggregation, *we mean a tuple* $(i_0, I_1, \ldots, I_k, C_1, \ldots, C_n)$, *where* $(i_0, I_1, \ldots, I_k)$ *is a partition, and for each i from 1 to n, $C_i : \mathbb{R} \to \mathbb{R}$ is a function of one variable such that the function F_{i_0} actually depends on x_{i_0}.*
- *By a* macrostate, *we mean a tuple* $y = (y_1, \ldots, y_k)$.
- *By a* microstate *corresponding to a macrostate y, we mean a state* $x = (x_1, \ldots, x_n)$ *in which for every index i, we have* $x_i = C_a(y_a)$, *where a is the cluster containing i* $(i \in I_a)$.
- *A microstate is called* modular *if it corresponds to some macrostate y.*

A conditional aggregation $(i_0, I_1, \ldots, I_k, C_1, \ldots, C_n)$ is said to be *consistent* with a dynamical system $(n, f_1, \ldots, f_n)$ if for every trajectory for which at some moment t, the microstate $x(t)$ is modular, it remains modular for all following moments of time $t' > t$. We say that a dynamical system is *conditionally k-aggregable* if it is consistent with some conditional aggregation $(i_0, I_1, \ldots, I_k, C_1, \ldots, C_n)$. Similarly, we can define when a system *conditionally $\leq k$-aggregable* and *conditionally aggregable*.

Example of conditional aggregation: phenotype-based description of an additive genetic trait. In general, the description of recombination and natural selection is a quadratic dynamical system [14–16]. Specifically, from one generation t to the next one $(t+1)$, the absolute frequency p_i (number of individuals with genotype i in a population) changes as follows:

$$p_z(t+1) = \sum_i \sum_j w_i \cdot w_j \cdot p_i(t) \cdot p_j(t) \cdot R_{ij \to z},$$

where w_i is the fitness of the i-th genotype (probability of survival multiplied by the number of offsprings), and $R_{ij \to z}$ is the recombination function that determines the probability that parental types i and j produce progeny z.

Let us assume that we have two alleles at each of g loci. In this case, each genotype i can be described as a binary string, i.e., a string consisting of 0s and 1s. Let a_i denote the number of 1s in the i-th string; then the number of 0s is $g - a_i$. A frequent simplifying assumption in quantitative genetics is that the contribution of each locus to phenotype is equal. In precise terms, this means that the fitness w_i depends only on the number of 1s in the corresponding binary string: $w_i = w_{a_i}$. A phenotype is formed by all the genotypes with a given number of 1s. In this case, since recombination at different loci are independent, the recombination function takes the form [1, 21] $R_{ij \to z} = R_{a_i a_j \to a_z}(L)$, where L is the number of common (overlapping) 1s between the binary sequences i and j: e.g., the sequences 1010 and 0011 have one overlapping 1 (in the 3rd place), and

$$R_{ab \to d}(L) = \left(\frac{1}{2}\right)^{a+b-2L} \binom{a+b-2L}{d-L}.$$

(See [1, 21] for derivation.)

In this situation, since different genotypes i within the same phenotype a have the same fitness, it is reasonable to assume that all these genotypes have the same frequency within each phenotype class $p_i = p_{a_i}$. It is easy to see that this this equal-frequency distribution is an equilibrium, i.e., that if we start with equal genotype frequencies within each phenotype $p_i(t) = p_{a_i}(t)$, then in the next generation, we also have equal genotype frequencies $p_i(t + 1) = p_{a_i}(t + 1)$. It was shown [1] that for many reasonable fitness functions w_a, this internal equilibrium solution is stable in the sense that if we apply a small deviation to this equilibrium, the system asymptotically returns to the equilibrium state.

In this case, the phenotype frequencies p_a are the macrovariables y_a, and each microvariable p_i is simply equal to the corresponding macrovariable $p_i = y_a$ (i.e., $F_i(y_a) = y_a$).

For the macrovariables p_a, the dynamic equations take the form

$$p_d(t + 1) = \sum_a \sum_b w_a \cdot w_b \cdot p_a(t) \cdot p_b(t) \cdot R_{ab \to d},$$

where

$$R_{ab \to d} = \sum_L P(L) \cdot R_{ab \to d}(L)$$

and

$$P(L) = \frac{\binom{i}{L}\binom{g-i}{j-L}}{\binom{g}{j}}$$

is the probability that in the equal-frequency state, the overlap is L.

Possibility of multi-parametric families of equilibrium states: a comment. It is worth mentioning that in some biologically important scenarios, we have multi-parametric families of equilibrium states. An example of such a situation is *linkage equilibrium* (see, e.g., [8–10, 20]), when to describe the equilibrium frequencies x_i of different genotypes i, it is sufficient to know the frequencies e_ℓ of alleles ℓ at different loci. For a genotype $i = \ell_1 \ldots, \ell_m$, the corresponding frequency is equal to the product of the frequencies of its alleles: $x_i = e_{\ell_1} \cdot \ldots \cdot e_{\ell_m}$.

If we have two alleles at each locus, then the sum of their frequencies is 1, so to describe the frequencies of these alleles, it is sufficient to describe one of these frequencies. In this case, for g loci with two alleles at each locus, there are $n = 2^g$ possible genotypes, so in general, we need 2^g different frequencies x_i to describe the state of this system. However, under the condition of linkage equilibrium, we only need g ($\ll 2^g$) frequencies $y_1, \ldots, y_g$ corresponding to g loci.

Such situations are not covered by our definitions and will require further analysis.

Conditional aggregation beyond equilibria. Our main motivation for the conditional aggregation was based on the assumption that within each each cluster, the state reaches an equilibrium. This assumption makes sense for quasi-equilibrium situations in which within-cluster interactions are much stronger than between-cluster interactions. However, this is not a necessary condition for aggregation. In situations where the between-cluster interaction is not weak, we can still have conditional aggregation – with microstates no longer in equilibrium within each cluster.

12.2.2 Conditional Aggregability: Linear Case

Definitions. The main idea behind conditional aggregation is that we only consider "modular" states, i.e., states in which an (quasi-)equilibrium is attained within each cluster. For linear systems, (quasi-)equilibrium means that for each cluster I_a, we have $x_i = y_a \cdot \beta_i$ for all $i \in I_a$. Here, β_i are the values which characterize a fixed quasi-equilibrium state, and y_a is a parameter describing the state of the a-th cluster.

Since in such modular states, the state of each cluster is uniquely characterized by the value y_a, this value y_a serves as a *macrovariable* characterizing this state. We thus arrive at the following definition.

Definition 9. *We say that a conditional aggregation*

$$(i_0, I_1, \ldots, I_k, C_1, \ldots, C_n)$$

is linear *if all the functions C_i are linear, i.e., if $C_i(y_a) = \beta_i \cdot y_a$ for all i.*

This definition can be reformulated in the following equivalent form. By a *linear conditional aggregation*, we mean a tuple $(i_0, I_1, \ldots, I_k, \beta)$, where $(i_0, I_1, \ldots, I_k)$ is a partition, and $\beta = (\beta_1, \ldots, \beta_n)$ is a tuple of real numbers for which $\beta_{i_0} \neq 0$. By a *microstate* corresponding to a macrostate

$y = (y_1, \ldots, y_k)$, we mean a state $x = (x_1, \ldots, x_n)$ in which for every index i, we have $x_i = y_a \cdot \beta_i$, where a is the cluster containing i ($i \in I_a$). A microstate is called *modular* if it corresponds to some macrostate y. We say that a dynamical system is *linearly conditionally k-aggregable* if it is consistent with some conditional linear aggregation $(i_0, I_1, \ldots, I_k, \beta)$.

We can similarly define linear conditional $\leq k$-aggregability and linear conditional aggregability.

Formulation of the problem. For every integer $k > 0$, we arrive at the following *linear conditional k-aggregability* problem:

- *given* a linear dynamical system;
- *check* whether the given system is linearly conditionally k-aggregable.

Given the existence of such an aggregation, it must be computed. Specifically, we must find the partition $I_1, \ldots, I_k$ and the weights β_i.

Discussion. The main motivation for discussing the notion of conditional aggregability is that the original notion of decomposable aggregability required decomposability for *all* possible states – and was, therefore, too restrictive. Instead, we require decomposability only for modular states, in which we have a quasi-equilibrium within each cluster. This part of the requirement of conditional aggregability is thus *weaker* than the corresponding condition of decomposable aggregability.

On the other hand, in decomposable aggregability, we are only concerned with the dynamics of macrostates, while in conditional aggregability, we also require that microstates also change accordingly (i.e., modular state are transformed into modular states). This part of the requirement of conditional aggregability is thus *stronger* than the corresponding condition of decomposable aggregability.

Since one part of the requirement is weaker and the other part of the requirement is stronger, it is reasonable to conjecture that the requirements themselves are of approximately equal strength. It turns out that, in fact, the two corresponding problems have the exact same computational complexity.

Main results. In this chapter, we prove the following two results:

Proposition 1. *For every $k \geq 2$, the linear conditional k-aggregability problem is NP-hard.*

Proposition 2. *There exists an efficient (polynomial-time) algorithm that, given a linear dynamical system (n, F) and a partition $(i_0, I_1, \ldots, I_k)$ under which the system is linearly conditionally aggregable, returns the corresponding weights β_i.*

The proof of both results is based on the following auxiliary statement. For every matrix F, let F^T denote a transposed matrix, with $F^T_{i,j} \stackrel{\text{def}}{=} F_{j,i}$.

Proposition 3. *A linear dynamical system (n, F) is linearly decomposably aggregable if and only if the system (n, F^T) is linearly conditionally aggregable (for the same partition).*

These results show that not only are two above statements true, but also that the problems of detecting linear decomposable aggregability and linear conditional aggregability have the exact same computational complexity. For example, if we can solve the problem of detecting linear decomposable aggregability, then we can apply this algorithm to the transposed matrix c^T and thus get an algorithm for detecting linear conditional aggregability. Vice versa, if we can solve the problem of detecting linear conditional aggregability, then we can apply this algorithm to the transposed matrix c^T and thus get an algorithm for detecting linear decomposable aggregability.

So, to prove Propositions 1 and 2, it is sufficient to prove the auxiliary Proposition 3.

Proof of Proposition 3. By definition, for a given partition $(i_0, I_1, \ldots, I_k)$, linear conditional aggregability means that for every macrostate $y = (y_1, \ldots, y_k)$, i.e., for all possible values $y_1, \ldots, y_k$, the equations of the dynamical system $x_i' = \sum_{j=1}^{n} F_{i,j} \cdot x_j$ transform the corresponding modular state $x_j = y_a \cdot \beta_j$ $(j \in I_a)$ into a modular state x_i'. In particular, for every cluster a $(1 \leq a \leq k)$, the corresponding modular state takes the form $x_j = \beta_j$ for $j \in I_a$ and $x_j = 0$ for all other j. For this modular state, the new state x_i' takes the form $x_i' = \sum_{j \in I_a} F_{i,j} \cdot \beta_j$.

This equation can be simplified if we use the notations that we introduced in our above analysis of linear dynamical systems. Specifically, we can represent each n-dimensional state vector x as a combination of vectors $x^{(a)}$ formed by the components x_i, $i \in I_a$. In these terms, the above equation takes the form $x'^{(b)} = F^{(a),(b)} \beta^{(a)}$ for all b. The new state x' must also be a modular state, so for every cluster b, the corresponding state $x'^{(b)}$ must be proportional to the fixed quasi-equilibrium state $\beta^{(b)}$ of this cluster: $x'^{(b)} = \lambda_{a,b} \beta^{(b)}$ for some constant $\lambda_{a,b}$. Thus, for every two clusters a and b, we must have

$$F^{(a),(b)} \beta^{(a)} = \lambda_{a,b} \beta^{(b)}. \tag{12.9}$$

Conversely, if this equation is satisfied, one can easily check that for every macrostate y, the corresponding modular state is also transformed into a new modular state.

Therefore, for a given partition $(i_0, I_1, \ldots, I_k)$, a linear dynamical system (n, F) is linearly conditionally aggregable if and only if there exist vectors $\beta^{(a)}$ for which the equations (12.9) hold for some values $\lambda_{a,b}$. A system (n, F) is linearly decomposably aggregable if and only if there exist vectors $\alpha^{(a)}$ for which the equations (12.9) hold for some values $\lambda_{a,b}$. The only difference between the equations (12.9) and (12.8) (apart from different names for $\alpha^{(a)}$ and $\beta^{(a)}$) is that in (12.9), we have the original matrix F, while in (12.8), we have the transposed matrix F^T. Thus, the linear system (n, F) is linearly conditionally aggregable if and only if the system (n, F^T) with a transposed matrix F^T is linearly decomposably aggregable. The proposition is proven.

Corollary. In the practically important case when the matrix F describing a linear dynamical system is symmetric $F = F^T$, the above Proposition 3 leads to the following interesting corollary:

Corollary 1. *A linear dynamical systems (n, F) with a symmetric matrix F is linearly conditionally aggregable if and only if it is linearly decomposably aggregable.*

Approximate aggregability: observation. One of the main cases of conditional aggregation is when we have clusters with strong interactions within a cluster and weak interactions between clusters. Due to the weakness of across-cluster interactions, it is reasonable to assume that the state of each cluster is *close* to the equilibrium. In the above text, we assumed that the clusters are *exactly* in the (quasi-)equilibrium states. In real life, such systems are only *approximately* conditionally aggregable.

Examples of approximately conditionally aggregable systems are given, e.g., in [24]. For an application to population genetics see [23].

Is detecting *approximate* linear conditional aggregability easier than detecting the (exact) linear conditional aggregability? In our auxiliary result, we have shown that the problem of detecting linear conditional aggregability is equivalent to a problem of detecting linear decomposable aggregability (for a related linear dynamical system). One can similarly show that approximate linear conditional aggregability is equivalent to approximate linear decomposable aggregability. In [6, 7], we have shown that detecting approximate linear decomposable aggregability is also NP-hard. Thus, detecting approximate linear conditional aggregability is NP-hard as well – i.e., the approximate character of aggregation does not make the corresponding computational problems simpler.

12.3 Identifying Aggregations in Lotka-Volterra Equations with Intraspecific Competition

12.3.1 Formulation of the Problem

Motivations. In the previous sections, we mentioned that in general, identifying aggregations is a computationally difficult (NP-hard) problem. This means that we cannot expect to have a feasible aggregations-identifying algorithm that is applicable to an *arbitrary* dynamical system. We can, however, hope to get such a feasible algorithm for specific classes of biology-related dynamical systems.

We start with possibly the most well-known dynamical system in biology: the Lotka-Volterra equations; see, e.g., [11, 12].

Lotka-Volterra equations. The standard Lotka-Volterra equations for competition between multiple species x_i exploiting the same resource in a community is

$$\frac{dx_i}{dt} = r_i \cdot x_i \cdot \left(1 - \frac{\sum\limits_{j} a_{ij} \cdot x_j}{K_i} \right) ; \tag{12.10}$$

where K_i is the *carrying capacity* of the i-th species, and a_{ij} is the effect of j-th species on the i-th species. In this equation:

- the terms a_{ij} corresponding to $i \neq j$ describe *interspecific* competition, i.e., competition between different species, while
- the term a_{ii} describes *intraspecific* competition, i.e., competitions between organisms of the same species.

In this chapter, we will only consider the case where there is an intraspecific competition, i.e., where $a_{ii} \neq 0$ for all i.

Known aggregation results about Lotka-Volterra equations. The known results about the aggregability of the Lotka-Volterra equations are described by Iwasa *et al.* in [4, 5]. Specifically, those papers analyze a simple case of aggregation when there are classes of competitors $I_1, \ldots, I_k$ such that:

- all the species i within the same class I_a have the same values of r_i and K_i;
- the interaction coefficients a_{ij} depend only on the classes I_a and I_b to which i and j belong, i.e., for all $i \in I_a$ and $j \in I_b$, the coefficient a_{ij} has the same value.

In this case, the actual aggregation of microvariables is simple and straightforward: we can have $y_a = \sum\limits_{i \in I_a} x_i$.

In [22, 23], it is shown that a similar "weighted" linear aggregation, with $y_a = \sum\limits_{i \in I_a} \alpha_i \cdot x_i$ and possible different weights α_i, is sometimes possible in situations when the values a_{ij} are not equal within classes – namely, it is possible when the values a_{ij} satisfy some symmetry properties. In this section, we will analyze the general problem of linear aggregability of such systems of equations.

Restriction to practically important cases. Before we present a precise mathematical formulation of our result, let us once again recall why this problem is practically useful. The main reason why aggregation is important is because aggregation simplifies the analysis of the complex large-size dynamical systems – by reducing them to simpler smaller-size ones, of size $k \ll n$. From this standpoint, the fewer classes we have, the simpler the reduced system, and the more important its practical impact.

The most interesting reduction is the one with the smallest possible number of classes. In other words, it is important to know whether we can subdivide the objects into 10 classes or less – but once we know that we can subdivide the objects into 7 classes, then the problem of checking whether we can also have a non-trivial subdivision into 9 classes sounds more academic.

In view of this observation, instead of checking whether a given system can be decomposed into exactly k classes, we study the possibility of checking whether it can be subdivided into $\leq k$ classes. Thus, we arrive at the following problem.

Exact formulation of the problem. For every integer $k > 0$, we arrive at the following *linear k-aggregability problem for Lotka-Volterra equations:*

- *given:* a Lotka-Volterra system, i.e., rational values n, r_i, K_i, $(1 \leq i \leq n)$, and a_{ij} $(1 \leq i \leq n,\ 1 \leq j \leq n)$;
- *check* whether the given system is linearly $\leq k$-aggregable.

When such an aggregation exists, the next task is to *compute* it, i.e., to find the partition $I_1, \ldots, I_k$ and the weights α_i which form the corresponding conditional aggregation.

12.3.2 Analysis of the Problem

Linearization seems to indicate that this problem is NP-hard. One can easily check that if a non-linear system $(n, f_1, \ldots, f_n)$ is k-aggregable, then for each state $x^{(0)} = (x_1^{(0)}, \ldots, x_n^{(0)})$ and for the deviations $\Delta x_i \stackrel{\text{def}}{=} x_i - x_i^{(0)}$, the corresponding linearized system

$$\Delta x_i' = f_i(x^{(0)}) + \sum_{j=1}^{n} \frac{\partial f_i}{\partial x_j} \cdot \Delta x_j \tag{12.11}$$

is also k-aggregable.

In particular, if the Lotka-Volterra equation is k-aggregable, then the corresponding linearized system

$$\Delta x_i' = \left(r_i - \sum_{j=1}^{n} r_i \cdot a_{ij} \cdot K_i^{-1} \cdot x_j^{(0)} \right) \cdot \Delta x_i - r_i \dot{x}_i^{(0)} \cdot K_i^{-1} \cdot \sum_{j=a}^{n} a_{ij} \cdot \Delta x_j \tag{12.12}$$

should also be k-aggregable. Since in the general Lotka-Volterra equations, we can have an arbitrary matrix a_{ij}, the corresponding linearized systems can have an arbitrary matrix $F_{i,j}$.

We already know that for general linear systems, the general problem of detecting linear k-aggregability for an arbitrary matrix $F_{i,j}$ is NP-hard. So, at first glance, it may seem like for Lotka-Volterra equations, the problem of detecting linear k-aggregability should also be NP-hard.

Why the above argument for NP-hardness is not a proof. In spite of the above argument, we will show that a feasible algorithm is possible for detecting k-aggregability of Lotka-Volterra equations. This means that the above argument in favor of NP-hardness cannot be transformed into a precise proof.

Indeed, the result about NP-hardness of the linear problem means that it is computationally difficult to check k-aggregability of a *single* linear system. On the other hand, k-aggregability of a non-linear system means, in general, that *several* different linear dynamic systems are k-aggregable – namely, the linearized systems (12.11) corresponding to all possible states $x^{(0)}$. So, even if for some state $x^{(0)}$, it is difficult to check k-aggregability, we may be able to avoid this computational difficulty if for other states $x^{(0)}$, the corresponding linear system is easily proven *not* to be k-aggregable.

12.3.3 Main Result

Result. The main result of this section is that for every $k > 0$, there exists a feasible (polynomial-time) algorithm for solving the above problem:

Proposition 4. *For every $k > 0$, there exists a polynomial-time algorithm for solving the linear k-aggregability problem for Lotka-Volterra equations.*

Corollary: how to compute the corresponding weights. As we will see from the proof, identifying the aggregating partition is feasible, albeit complicated.

However, as we'll see from the same proof, once we know the aggregating partition $I_1, \ldots, I_k$, we have a straightforward formula for determining the wights α_i of the corresponding macrovariables $y_a = \sum_{i \in I_a} \alpha_i \cdot x_i$: namely, we can take $\alpha_i = r_i \cdot a_{ii} \cdot K_i^{-1}$.

Discussion. It may be worth mentioning that the approach behind our algorithm will not work for a general recombination system (as described above). Specifically, in our algorithm, we essentially used the fact that in the Lotka-Volterra equations, all the quadratic terms in the expression for the new value x_i' are proportional to the previous value x_i of the same quantity. In contrast, in the recombination system, this is not necessarily the case, because a genotype z need not be a progeny of z and some other genotype.

12.3.4 Proof

Reduction to minimal aggregability. According to the precise formulation of our problem, we want to know, for a given $k > 0$, whether there exists a linear ℓ-aggregation for some $\ell \le k$. If such a linear aggregation exists, then among all such aggregations we can select a *minimal* one, i.e., a linear aggregation for which no linear aggregation with fewer classes is possible. Thus, to check whether a system is linearly ℓ-aggregable for some $\ell \leq k$, it is sufficient to check whether it is minimally linearly ℓ-aggregable for some $\ell \leq k$.

Once we have feasible algorithms for checking minimal linear ℓ-aggregability for different ℓ, we can then apply these algorithms for $\ell = 1, 2, \ldots, k$ and thus decide whether the original system is $\leq k$-aggregable. For every given k, we have a finite sequence of feasible (polynomial-time) algorithms. The computation time for each of these algorithms is bounded by a polynomial of the size of the input. Thus, the total computation time taken by this sequence is bounded by the sum of finitely many polynomials, which is itself a polynomial.

In view of this observation, in the following text, we will design, for a given integer $k > 0$, an algorithm for detecting minimal linear k-aggregability of a given Lotka-Volterra equation.

Simplification of the Lotka-Volterra equation. In order to describe the desired algorithm, let us first reformulate the Lotka-Volterra equations in a simplified form $x'_i = r_i \cdot x_i - \sum_j b_{ij} \cdot x_j$, where $b_{ij} \overset{\text{def}}{=} r_i \cdot a_{ij} \cdot K_i^{-1}$.

Linear aggregability: reminder. Linear k-aggregability means that for the macrovariables $y_a = \sum_{i \in I_a} \alpha_i \cdot x_i$, their changes $y'_a = \sum_{i \in I_a} \alpha_i \cdot x'_i$ are uniquely determined by the previous values $y_1, \ldots, y_k$. Substituting the expression for x'_i into the formula for y'_a, we conclude that

$$y'_a = \sum_{i \in I_a} \alpha_i \cdot r_i \cdot x_i - \sum_{i \in I_a} \sum_{j=1}^{n} \alpha_i \cdot b_{ij} \cdot x_i \cdot x_j. \qquad (12.13)$$

Dividing the sum over all j into sums corresponding to different classes a, we conclude that

$$y'_a = \sum_{i \in I_a} \alpha_i \cdot r_i \cdot x_i - \sum_{i \in I_a} \sum_{j \in I_a} \alpha_i \cdot b_{ij} \cdot x_i \cdot x_j - \sum_{b \neq a} \sum_{i \in I_a} \sum_{j \in I_b} \alpha_i \cdot b_{ij} \cdot x_i \cdot x_j. \qquad (12.14)$$

This expression must depend only on the values $y_1, \ldots, y_k$. Since the expression for y'_a in terms of microvariables x_i is quadratic, and $y_1, \ldots, y_k$ are linear functions of the microvariables, the dependence of y'_a on $y_1, \ldots, y_k$ must also be quadratic.

Since y'_a depends only on the variables x_i for $i \in I_a$, we can only have a linear term proportional to y_a. Similarly, since quadratic terms are proportional to x_i for $i \in I_a$, quadratic terms in the expression for y'_a must be proportional to y_a. So, we arrive at the following expression:

$$y'_a = R_a \cdot y_a + B_{aa} \cdot y_a^2 + \sum_{b \neq a} B_{ab} \cdot y_a \cdot y_b. \qquad (12.15)$$

Substituting the expressions $y_a = \sum_{i \in I_a} \alpha_i \cdot x_i$ into the right-hand side of the formula (12.15), we conclude that

$$y'_a = R_a \cdot \sum_{i \in I_a} \alpha_i \cdot x_i + B_{aa} \cdot \left(\sum_{i \in I_a} \alpha_i \cdot x_i \right)^2 +$$

$$\sum_{b \neq a} B_{ab} \cdot \left(\sum_{i \in I_a} \alpha_i \cdot x_i \right) \cdot \left(\sum_{j \in I_b} \alpha_j \cdot x_j \right). \qquad (12.16)$$

Aggregability means that the right-hand sides of the expressions (12.15) and (12.16) must coincide for all possible values of the microvariables x_i. Both expressions are quadratic functions of x_i. For the quadratic functions to coincide, they must have the exact same coefficients at x_i and the exact same coefficients at all the products $x_i \cdot x_j$. Let us see what we can conclude about the system from this condition.

Possibility of zero weights: analysis of the degenerate case. Let us first take into account that, in general, it is possible that the weight α_j of some variables is 0; our only restriction is that $\alpha_{i_0} \neq 0$ for a fixed microvariable i_0.

By the definition of linear aggregation, the fact that $\alpha_j = 0$ for some j means that none of the macrovariables $y_1, \ldots, y_k$ depend on the corresponding microvariable x_j and thus, the expression y'_a also cannot depend on x_j.

From the above expression for y'_a, we can thus conclude that for every i for which $\alpha_i \neq 0$, we must have $b_{ij} = 0$. Thus, if $\alpha_i \neq 0$ and $b_{ij} \neq 0$, then we must have $\alpha_j \neq 0$.

As we have just mentioned, we have $\alpha_{i_0} \neq 0$. So, if $b_{i_0 j} \neq 0$, we must have $\alpha_j \neq 0$; if for such j, we have $b_{jk} \neq 0$, then we must have $\alpha_k \neq 0$, etc. This fact can be described in graph terms if we form a directed graph with the microvariables $1, \ldots, n$ as vertices, and a connection $i \to j$ if and only if $b_{ij} \neq 0$. In terms of this graph, if there is a path (sequence of connections) leading from i_0 to j, then $\alpha_j \neq 0$.

It is known that in polynomial time, we can find out whether every vertex can be reached; see, e.g., [2]. For this, we first mark i_0 as reachable. At each stage, we take all marked vertices, take all edges starting with them, and mark their endpoints. Once there are no new vertices to mark, we are done: if all vertices are marked, this means that all vertices are reachable, otherwise this means that some vertices are not reachable.

At each stage except for the last one, we add at least one vertex to the marked list; thus, the number of steps cannot exceed the number n of vertices. Each step requires polynomial time; thus, overall, this graph algorithm takes polynomial time.

If all states are reachable from i_0, this means that in every aggregation, we must have $\alpha_i \neq 0$. If some states are not reachable, then for these states, we can set $\alpha_i = 0$ and keep the aggregation.

Reduction to non-degenerate case. In view of the above, to check for the existence of a linear aggregation, it is sufficient to first mark all reachable vertices and then to restrict ourselves only to reachable vertices.

For these vertices, $\alpha_i \neq 0$. So, in the following text, we will assume that all the vertices are reachable and all the weights α_i are non-zeros – i.e., that we have a "non-degenerate" situation.

For this non-degenerate situation, let us make conclusions from the equality of the coefficients at x_i and at $x_i \cdot x_j$ in the right-hand sides of the formulas (12.15) and (12.16).

Comparing coefficients at x_i. Comparing coefficients at x_i, we get $\alpha_i \cdot r_i = R_a \cdot \alpha_i$. Since $\alpha_i \neq 0$, we can divide both sides of this equality by α_i and conclude that $r_i = R_a$, i.e., that for all i from the same class $i \in I_a$, we have the same value r_i.

Comparing coefficients at x_i^2. Comparing coefficients at x_i^2, we get $\alpha_i \cdot b_{ii} = B_{aa} \cdot \alpha_i^2$. Since $\alpha_i \neq 0$, we conclude that $b_{ii} = B_{aa} \cdot \alpha_i$. Since we only consider the situations with intraspecific competition $b_{ii} \neq 0$, and we know that $\alpha_i \neq 0$, we thus conclude that $B_{aa} \neq 0$ for all a.

Let us use non-uniqueness in y_a to further simplify the formulas. The macrovariables y_a are not uniquely determined. In principle, instead of the original macrovariables y_a, we can consider new macrovariables $\widetilde{y}_a = k_a \cdot y_a$ for arbitrary constants $k_a \neq 0$. Let us use this non-uniqueness to further simplify our equations.

Specifically, we will consider the new macrovariables $\widetilde{y}_a = B_{aa} \cdot y_a$. From the original equation

$$y_a' = R_a \cdot y_a + B_{aa} \cdot y_a^2 + \sum_{b \neq a} B_{ab} \cdot y_a \cdot y_b,$$

we conclude that

$$\widetilde{y}_a' = B_{aa} \cdot y_a' = B_{aa} \cdot R_a \cdot y_a + B_{aa}^2 \cdot y_a^2 + \sum_{b \neq a} B_{aa} \cdot B_{ab} \cdot y_a \cdot y_b.$$

Representing the values y_a and y_b in the right-hand side in terms of the new macrovariables $\widetilde{y}_a$ and $\widetilde{y}_b$, as $y_a = \dfrac{\widetilde{y}_a}{B_{aa}}$ and $y_b = \dfrac{\widetilde{y}_b}{B_{bb}}$, we conclude that

$$\widetilde{y}_a' = R_a \cdot \widetilde{y}_a + \widetilde{y}_a^2 + \sum \widetilde{B}_{ab} \cdot \widetilde{y}_a \cdot \widetilde{y}_b,$$

where $\widetilde{B}_{ab} \stackrel{\text{def}}{=} \dfrac{B_{ab}}{B_{bb}}$. For these new macrovariables, $\widetilde{B}_{aa} = 1$.

Thus, without loss of generality, we can conclude that $B_{aa} = 1$ for all a. In this case, the above conclusion $b_{ii} = B_{aa} \cdot \alpha_i$ takes a simplified form $\alpha_i = b_{ii}$.

Comparing coefficients at $x_i \cdot x_j$ when i and j are in different classes. When $i \in I_a$ and $j \in I_b$ ($a \neq b$), comparing coefficients at $x_i \cdot x_j$ leads to $\alpha_i \cdot b_{ij} = B_{ab} \cdot \alpha_i \cdot \alpha_j$. Since $\alpha_i \neq 0$, this results in $b_{ij} = B_{ab} \cdot \alpha_j$. We already know that $\alpha_j = b_{jj}$, so we can conclude that for every i and j from different classes $a \neq b$, the ratio

$$r_{ij} \stackrel{\text{def}}{=} \dfrac{b_{ij}}{b_{jj}}$$

takes the same value B_{ab}, irrespective of the choice of $i \in I_a$ and $j \in I_b$.

Comparing coefficients at $x_i \cdot x_j$ when i and j are in the same class.
When $i, j \in I_a$, comparing coefficients at $x_i \cdot x_j$ (and at the same term $x_j \cdot x_i$)
and using the fact that $B_{aa} = 1$ leads to the equation

$$\alpha_i \cdot b_{ij} + \alpha_j \cdot b_{ji} = 2\alpha_i \cdot \alpha_j.$$

Dividing both sides of this equality by $\alpha_i = b_{ii}$ and $\alpha_j = b_{jj}$, we conclude
that
$$\frac{b_{ij}}{b_{jj}} + \frac{b_{ji}}{b_{ii}} = 2.$$

Using the notation r_{ij} that we introduced in the previous section, we conclude
that $r_{ij} + r_{ji} = 2$.

Summarizing the analysis. Combining the analysis of all linear and
quadratic terms, we conclude that for the aggregating partition into classes
$I_1, \ldots, I_k$, the following must be true:

- for all i within each class I_a, the values r_i are the same: $r_i = R_a$ (for some
  value R_a);
- for all $i, j \in I_a$, we have $r_{ij} + r_{ji} = 2$;
- for every $a \neq b$, for all $i \in I_a$ and $j \in I_b$, the ratios r_{ij} are the same:
  $r_{ij} = B_{ab}$ (for some value B_{ab}).

Vice versa, if we have a partition for which these properties are satisfied, then,
as one can easily see, we have an aggregation.

Taking minimality into account. As we have mentioned in the beginning
of this proof, we are looking for a *minimal* aggregation. This means, in particular, that if we simply combine two classes $a \neq b$ into a single one, we will
no longer get an aggregation. This means, in turn, that one of the three above
conditions is not satisfied for the new class, i.e., that (at least) one of the
following three things is happening:

- either $R_a \neq R_b$;
- or $B_{ab} + B_{ba} \neq 2$;
- or for some $d \neq a$, $d \neq b$, we have $B_{ad} \neq B_{bd}$ or $B_{da} \neq B_{db}$.

Towards an algorithm for distinguishing $i \notin I_a$ versus $i \notin I_b$. To exploit
this consequence of minimality, let us select a point s_a in each class I_a. Let us
show that once we know these points, we can use the above property to tell,
for every two classes $a \neq b$ and for each i, whether $i \notin I_a$ or $i \notin I_b$.

Indeed, at least one of the above three properties holds for $a \neq b$. If
this property is $R_a \neq R_b$, then we cannot have both $r_i = R_a = r_{s_a}$ and
$r_i = R_b = r_{s_b}$. So:

- if $r_i \neq r_{s_a}$, we have $i \notin I_a$;
- if $r_i \neq r_{s_b}$, we have $i \notin I_b$.

If this property is $B_{ab} + B_{ba} \neq 2$, this means that:

- for $i \in I_a$, we have $r_{is_a} + r_{s_a i} = 2$ but $r_{is_b} + r_{s_b i} = B_{ab} + B_{ba} \neq 2$;
- for $i \in I_b$, we have $r_{is_b} + r_{s_b i} = 2$ but $r_{is_a} + r_{s_a i} = B_{ab} + B_{ba} \neq 2$.

Thus:

- if $r_{is_a} + r_{s_a i} \neq 2$, we have $i \notin I_a$;
- if $r_{is_b} + r_{s_b i} \neq 2$, we have $i \notin I_b$.

If this property is $B_{ad} \neq B_{bd}$, this means that for $i \in I_a$, we have $r_{is_d} = B_{ad} \neq B_{bd}$, while for $i \in I_b$, we have $r_{is_d} = B_{bd} \neq B_{ad}$. Thus:

- if $r_{is_d} \neq r_{s_a s_d} = B_{ad}$, we have $i \notin I_a$;
- if $r_{is_d} \neq r_{s_b s_d} = B_{bd}$, we have $i \notin I_b$.

As a result, we arrive at the following auxiliary algorithm.

Auxiliary algorithm. In this algorithm, we assume that we have selected a representative s_a from each class I_a. This algorithm enables us, given $a \neq b$ and i, to check whether $i \notin I_a$ or $i \notin I_b$. This algorithm works as follows.

On the first stage of this algorithm, we compare r_i with r_{s_a} and r_{s_b}:

- if $r_i \neq r_{s_a}$, we conclude that $i \notin I_a$ (and stop);
- if $r_i \neq r_{s_b}$, we conclude $i \notin I_b$ (and stop);
- otherwise (i.e., if $r_i = r_{s_a} = r_{s_b}$), we go to the next stage.

On the second stage, we do the following:

- if $r_{is_a} + r_{s_a i} \neq 2$, we conclude that $i \notin I_a$ (and stop);
- if $r_{is_b} + r_{s_b i} \neq 2$, we conclude that $i \notin I_b$ (and stop);
- otherwise (i.e., if $r_{is_a} + r_{s_a i} = r_{is_b} + r_{s_b i} = 2$), we go to the next stage.

On the third stage, for all $c \neq a, b$, we compute the values r_{is_c}, $r_{s_c i}$, $r_{s_a s_c}$, $r_{s_c s_a}$, $r_{s_b s_c}$, $r_{s_c s_b}$.

- If for some d, we get $r_{is_d} \neq r_{s_a s_d}$ or $r_{s_d i} \neq r_{s_d s_a}$, we conclude that $i \notin I_a$.
- If for some d, we get $r_{is_d} \neq r_{s_b s_d}$ or $r_{s_d i} \neq r_{s_d s_b}$, we conclude that $i \notin I_b$.

(Due to the above minimality property, this algorithm always decides whether $i \notin I_a$ or $i \notin I_b$.)

For every i, a, and b, this algorithm requires that we compute at most 6 values r_{xs_d} or $r_{s_d x}$ for each of k classes d, to the total of $\leq 6k$ computational steps.

Once we know representatives $s_1, \ldots, s_k$, we can determine the partition $(I_1, \ldots, I_k)$. Let us now show that once we know the representatives $s_1, \ldots, s_k$, we can assign each element i to the appropriate class I_a as follows.

In the beginning, we only know that i belongs to one of the classes I_a, where a belongs to the k-element set $S = \{1, \ldots, k\}$. We will show how we can sequentially decrease this set until we get one consisting of a single element.

If the set S of possible classes containing i contains at least two different classes $a \neq b$, then we can use the above algorithm to check whether $i \notin I_a$ or $i \notin I_b$. Whichever of the two conclusions we make, in both cases we delete one element from the set S. So, after $k-1$ steps, we get a set S consisting of a single class a. Thus, we have computed the class to which i belongs.

This computation takes $k-1$ applications of the above auxiliary algorithm. So, overall, it takes $(k-1) \cdot 6k = O(k^2)$ steps. For a given k, this is simply a constant.

Once we know a partition, we can check whether it leads to the aggregation. In accordance with the above characterization of the aggregating partition, once we know a partition $I_1, \dots, I_k$, in order to determine whether it leads to an aggregation, we need to check the following conditions:

- for all i within each class I_a, the values r_i are the same: $r_i = R_a$ (for some value R_a);
- for all $i, j \in I_a$, we have $r_{ij} + r_{ji} = 2$;
- for every $a \neq b$, for all $i \in I_a$ and $j \in I_b$, the ratios r_{ij} are the same: $r_{ij} = B_{ab}$ (for some value B_{ab}).

This requires checking all pairs (i, j), $1 \leq i, j \leq n$, which takes $O(n^2)$ computational steps.

Final algorithm. For a given k, to check k-aggregability of a given Lotka-Volterra system, we try all possible combinations of points $s_1, \dots, s_k$ ($1 \leq s_a \leq n$). For each of these combinations, we find the corresponding partition and check if it leads to an aggregation.

If one of these partitions leads to an aggregation, the system is aggregable. In the process, we have computed the partition, and we know the weights $\alpha_i = b_{ii}$.

If none of the partitions leads to an aggregation, this means that the original Lotka-Volterra system is not linearly k-aggregable.

Computation time. For each class a, there are n values choices of s_a. We need to make this choice for k different classes, so we test n^k possible tuples $(s_1, \dots, s_k)$. For each tuple, we take $O(n^2)$ time, so the overall computation time is $n^k \cdot O(n^2) = O(n^{k+2})$.

For a fixed k, this is polynomial time. The proposition is proven.

Comment. It is important to emphasize that while for every given k, the algorithm is polynomial, but its computation time grows exponentially with k. It is not clear whether it is possible to have an algorithm whose computation time grows polynomially with k as well.

Conclusions and Open Problems

Aggregability is an important property of biological systems, a property that simplifies their analysis. In view of this importance, it is desirable to be able to detect aggregability of a given system.

In our previous papers [6, 7], we analyzed the problem of detecting and identifying aggregability for linear systems. We showed that this problem is, in general, computationally difficult (NP-hard). We also showed that, once an aggregating partition of microvariables $x_1, \ldots, x_n$ into classes $I_1, \ldots, I_k$ is identified, we can efficiently compute the weights α_i describing the corresponding macrovariables $y_a = \sum_{i \in I_a} \alpha_i \cdot x_i$.

In this chapter, we extend our analysis in two different directions. First, we consider *conditional aggregability*, i.e., aggregability of modular states. For linear systems, we get results similar to general (unconditional) aggregability: the problem of identifying conditional aggregability is, in general, NP-hard, but once a partition is identified, we can efficiently compute the corresponding weights.

Second, we consider a biologically important case of non-linear systems: Lotka-Volterra systems with interspecific competition. For such systems, we have designed an efficient (polynomial-time) algorithm for identifying aggregability and computing the corresponding weights. There is a great deal of data about interspecific competition in biological populations, so the algorithm developed here can be applied to identify clusters in such systems.

For conditional aggregability, it would be of interest to extend our results to situations like linkage equilibrium, when we have a non-linear relation dependence of microvariables on the macrovariables. For non-linear systems, it is also desirable to extend our non-linear results to Lotka-Volterra systems without intraspecific competition, and to other biologically relevant classes of non-linear systems (such as predator-prey or parasite-host systems). Finally, we would like to be able to generalize our results to aggregations in which blocks I_a are allowed to overlap but remain smaller than the set of all the microvariables.

Acknowledgments.

This work was supported in part by NSF grants HRD-0734825, EAR-0225670, and EIA-0080940, and by Texas Department of Transportation grant No. 0-5453. The authors are thankful to the anonymous referees for valuable suggestions.

References

1. Barton NH, Shpak M (2000) The stability of symmetric solutions to polygenic models. Theoretical Population Biology 57:249–263
2. Cormen T, Leiserson CE, Rivest RL, Stein C (2001), Introduction to Algorithms, MIT Press, Cambridge, MA
3. Courtois PJ (1977) Decomposability: queueing and computer system applications. Academic Press, New York
4. Iwasa Y, Andreasen V, Levin SA (1987) Aggregation in model ecosystems. I. Perfect aggregation. Ecological Modelling 37:287–302

5. Iwasa Y, Levin SA, Andreasen V (1989) Aggregation in model ecosystems. II. Approximate aggregation. IMA Journal of Mathematics Applied in Medicine and Biology 6:1–23

6. Kreinovich V, Shpak M (2006) Aggregability is NP-hard. ACM SIGACT, 37(3):97–104

7. Kreinovich V, Shpak M (2007) Decomposable aggregability in population genetics and evolutionary computations: algorithms and computational complexity. In: Kelemen, A (ed.), Computational Intelligence in Medical Informatics, Springer-Verlag (to appear).

8. Laubichler MD, Wagner GP (2000) Organism and character decomposition: Steps towards an integrative theory of biology. Philosophy of Science 67:289–300.

9. Lewontin RC (1974) The Genetic Basis of Evolutionary Change, Columbia University Press, New York

10. Lewontin RC, Kojima K-i (1960) The evolutionary dynamics of complex polymorphisms. Evolution 14(4):485–472

11. MacArthur R, Levins R (1967) The limiting similarity, convergence and divergence of coexisting species. American Naturalist 101:377-385.

12. May RM (1973) Stability and Complexity in Model Ecosystems, Princeton University Press, Princeton, NJ

13. Moey CCJ, Rowe JE (2004). Population aggregation based on fitness. Natural Computing 3(1):5–19

14. Rabani Y, Rabinovich Y, Sinclair A (1995) A computational view of population genetics. Proceedings of the 1995 Annual ACM Symposium on Theory of Computing, Las Vegas, Nevada, 83–92

15. Rabani Y, Rabinovich Y, Sinclair A (1998) A computational view of population genetics. Random Structures & Algorithms 12(4):313–334

16. Rabinovich Y, Sinclair A, Wigderson A (1992) Quadratic dynamical systems. Proc. 33rd Annual Symp. on Foundations of Computer Science, IEEE Computer Society Press, Los Alamitos, California, 304–313

17. Rowe J (1998) Population fixed-points for functions of unitation. In: Reeves C, Banzhaf W (Eds), Foundations of Genetic Algorithms, Morgan Kaufmann Publishers, Vol. 5

18. Rowe JE, Vose MD, Wright AH (2005). Coarse graining selection and mutation. In: Proceedings of the 8th International Workshop on Foundations of Genetic Algorithms FOGA'2005, Aizu-Wakamatsu City, Japan, January 5–9, 2005, Springer Lecture Notes in Computer Science, Vol. 3469, pp. 176–191

19. Rowe JE, Vose MD, Wright AH (2005) State aggregation and population dynamics in linear systems. Artificial Life 11(4):473–492

20. Shpak M, Gavrilets S (2006) Population genetics: multilocus. In: Encyclopedia of Life Sciences, Wiley, Chichester, 2005, http://www.els.net/

21. Shpak M, Kondrashov AS (1999) Applicability of the hypergeometric phenotypic model to haploid and diploid production. Evolution 53(2):600–604

22. Shpak M, Stadler PF, Wagner GP, Hermisson J (2004) Aggregation of variables and system decomposition: application to fitness landscape analysis. Theory in Biosciences 123:33–68

23. Shpak M, Stadler PF, Wagner GP, Altenberg L (2004). Simon-Ando decomposability and mutation-selection dynamics. Theory in Biosciences 123:139–180

24. Simon H, Ando F (1961) Aggregation of variables in dynamical systems. Econometrica 29:111–138

13

Conceptual Biology Research Supporting Platform: Current Design and Future Directions

Ying Xie[1], Jayasimha Katukuri[2], Vijay V. Raghavan[2], and Tony Presti[3]

[1] Department of Computer Science and Information Systems, Kennesaw State University, Kennesaw, GA 30144, USA
yxie2@kennesaw.edu
[2] Center for Advanced Computer Studies, University of Louisiana at Lafayette, Lafayette, LA 70503, USA
{jrk8907,Raghavan}@cacs.louisiana.edu
[3] Araicom Research, LLC.
tony.presti@araicom.com

Summary. Conceptual biology utilizes a vast amount of published biomedical data to enhance and speed up biomedical research. Current computational study on conceptual biology focuses on hypothesis generation from biomedical literature. Most of the algorithms for hypothesis generation are dedicated to produce one type of hypothesis called pairwise relation by interacting with certain search engines such as PubMed. In order to fully realize the potential of conceptual biology, we designed and implemented a conceptual biology research support platform that consists of a set of interrelated information extraction, mining, reasoning, and visualizing technologies to automatically generate several types of biomedical hypotheses and to facilitate researchers in validating generated hypotheses. In this chapter, we provide detailed descriptions of the platform architecture, the algorithms for generating novel hypotheses, and the technologies for visualizing generated hypotheses. Furthermore, we propose a set of computational procedures and measures for evaluating generated hypotheses. The experimental analysis of the proposed hypothesis generation algorithms is also presented.

13.1 Introduction

With the exponential growth of the accumulated biomedical facts in various databases, the rapid advance of data mining techniques, and the further development of biomedical ontology, conceptual biology is anticipated to take its

---

[3] The technologies described in this paper are patent-pending properties assigned to Araicom Research, LLC. Please contact Tony Presti at tony.presti@araicom.com for licensing and related information.

Y. Xie et al.: *Conceptual Biology Research Supporting Platform: Current Design and Future Directions*, Studies in Computational Intelligence (SCI) **122**, 307–324 (2008)
www.springerlink.com   © Springer-Verlag Berlin Heidelberg 2008

place as an essential component in biomedical research [12]. By following the hypothesis-driven, experimental research paradigm, conceptual biology utilizes vast amount of published data as sources to generate and test hypotheses at the conceptual level. Compared with labor-based biological research, conceptual biology is expected to be more efficient, cost-effective, and knowing no bounds across fields [12].

Current research on conceptual biology focuses on hypothesis generation from biomedical literature. Most of these algorithms are dedicated to produce one type of hypothesis called pairwise relation by interacting with certain search engines such as PubMed [16]. However, in order to fully implement its potential, we see the need for constructing a comprehensive conceptual biology research supporting platform, which supports generating and conceptually testing multiple types of biomedical hypotheses.

Guided by this vision, we developed a novel integrated architecture that encapsulates a suit of interrelated data structures and algorithms which support 1) automatically revealing multiple types of potential relations among biomedical entities embedded in enormous amount of literature data sets without any users input; 2) visualizing multiple types of hypotheses; 3) facilitating validation of generated hypotheses based on heterogeneous repositories; 4) tracking research history of a biological discovery; 5) providing Application Programming Interface (API) that allow users to develop their own types of hypothesis generation and testing approaches. In this chapter, we will discuss the architecture and some major components/algorithms of the conceptual research support platform. Future research and design issues will also be discussed.

13.2 Reviewing Works in Conceptual Biology

The research activities in the domain of conceptual biology can be traced back to Swanson's work in 1986 that discovered the novel connection between Raynaud disease and fish oils by examining two disjointed biomedical literature sets [2]. The hypothesis of the beneficial effect of fish oils on Raynaud disease was proved by an independent clinical trial two years later, which demonstrated the value of literature as a potential source of new knowledge. Swansons hypothesizing model can be simply described as "A relates to B, B relates to C, therefore A may relate to C", so called Swansons ABC model [3, 18]. In his follow-up work in 1990, Swanson suggested a trial-and-error search strategy, by which the ABC model guides a manual online search for identifying logically related non-interactive literature [3]. By applying this strategy for citation analysis, Swanson discovered some other novel biomedical hypotheses, such as the implicit connection between the blood levels of Somatomedin C and dietary amino acids arginine [3, 4], and hidden linkage between the mineral magnesium and a medical problem migraine [3].

Along with the advance in the text retrieval and mining techniques, researchers have made all efforts to partially automate Swansons ABC model for hypothesizing. Stegmann and Grohman proposed a way to guide a researcher to identify a set of promising B terms by conducting clustering analyses of terms on both the retrieval result set of topic A and the retrieval result set of topic C [10]. Their work used measures called centrality and density to evaluate the goodness of term clusters and showed that the promising B terms that link disjoined literature for A and C tend to appear in clusters of low centrality and density. Srinivasans approach to identify promising B terms starts with building two profiles for both topic A and topic C from the retrieval result sets of A and C respectively [14]. In her work, the profile of a topic consists of terms that have high frequency in the retrieval result set of that topic and belong to semantic types interesting to the user. Then the intersection of As profile and Cs profile generates candidate B terms. Identifying B terms from given topics A and C is called close discovery. In her work, Srinivansan also applies the topic profile idea to conduct open discovery, which identifies both B terms and C terms given only topic A. Srinivansans open discovery algorithm can be simply described as follows: Top-ranking B terms are selected from the profile of topic A. Then, a profile for each selected B term is created from the retrieval result set of that B term. The top-ranking terms in a B terms profile form candidate C terms. If topic As retrieval result set is disjointed from a candidate C terms retrieval result set, then this candidate C term is reported as having potential relation with topic A via term B. Slightly different from Srinivansans topic profile approach, Pratt and Yildiz directly applied association mining on the retrieval result set of topic A to conduct open discovery [20]. In their work, the logic reference based on two association rules $A \rightarrow B$, $B \rightarrow C$ leads to the finding a candidate C term.

One of the problems that almost all the hypothesizing approaches face is the large amount of spurious hypotheses generated in the process of automating Swansons ABC model. In order to eliminate those noisy ones as much as possible, different components of the biomedical ontology system UMLS (United Medical Language System) [19] have been utilized. Weeber and colleagues used Metathesaurus of the UMLS to extract biomedical phrases and further limited the desired phrases by using the semantic types of the UMLS as an additional filter [21]. Similar strategies are widely used by most of the follow-up works [9, 14, 20]. Hu and his colleagues took advantage of the semantic networks, another UMLS component that specifies possible relations among different semantic types, to restrict the association rules generated from the retrieval result set of topic A in the process of open discovery [9]. Besides utilizing biomedical ontology system, we envision that cross-repository validation may be another effective addition for eliminating noisy hypotheses.

Although the pioneering work on biomedical hypothesizing is traced back two decades ago, the phrase "conceptual biology" was first coined in the paper "Unearthing the gems" published in Nature in 2002 [12]. This article delineated conceptual biology as scientific investigation based on accumulated biological facts themselves. The key activities involved in this type of investigation include formulating hypothesis by connecting retrievable facts, and searching experiments from literature that are crucial for testing the hypothesis. This article envisioned conceptual biology as an essential component of biological research that is able to meet the challenges brought by exponentially increasing amounts of information and artificial separation of biology into different disciplines.

Despite of its great potential, conceptual biology is still in its early stage of development. No matter whether designed for close discovery or open discovery, existing works are still constrained in the category of automating and refining Swansons ABC hypothesizing model. Furthermore, all the approaches are based on retrieval result set of one or two initial topics provided by a user, instead of being able to scale up to the whole set of literature database for the purpose of discovering real novel and crossing fields biomedical hypotheses. Last but not least, there is a lack of a comprehensive platform that really supports various activities in conceptual biology. Our work, therefore, aims to extend the realm of existing endeavors in conceptual biology to further increase its potential.

13.3 The Architecture of the Conceptual Biology Research Supporting Platform

Conceptual biology utilizes online literature and various facts databases as sources to generate and test hypotheses at the conceptual level. By following a typical research paradigm, we divide the conceptual biology research process into the following phases: Idea generating phase, problem definition phase, procedure design phase, conceptual experimentation phase, data analysis phase, and results interpretation phase. The activities involved in each phase was described as follows in [22]:

- Idea Generating Phase
  - Extracting facts from literature and other databases.
  - Conducting inference on the facts to generate hypotheses.
- Problem Definition Phase
  - Precisely and clearly defining and formulating hypotheses.
- Procedure Design Phase
  - Identifying expected findings, intermediary results, and supporting experiments.
- Conceptual Experimentation Phase
  - Retrieving and extracting relevant data, findings, and experiments from literature.
- Data Analysis Phase
  - Conducting analysis and inference based on collected data, findings, and supporting experiments.
- Results Interpretation Phase

– Building rational models and theories that explain the results from
  data analysis phase
– Hypothesis evaluation and testing

We envision the ideal conceptual biology research supporting platform
being able to automate and/or facilitate each phase of conceptual biology
research activities, as shown in figure 13.1.

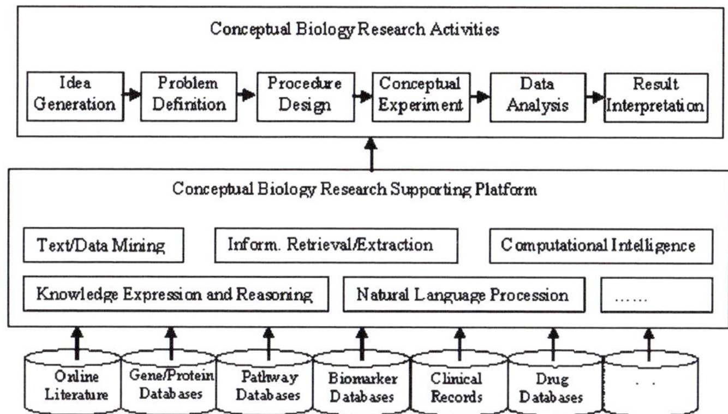

Fig. 13.1. Ideal Conceptual Biology Research Supporting Platform

As a step towards the ideal platform, the current version of our conceptual
biology research supporting platform focuses on automating the generation
of several types of biomedical hypothesis and facilitating the validation of
generated hypotheses. The architecture of the platform is shown in figure
13.2.

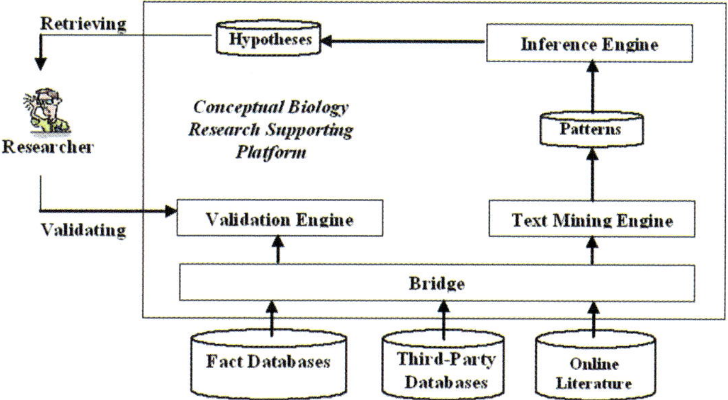

Fig. 13.2. Current Architecture of Conceptual Biology Research Supporting Platform

The working mechanism of the platform can be briefly described as follows. The text-mining engine first extracts biomedical entities and co-occurrence relations between entities from biomedical literature databases (currently we use the whole MedLine XML data set). Then, the extracted entity pairs are further categorized by referring to the semantic networks provided by the UMLS. Finally, the text-mining engine stores the categorized entity co-occurrence pairs into a database called "patterns". When the patterns database is populated, the inference engine starts to scan the whole database and generate the following types of biomedical hypotheses: pairwise, substitution, and chaining. The algorithms used by the inference engine to generate hypotheses will be described in section 4. The text-mining engine and the inference engine work in an off-line mode to populate the hypotheses database. Nevertheless, these two engines can periodically incorporate new contents added to the literature data set and modify the patterns database and hypotheses database accordingly.

After the hypotheses database is ready, the researcher can retrieve or browse hypotheses in real time. As you may notice, by our platform, the hypotheses are generated in a real mining mode, instead of the retrieval mode used by other cited works. By retrieval mode, the researcher has to initiate a term or phrase to begin with and interact with certain search engine such as PubMed. Hypotheses generated in retrieval mode are constrained by the researchers initial thinking. On the contrary, hypotheses generated by our platform can be really novel with high degree of surprisingness.

Furthermore, for each hypothesis selected by the researcher, the validation engine of the platform dynamically pulls relevant information from various fact databases for the researchers reference in real time. The details of validation engine will be described in section 5.

Finally, the bridge of the platform is an integration of both ETL (Extraction, Transformation, and Loading) technology and SELEGO (Search Engine Lego) technology [8, 17]. As an ETL-like module, the bridge can extract and transform contents from any third-party OLE databases, such as clinical records and private drug database, to feed the text-mining engine. While as an SELEGO-like module, it allows the user to incorporate the search page of any biomedical facts database for validation purpose.

13.4 Biomedical Hypothesis Generation Algorithms

Patterns can be viewed as existing relationships among biomedical concepts extracted from biomedical literature by the text mining engine of the platform. After the patterns database is populated, the inference engine starts to scan all the extracted patterns and generates different types of hypotheses, which are potential but not existing relationship among biomedical concepts. One of the types of hypothesis generated by the platform is called pairwise, which is based on Swansons ABC model. Nevertheless, our platform generates this type of hypotheses in a batch/mining mode rather than retrieval mode adopted by other works. In other words, our platform is able to generate all pairwise hypotheses embedded in the underlying literature database, instead of generating one or several hypotheses at one time based on the users query. The advantage of using batch/mining mode is that it may generate hypotheses that are really novel and crossing fields, rather than being bounded by the users initial thinking. The abstract algorithm for generating pairwise hypotheses in a batch mode can be described in Algorithm 2.

Procedure 1 Creating concept co-occurence matrix and binary association matrix

---

1: Create a symmetric matrix L such that $l_{C_iC_j}$ represents the co-occurence frequency of the concept C_i represented by the *ith* row of L and the concept C_j represented by the *jth* column of L. {for any matrix P, we have $rowSize_P$ and $colSize_P$ represent the number of rows and the number of columns in the matrix P respectively.}

2: Create a matrix M with size $rowSize_L \times colSize_L$.

3: **for** $(i = 0; i < rowSize_L; i + +)$ **do**

4:    **for** $(j = 0; j < colSize_L; j + +)$ **do**

5:       **if** $l_{C_iC_j} < minSupport$ or $(l_{C_iC_j} / \sum_j l_{C_iC_j}) < minConfidence$ **then**

6:          $m_{C_iC_j} = 0$

7:       **else**

8:          $m_{C_iC_j} = 1$

9:       **end if**

10:    **end for**

11: **end for**

Algorithm 2 Pairwise Hypothesis Generation

Input: the concept co-occurence matrix L and the binary association matrix M generated in Procedure 1

Output: the set of candidate pairwise hypotheses.

1: Create a matrix N such that $N = M^T M$

2: **for** $(i = 0; i < rowSize_N; i + +)$ **do**

3:    **for** $(j = 0; j < colSize_N; j + +)$ **do**

4:       **if** $n_{C_iC_j} > minSupport2rd$ and $l_{C_iC_j} < minSupport$ **then**

5:          Output $C_i \rightarrow C_j$ as a candidate pairwise hypothesis.

6:       **end if**

7:    **end for**

8: **end for**

As one may see, a non-zero value $m_{C_iC_j}$ of the matrix M actually represents an association rule $C_i \rightarrow C_j$ whose support is greater than or equal to *minSupport* and confidence is greater than or equal to *minConfidence*. We call the matrix M *binary association matrix*. Then the matrix N is the *second order association matrix*. Therefore, a candidate pairwise hypothesis is a pair of concept that has strong relation in the second order association matrix N, while has weak relation in the co-occurence matrix L. In order to be able to scale up to large volume of data and reduce the number of noisy hypotheses, the actual algorithm is further optimized on the following aspects: 1) only the necessary parts of the matrix L, M and N are actually calculated and stored on demand; 2) only those concepts that are located at the leaf levels at the concept hierarchy trees provided by the UMLS Metathesaurus are involved in calculation (i.e., concepts having broad meaning are excluded); 3) only concepts that belong to desired semantic types are considered at certain steps of calculations; 4) only concept pairs that belong to desired semantic relations are considered at certain steps of calculations. An example pairwise hypothesis generated by the platform is shown in figure 13.3. The above optimization steps are not included in the algorithm description in order to keep the major logic flow of the algorithm clear.

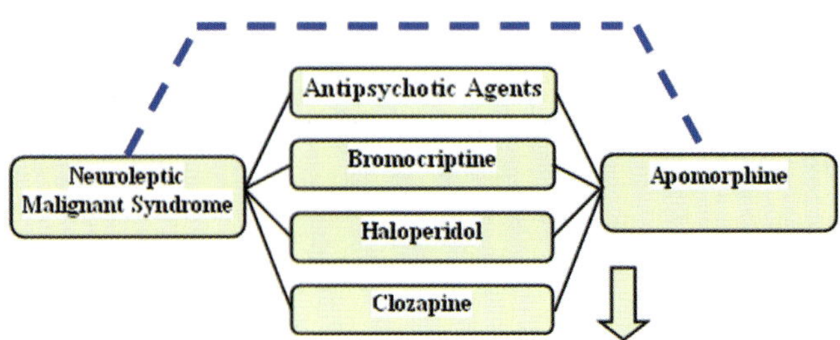

Fig. 13.3. Example Pairwise Hypothesis

Our platform also automatically generates another new type of hypotheses that is called "chaining". The basic model for chaining is that "If concept A relates to B, B relates to C, and A relates to C, then A, B, and C may be related altogether". This type of hypothesis may help identify chaining relation among chemical compounds, predict biologic pathways, and analyze combinational effects of drugs. One way to generate this type of chaining hypotheses can be described in Algorithm 3.

As one may see that a candidate chaining hypothesis consists of three concepts belonging to the same desired semantic type. From these three concepts, three pairwise association rules can be obtained to form a chain, but the co-occurence frequency of these three altogether is weak. The actual algorithm for chaining hypothesizing is further optimized on the following aspects (The optimization steps are not included in the algorithm description in order to keep the major logic flow of the algorithm clear): 1) only the necessary parts of the matrix L and M are actually calculated and stored on demand; 2) only those concepts that are located at the leaf levels at the concept hierarchy trees provided by the UMLS Metathesaurus are involved in calculation.

Algorithm 3 Chaining Hypothesis Generation

Input: the binary association matrix M generated in Procedure 1.
Output: the set of candidate chaining hypotheses.
1: **for** $(i = 0; i < rowSize_M; i + +)$ **do**
2:    **for** $(j = 0; j < colSize_M; j + +)$ **do**
3:      **if** $m_{C_i C_j} == 1$ and C_i, C_j belong to the same desired semantic type **then**
4:        **for** $(k = 0; k < colSize_M; k + +)$ **do**
5:          **if** $m_{C_j C_k} == 1$ and C_j, C_k belong to the same desired semantic type
          then
6:            **if** $m_{C_i C_k} > 0$ **then**
7:              Calculate the co-occurence frequency for C_i, C_j,and C_k altoghter
              and denote it as $freq_{C_i,C_j,C_k}$.
8:              **if** $freq_{C_i,C_j,C_k} < minSupportfor3$ **then**
9:                Output C_i, C_j, C_k as a candidate chaining hypothesis.
10:              **end if**
11:            **end if**
12:          **end if**
13:        **end for**
14:      **end if**
15:    **end for**
16: **end for**

An example chaining hypothesis generated by our platform is shown as below. In this example, Vincristine and Carmustine are chemotherapy drugs given as a treatment for some types of cancer. Articles also show that Carmustine and Vincristine are used in combination chemotherapy. Semustine is an investigational chemotherapy drug and articles show its use in combination with Vincristine. However there has not been evidence of all three drugs used in combination and thus our platform suggests this hypothesis. Therefore, the combination of these three is reported as a chaining hypothesis by the platform.

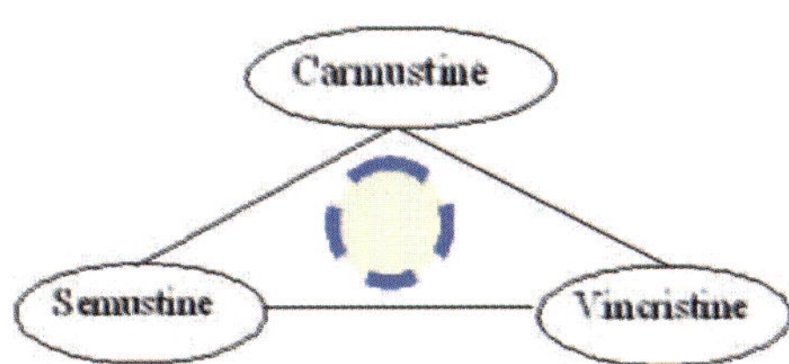

Fig. 13.4. Example Chaining Hypothesis

Another new type of hypotheses generated by our platform is called "substitution". The basic model for substituion can be described as "If concept A is similar with C, and A is strongly related to B, then C may be also related to B". One way to generate this type of chaining hypotheses can be described in Algorithm 4. In this algorithm, the similarity of two concepts is evaluated by a similarity measure between these two concepts' contexts. The context of a concept is represented by a vector of confidence values with which this concept is associated with other concepts.

Algorithm 4 Substitution Hypothesis Generation

Input: the concept co-occurence matrix L and the binary association matrix M generated in Procedure 1.

Output: the set of candidate substitution hypotheses.

1: **for** $(i = 0; i < rowSize_M; i + +)$ **do**
2:    **for** $(j = 0; j < colSize_M; j + +)$ **do**
3:       **if** $m_{C_i C_j} == 1$ and C_i, C_j belong to the same desired semantic type such as drugs **then**
4:          Call procedure 5 to calculate the similarity between concept C_i and C_j that is denoted as $sim(C_i, C_j)$.
5:          **if** $sim(C_i, C_j) > minSim$ **then**
6:             Create a set of concepts called $C_i Assoc$, such that for the kth concept in $C_i Assoc$, denoted as C_k, we have $m_{C_i C_k} = 1$.
7:             Create a set of concepts called $C_j Assoc$, such that for the kth concept in $C_j Assoc$, denoted as C_k, we have $m_{C_j C_k} = 1$.
8:             let a set of concepts $C_i Assoc\_minus\_C_j Assoc = C_i Assoc - C_j Assoc$.
9:             **for** each concept C_k in $C_i Assoc\_minus\_C_j Assoc$ **do**
10:               **if** C_k belongs to desired semantic type and $l_{C_j C_k} < minSupport$ **then**
11:                  Output concept C_j as a candidate substitution of concept C_i on concept C_k.
12:               **end if**
13:             **end for**
14:          **end if**
15:       **end if**
16:    **end for**
17: **end for**

One substitution hypothesis generated by our platform is described as follows. Both concept "Vinblastine" and "Prednisone" have strong relations with quite a few common concepts including "Methotrexate", "Breast Neoplasms", "Hodgkin Disease", "Cancer Staging", "Cyclophosphamide", and many others. In addition, both these two concepts belong to category "Drugs". Therefore, the platform views these two concepts are similar. Then the platform is looking for concepts that are only associated with "Prednisone", but not "Vinblastine" and vice versa. For instance, "Asthma" is one of the concepts that are only associated with "Prednisone" but not with "Vinblastine". Furthermore, "Asthma" belongs to one of the desired semantic types, which is "Disease". Therefore, the platform hypothesizing that "Vinblatine" may be a substitution for "Prednisone" on "Asthma". This hypothesis is visualized in

---

Procedure 5 Calculate the similarity between two concepts A and B

Input: the concept co-occurence matrix L and the binary association matrix M
   generated in Procedure 1; concepts A and B
Output: $sim(A, B)$
   1: Create a set of concepts called $AAssoc$, such that for the ith concept in $AAssoc$,
      denoted as C_i, we have $m_{AC_i} = 1$.
   2: Create a set of concepts called $BAssoc$, such that for the ith concept in $BAssoc$,
      denoted as C_i, we have $m_{BC_i} = 1$.
   3: $ABAssoc = AAssoc \cap BAssoc$.
   4: Create a vector called $AContext$, such that the ith number in $AContext$ equals
      to $l_{AC_i} / \sum_j l_{C_i j}$, where C_i is the ith concept in $ABAssoc$.
   5: Create a vector called $BContext$, such that the ith number in $BContext$ equals
      to $l_{BC_i} / \sum_j l_{C_i j}$, where C_i is the ith concept in $ABAssoc$.
   6: Output $sim(A, B) =< AContext, BContext >$.

---

figure 13.5. On the right side of the middle column, it shows those concepts
that both "Prednisone" and "Vinblatine" have strong relations with; while
on the left side, it shows concepts that have strong relation only with "Prednisone" but not "Vinblatine". The blue dot line represents potential relation
with two concepts.

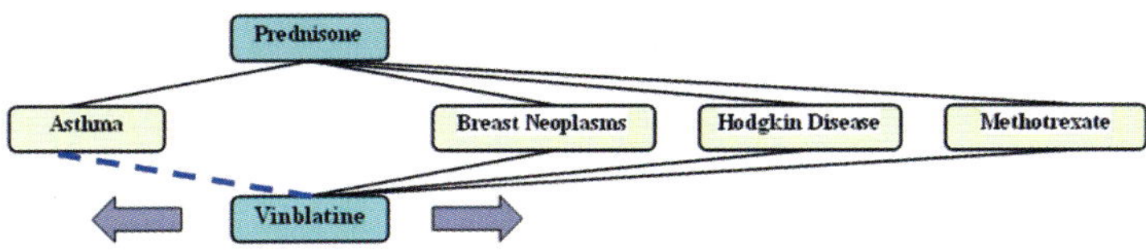

Fig. 13.5. Example Substitution Hypothesis

Finally, we would like to introduce another type of hypotheses that our
platform generates specifically for pathway predication. Given a set of concepts $P = P_1, P_2, , P_n$ which contains known elements involved in a target
pathway, the platform looks for concept A that has strong relation with at
least certain percentage of P concepts. If A belongs to desired semantic type,
then the platform hypothesizes that A may be involve in the target pathway.

13.5 Computational Evaluation of the Hypothesis Generation Algorithms

Evaluating the effectiveness of a hypothesis generation algorithm is a big
challenge, due to the fact that hypotheses describe the non-existing relations
among concepts. Cohen and Hersch [1] noted that most work uses the following
two approaches for evaluation.

- Attempting to recreate Swansons discoveries. Gordon and Lindsay [11], Weeber et al. [21] and Srinivasan [14] followed this approach.
- Manually reviewing the literature supporting the extracted hypothesis for scientific plausibility and relevance.

Both these approaches are useful to demonstrate the potential of the literature based discovery, but a more systematic and automatic approach is required to work as a standard experiemental method to examine different hypothesis generation algorithms. Bekhuis [18] mentioned the way of mining hypotheses from a document collection up to a cutoff date and verifying the hypotheses using a later document collection. However, to the best of our knowledge there has not been any detailed work in this regard. In this chapter, we propose a set of evaluation procedures and measures based on this methodology and use it to conduct experimental analysis for the proposed hypothesis generation algorithms.

In our experiments, the Medline XML collection from year 1965 to 1990 is used to generate three types of hypotheses, while the Medline collection from year 1991 to 2000 is used as test collection to validate the generated hypotheses. We calculate the *precision at each support level* to evaluate the performance of a hypothesis generation algorithm. The *precision* measure is defined as the percentage of generated hypotheses are "validated" in the test collection. A hypothesis is called "validated" if the described relation appears in minimum number of documents in the test collection. The minimum number of documents in the test collection that is required to validate a hypothesis is called the *support level*. The following subsections show the experiemental results for each hypothesis generation algorithm. For all experiements, we set *minSpport* and *minConfidence* described in Procedure 1 to be 30 and 0.02 respectively.

13.5.1 Evaluation of the Pairwise Hypothesis Generation Algorithm

Recall that the basic model for the substitution hypothesis is "if A relates to B, B relates to C, A may relate to C" We set two cases for this experiment. The first case, denoted as *case A*, requires that the semantic type of A to be diseases; while the second case, denoted as *case C*, requires that the semantic type of C to be deseases. The total numbers of hypotheses generated for *case A* and *case C* are 127 and 500 respectively. The precisions at each support level for these two cases are shown in Figure 13.6. As one can see, the precision value is higher for *case A* compared to *case C* for at all support levels.

Here is an example pairwise hypothesis that is validated in the test collection. Thiazolidinediones is now used as an antidiabetic agent [13, 23]. There are only three publications that contain both "Thiazolidinediones" and "Obesity" in Diabetes before 1990. None of them stated Thiazolidinediones as a treatment for Obesity in Diabetes. The pairwise hypotheses generation

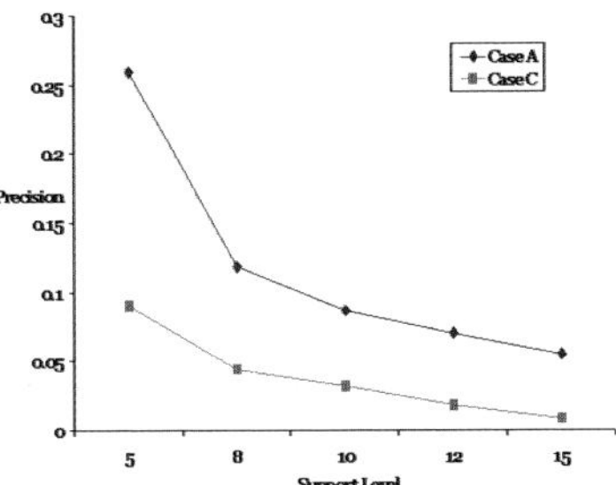

Fig. 13.6. Precision at each support level for the pairwise hypothesis generation algorithm

algorithm mines this hidden connection using the common term Hypoglycemic Agents.

13.5.2 Evaluation of the Chaining Hypothesis Generation Algorithm

Recall that the basic model for the substitution hypothesis is "If concept A relates to B, B relates to C, and A relates to C, then A, B, and C may be related altogether". In this experiement, we require that the semantic type of A, B, and C should be the same and one of the followings: drugs, chemical compounds, or genes. The parameter $minSupportfor3$ as described in Algorithm 3 is set to be 5. The precisions at each support level curve is shown in Figure 13.7.

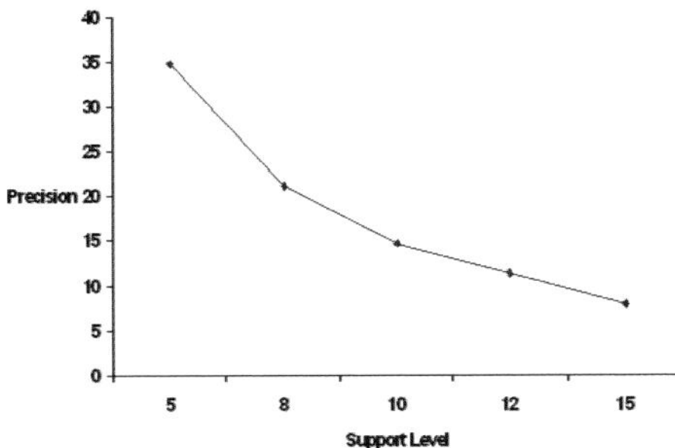

Fig. 13.7. Precision at each support level for the chaining hypothesis generation algorithm

13.5.3 Evaluation of the Substitution Hypothesis Generation algorithm

Recall that the basic model for the substitution hypothesis is "If concept A is similar with C, and A is strongly related to B, then C may be also related to B". In this experiment, we require that the semantic type for both A and C be drugs, and the semantic type for B be diseases. Figure 13.8 shows the precision at each support level curves with different $minSim$ values as described in Algorithm 4. A precision of 40% is obtained when $minSim = 0.2$ at the support level of 5.

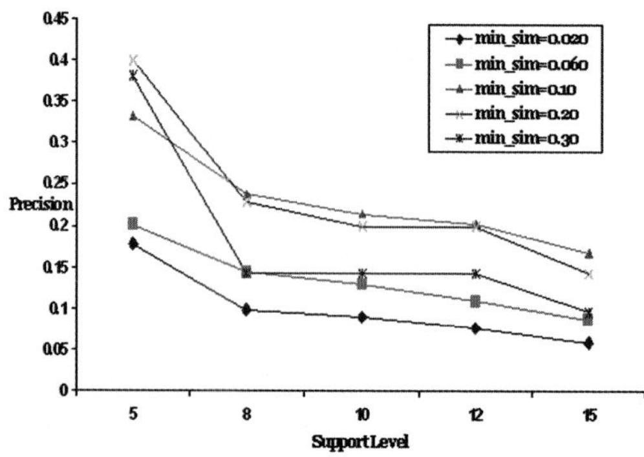

Fig. 13.8. Precision at each support level for substitution hypotheses

Here is an example substitution hypothesis that is validated in the test collection. Flutamide is now an accepted treatment for the Polycystic Ovary Syndrome [5]. Before 1990, there was only one publication in the MedLine that contains both "Flutamide" and "Polycystic Ovary Syndrome". This publication did not mention Flutamide as a treatment for Polycystic Ovary Syndrome. The substitution hypotheses algorithm generated this hypothesis using the document collection from 1965 to 1990. The hypotheses algorithm found that Flutamide is similar to drug Buserelin and Buserelin is strongly connected to Polycystic Ovary Syndrome and thus predicted that Flutamide could be used for Polycystic Ovary Syndrome.

13.5.4 Discussion

Strictly speaking, the precision measure is actually not a precise one to evaluate the performance of a hypothesis generation algorithm. A hypothesis, which describes a non-existing relation among concepts, may be proven to be meaningful in the future even though it is unable to be validated in the test collection. Nevertheless, the experiemental results shown above do give us confidence that at least a portion of the hypotheses generated by our platform have the value to provide research clues or suggest research directions. Another issue that needs to be mentioned is whether a hypothesis generated by our algorithms really describes a non-existing relation among concepts. Recall that the proposed algorithms require that the co-occurence frequence of the concepts invovled in a hypothesis be less than certain threshold. But this constrain does not guarantee the relation suggested by a hypothesis never being desribed in any existing article. However, even if this is the case, we can still say that a hypothesis suggests a relation that is supposed to have more acticles to support than it has already had.

13.6 Facilitating Hypothesis Validation

The validation engine of our platform provides multi-faced information from various related databases to facilitate researcher in validating interesting hypotheses. Figure 13.9 illustrates the way the platform facilitates the validation of the chaining hypothesis we mentioned as an example in previous section. When the researcher clicks the link between Carmustine and Semustine, or the icon above adjacent to the link, the platform pulls citations that discuss both these two concepts from PubMed [16]. The same for other two links. When each concept or the icon adjacent to it is clicked, the platform shows the information related this drug from drugbank [6]. In this example, since "Semustine" does not have entry in drugbank, so there is no icon attached to it.

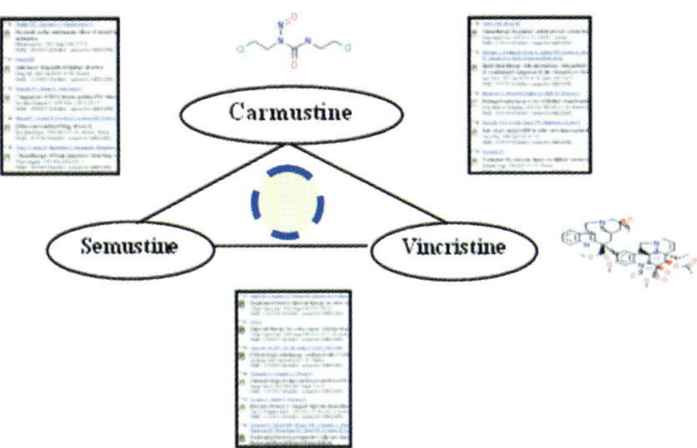

Fig. 13.9. Example Substitution Hypothesis

Depending upon the semantic type of a concept involved in a hypothesis, the platform may show multiple icons adjacent to it. For example, if a concept is a protein, then two icons will be put adjacent to it, one representing protein function and sequence pulled from ExPASy [7], and the second representing protein 3D structure pulled from pdb [15].

The bridge component of our platform allows users to register other online repositories to the platform. Each repository maps to one or more semantic types. When registered, a wrapper that allows the platform to retrieve and extract information from the corresponding repository is automatically created for this repository [8, 17]. When a concept of certain semantic type is presented in a hypothesis, the platform will automatically retrieve and extract information through corresponding wrappers from those repositories that map to the semantic type the concept belongs to.

13.7 Future Research and Development

Our future research on conceptual biology research support platform will focus on algorithms that generate new types of hypotheses supporting biomarker discovery, drug target discovery and other types of biomedical research. The generation of these types of hypotheses requires 1) designing a set of more comprehensive information extraction and mining technologies that are able to extract complex patterns crossing multiple heterogeneous data repositories such as literature database, microarray database, gene/protein sequence and structure databases, pathway databases, and so on; 2) more advanced inference mechanisms that are able to connect different type of patterns extracted from different repositories to form different types of hypotheses.

Another research project that we are planning is to enhance the platforms capability to facilitate hypothesis validation. More specifically, we would like the platform to retrieve relevant experimental results from literature to support generated hypotheses. In order to achieve this goal, we need to 1) design templates to describe types of facts that experiments support; 2) index and categorize various experimental descriptions extracted from literature based on the identified templates; 3) identify possible validation paths for each type of hypothesis; 4) design formal language to categorize and describe each identified validation paths; 5) design advanced retrieval model to match the facts that experiments support with sections of possible validation paths of a given hypothesis.

Furthermore, we would like to expose as much APIs of the platform as possible to the public as web/grid services, so that third parties can incrementally contribute to the platform by utilizing the exposed APIs of the platform to extract novel patterns, generate new types of hypotheses, design fresh visualizing schemes, enhance validation approach, and build special applications.

Finally and most importantly, we would like to involve biomedical subject experts to evaluate hypotheses generated by the platform to fine tune the hypothesis models and generation algorithms.

13.8 Conclusion

Conceptual biology research aims to take advantage of the vast amount of published data to enhance and speed up biomedical research. In order to fully realize the potential of conceptual biology, we designed and implemented a conceptual biology research support platform. This platform consists of a set of interrelated information extraction, mining, reasoning, and visualizing technologies to automatically generate several types of biomedical hypotheses and facilitate the researchers in validating the generated hypotheses. To the best of our knowledge, this platform is the first one that is able to generate multiple types of hypotheses in a real mining mode based on a whole set of biomedical literature database such as MedLine, instead of generating a single type of hypotheses from result lists returned by a search engine in response to a users certain query. Furthermore, the platform provides intuitive visualization for a users selected hypothesis, and retrieves and extracts information from various heterogeneous repositories to facilitate the validation of the hypothesis. Several future research and development projects are planed for further improving the platform.

References

1. Cohen AM, Hersh WR (2005) A survey of current work in biomedical text mining. Briefings in Bioinformatics. 6:57–71.
2. Don RS (1986) Raynaud's syndrome, and undiscovered public knowledge. Perspective in Biology and Medicine 30(1):7–18.
3. Don RS (1990) Medical literature as a potential source of new knowledge. Bulletin of the Medical Library Association 78(1):29–37
4. Don RS (1990) Somatomedin C and arginine: implicit connection between mutually-isolated literatures. Perspective in Biology and Medicine 33(2):157–186.
5. Gambineri A, Patton L, Vaccina A, Cacciari M, Morselli-Labate AM, Cavazza C, Pagotto U, Pasquali R (2006) Treatment with flutamide, metformin, and their combination added to a hypocaloric diet in overweight-obese women with polycystic ovary syndrome: a randomized, 12-month, placebo-controlled study. Journal of Clinical Endocrinology & Metabolism 91(10):3970–3980.
6. Drugbank. http://redpoll.pharmacy.ualberta.ca/drugbank, as of October, 2007.
7. ExPASy Proteomics Server (ExPASy). http://ca.ExPASy.org/, as of October, 2007.
8. Hongkun Z, Weiyi M, Zonghuan W, Vijay VR, Clement TY (2005) Fully automatic wrapper generation for search engines. Proceedings of the 14th International World Wide Web Conference (Chiba, Japan) 66–75.
9. Hu X, Zhang X, Li G, Yoo I, Zhou X, Wu D (to be published) Mining Hidden Connections among Biomedical Concepts from Disjoint Biomedical Literature Sets through Semantic-based Association Rule. International Journal of Intelligent Systems.

10. Johannes S, Guenter G (2003) Hypothesis generation guided by co-word clustering. Scientometrics 56(1):111–135.
11. Michael DG, Robert KL (1996) Toward Discovery Support Systems: A Replication, Re-Examination, and Extension of Swanson's Work on Literature-Based Discovery. Journal of the American Society for Information Science 47(2):116–128.
12. Mikhail VB, Arthur BP (2002) Unearthing the gems. Nature 416:373.
13. Ohtomo S, Izuhara Y, Takizawa S, Yamada N, Kakuta T, van Ypersele de Strihou C, Miyata T (2007) Thiazolidinediones provide better renoprotection than insulin in an obese, hypertensive type II diabetic rat model. Kidney international.
14. Padmini S (2004) Text mining: generating hypotheses from MedLine. Journal of American Society for Information Science and Technology 55(5):396–413.
15. The RCSB Protein Data Bank (PDB). http://www.rcsb.org/pdb/home/home.do, as of October, 2007
16. PubMed. http://www.ncbi.nlm.nih.gov/sites/entrez?db=PubMed, as of October, 2007.
17. Selego. http://www.selego.com, as of Octorber, 2007.
18. Tanja B(2006) Conceptual biology, hypothesis discovery, and text mining: Swanson's legacy. Biomedical Digital Library 3(2).
19. Unified Medical Language System (UMLS): http://umlsinfo.nlm.nih.gov/, as of October, 2007.
20. Wanda P, Meliha Y(2003) LitLinker: capturing connections across the biomedical literature. Proceedings of the 2nd international conference on Knowledge capture (New York, USA) 105–112.
21. Weeber M, Klein H, Aronson AR, Mork JG, JongVan Den Berg L, Vos R(2000) Text-based discovery in biomedicine: the architecture of the DAD-system. Proceedings of the AMIA Annual FALL Symposium(Philadelphia, USA) 903–907.
22. Yiyu Y, A Framework for Web-based Research Support Systems. Proceedings of 27th Annual International Computer Software and Applications Conference (Dallas, USA) 601–606.
23. 1999 The Diabetes Prevention Program. Design and methods for a clinical trial in the prevention of type 2 diabetes. Diabetes Care 22(4):623–634.

Computational Intelligence in Electrophysiology: Trends and Open Problems

Cengiz Günay[1], Tomasz G. Smolinski[1], William W. Lytton[2], Thomas M. Morse[3], Padraig Gleeson[4], Sharon Crook[5], Volker Steuber[4,6], Angus Silver[4], Horatiu Voicu[7], Peter Andrews[8], Hemant Bokil[8], Hiren Maniar[8], Catherine Loader[9], Samar Mehta[10], David Kleinfeld[11], David Thomson[12], Partha P. Mitra[8], Gloster Aaron[13], and Jean-Marc Fellous[14]

[1] Dept. of Biology, Emory University, Atlanta, Georgia 30322, USA
[2] Depts of Physiology/Pharmacology and Neurology, State University of New York - Downstate, Brooklyn, New York 11203, USA
[3] Dept. of Neurobiology, Yale University, New Haven, CT 06510, USA
[4] Dept. of Physiology, University College London, London, UK
[5] Dept. of Mathematics and Statistics, Arizona State University, Tempe, Arizona, USA
[6] School of Computer Science, University of Hertfordshire, Hatfield, Herts, UK
[7] Dept. of Neurobiology and Anatomy, University of Texas Health Science Center, Houston, TX 77030, USA
[8] Cold Spring Harbor Laboratory, Cold Spring Harbor, NY 11724, USA
[9] Dept. of Statistics, University of Auckland, Auckland 1142, New Zealand
[10] School of Medicine, State University of New York - Downstate, Brooklyn, New York 11203, USA
[11] Dept. of Physics, University of California, San Diego, La Jolla, CA 92093, USA
[12] Dept. of Mathematics and Statistics, Queen's University, Kingston, ON, Canada
[13] Dept. of Biology, Wesleyan University, Middletown, CT 06459, USA
[14] Dept. of Psychology, University of Arizona, Tucson, AZ 85721, USA

Summary. This chapter constitutes mini-proceedings of the Workshop on Physiology Databases and Analysis Software that was a part of the Annual Computational Neuroscience Meeting CNS*2007 that took place in July 2007 in Toronto, Canada (`http://www.cnsorg.org`). The main aim of the workshop was to bring together researchers interested in developing and using automated analysis tools and database systems for electrophysiological data. Selected discussed topics, including the review of some current and potential applications of Computational Intelligence (CI) in electrophysiology, database and electrophysiological data exchange platforms, languages, and formats, as well as exemplary analysis problems, are presented in this chapter. The authors hope that the chapter will be useful not only to those already involved in the field of electrophysiology, but also to CI researchers, whose interest will be sparked by its contents.

C. Günay et al.: *Computational Intelligence in Electrophysiology: Trends and Open Problems*, Studies in Computational Intelligence (SCI) **122**, 325–359 (2008)
www.springerlink.com   © Springer-Verlag Berlin Heidelberg 2008

14.1 Introduction

Recording and simulation in electrophysiology result in ever growing amounts of data, making it harder for conventional manual sorting and analysis methods to keep pace. The amount of electrophysiological data is increasing as more channels can be sampled and recording quality improves, while rapid advances in computing speed and capacity (e.g., in grid computing) have enabled researchers to generate massive amounts of simulation data in very short times. As a result, the need for automated analysis tools, with emphasis on Computational Intelligence-based techniques, and database systems has become widespread. The workshop on "Developing Databases and Analysis Software for Electrophysiology: Design, Application, and Visualization," organized by Cengiz Günay, Tomasz G. Smolinski, and William W. Lytton in conjunction with the 16th Annual Computational Neuroscience Meeting CNS*2007 (http://www.cnsorg.org), provided a venue for researchers interested in developing and using such tools to exchange knowledge and review currently available technologies and to discuss open problems.

This chapter constitutes mini-proceedings of the workshop and comprises of several selected contributions provided by the participants. In Section 14.2, Thomas M. Morse discusses the current uses and potential applications of CI for electrophysiological databases (EPDBs). Sections 14.3 by Padraig Gleeson et al., 14.4 by Horatiu Voicu, 14.5 by Cengiz Günay, and 14.6 by Peter Andrews et al., describe some currently available data-exchange and analysis platforms and implementations. Finally, Sections 14.7 by Gloster Aaron and 14.8 by Jean-Marc Fellous present some interesting open problems in electrophysiology with examples of analysis techniques, including CI-motivated approaches.

14.2 Computational Intelligence (CI) in electrophysiology: A review[1,2]

14.2.1 Introduction

There are 176 neuroscience databases listed in the Neuroscience Database Gateway [7]. Only one of these, Neurodatabase [32], currently has electrical recordings available, indicating that electrophysiology datasets are currently rarely publicly available. We review potential applications of electrophysiology databases (EPDBs) in hopes to motivate neuroscience, computer intelligence, computer science, and other investigators to collaborate to create and contribute datasets to EPDBs. We hope that some of these applications will

---

[1] Contribution by T.M. Morse.

[2] This work was supported in part by the NIH Grant 5P01DC004732-07 and 5R01NS011613-31.

inspire computational intelligence (CI) investigators to work on real (from future EPDBs) and simulated datasets that may then be immediately applicable to electrophysiology investigations and clinical applications. Currently, there are many tools available online [55] for the processing of fMRI data, ranging from simple statistical calculations to CI methods. There are many published tools available for the electrophysiologist (see below) but as yet there are no tool databases specifically for electrophysiologists (although the laboratory developing Neurodatabase also has plans to develop a tool database [33]), so we advocate tool database construction.

We describe the background for, and present open questions of EPDB CI tools, first in neuronal networks, then single cells, and finally ion channel/receptors recordings and analysis.

14.2.2 Neuronal Network recordings: Spike-Sorting

Spike sorting (the process of identifying and classifying spikes recorded in one or more electrodes as having been produced by particular neurons, by only using the recorded data) tools have been published since 1996 [27] and CI methods began to appear in 2002 [49]. Limited comparisons between tools have been made, for example recordings from an in vitro 2D network were used to compare wavelet packets decomposition with a popular principle components analysis method and an ordinary wavelet transform [49]. Spike sort method comparison studies have been limited by available data and by the subsequent availability of the data for further study.

The field needs a comprehensive review of spike sorting methods. Such a review would only be possible if many sets of data (from different laboratories) were available to use as the input for these methods. Different types of electrodes and preparations from different cell types and brain regions, species, etc. produce signals with different characteristics (neuronal population density, levels and types of activity, and levels and shape of noise). Having the traces available in publicly accessible EPDBs would then allow the methods to have their domains of applicability tested, as well as noting strengths and weaknesses of particular methods in particular domains.

The absence of publicly available extracellular recordings likely lead to the use of neural network model output to compare spike sorting routines, see for example [66]. ModelDB [47] is an additional source for network models, which if used as a resource for spike sorting tool comparison's could extend the testing with simulated epilepsy, trauma, and sleep activity [89, 101]. Comparison studies would be easier to create if spike-sorting tools were assembled into a tools database (as in for example those for MRI analysis [55]).

One important application of spike sorting methods are their anticipated role in neuroprosthetic devices. It has been shown that sorting spikes (which separates neurons that are being used for detection of neuronal patterns) increases the accuracy of reaching to a target prediction (a common task in neuroprosthetic research) between 3.6 and 6.4% [90].

It is a reasonable conjecture that spike sorting may also play a role in future understanding and/or treatment of Epilepsy [56]. One paper [87] references 10 off-line and only 3 online (real time) spike sorting technique papers and stresses the need for more research in online spike sorting methods. The authors of [87] had access to recordings from a chronic electrode implanted in the temporal lobe of an epileptic patient to test their methods.

The increasing size of recordings is another technical difficulty that electrophysiologists are facing. It is now possible to simultaneously record on the order of a hundred cells with multi-electrode arrays and on the order of a thousand cells with optical methods (up to 1300 cells in [53]). We suggest the online (real time) spike sorting algorithms are needed here to reduce the raw voltage or optical time series data to event times, thus vastly reducing the storage requirements for the data.

There is a pressing need for samples of all types of electrical recordings applicable to comparing spike-sorting methods to be deposited in a publicly accessible EPDB. If that data was available the following open questions in CI could be addressed. Does a single spike sorting algorithm outperform others in all recorded preparations?

If not which spike sorting methods work better in which preparations? Is there a substantial difference in thoroughness or accuracy between online (real time) and off-line (unrestricted processing time) spike sorting methods?

14.2.3 CI in single cell recordings

CI methods have been applied since the middle 1990's to single cell recordings to extract model parameters for models which describe the input output electrical function of the cell or channels or receptors within the cell (see for example the review in [102]).

Their appearance was a natural evolution in sophistication of early electrophysiology parameter extraction methods such as those in the famous Hodgkin Huxley (HH) 1952 paper [48]. In HH the authors use chemical species substitution and voltage clamp to isolate the voltage gated sodium and potassium channels. They created parameterized functions to describe the voltage and time dependence of the channels and extracted from their data the best fit for these functions.

Pharmacological isolation of additional channels, and morphological measurements guided by cable theory [81] enhanced the biological realism and types of channels and cells these HH-style models described. Passive parameter (membrane resistance, capacitance) extraction is commonly practiced, see for example [85, 94]. Pre-CI methods were provided by papers which used either brute force and conjugate descent methods [6] or transformations of gradient descent into a series of single parameter optimizations or a Chebyshev approximation [100].

Early CI methods in single cell model optimizations to electrophysiology traces are reviewed in [102]. This paper compared gradient descent, genetic algorithms, simulated annealing, and stochastic search methods. They found that simulated annealing was the best overall method for simple models with a small number of parameters, however genetic algorithms became equally effective for more complex models with larger numbers of parameters. A more recent paper incorporated simulated annealing search methods into an optimization of maximum conductances and calcium diffusion parameters with a hybrid fitness function and a revised (modified from the version in [102]) boundary condition handling method [104].

As the number of model parameters (the dimension of the parameter space) being determined increased, the complexity of the solution (the set of points in parameter space for which the error function is small enough or the fitness function large enough) also increased. Papers [39] have examined the parameter space that defines real cells or models by criteria of either a successful fit, or of the activity patterns of the cell [79], and have found that these spaces have non-intuitive properties due to the shapes of the solution spaces (see also Figure 2 of [6] and see also the call to investigate ways of reducing the parameter size and other issues in discussion in [102]). Tools to examine or to help describe these high dimensional model parameter spaces are important open CI research areas, because the (biological) cells parameters are traversing these spaces throughout the cells life. This is relevant to EPDBs because the model parameters extracted out of many collected recordings over different cell ages (embryonic, juvenile, adult) and environments would then be representative of the solution space of the cell, and hence the desire to view or to be able to describe that space.

Several of the previously cited papers use model current clamp data as targets for the search methods [104], single cell electrophysiology current clamp data [6], or both [102]. In addition, pre-CI paper methods which used model target data (for example [100]) could also be tested with electrophysiology data. The public availability of real training data would allow comparing the reliability and efficiency of model parameter extraction methods from electrophysiology data. The best extraction methods and associated training protocols would likely be determined per cell-type which would then determine which electrophysiology protocols would be performed and uploaded to the EPDBs in iterative improvements.

Voltage clamp caveat and another dataset

Voltage clamp protocols when performed in some cells exhibits what is called a space clamp error. The voltage clamp method attempts to maintain the cell at a voltage (a constant determined by the experimenter) and measures the (usually time-varying) injected current that is required to do so. The error arises when the voltage in parts of the cell that are distant from the electrode(s) have more influence from local currents than the electrode(s) due to either large local currents, small neuronal processes (reducing current flow from the electrode(s)), or being at a large distance from the electrode(s). In these cases membrane voltages drift from the command voltage and this is called "space-clamp error." A 2003 paper [91] was able to incorporate this phenomenon for measuring (the restricted case of) pharmacologically isolated hyperpolarizing channel densities, however their method is general enough in this domain to measure heterogeneous conductance density distributions (for those hyperpolarizing channels). We suggest that the currents recorded in [91, 92] would be helpful practice datasets if available in an EPDB for other electrophysiologists learning their method.

In addition to the above mentioned uses of EPDBs we present a speculative open question in CI; initial work could be done with models, however, it would only be through the availability of single cell recordings in EPDBs that the methods could be confirmed. Is it possible to characterize ion channels in (a well space clamped) experimental cell from a voltage clamp protocol that was the result of an optimized protocol developed and tested in models? Such a procedure would again only apply to well space-clamped cells. By "characterize" we mean to discover conductance densities, reversal potentials, and/or kinetics.

14.2.4 Single channels/receptors

Ion channel kinetic analysis preparations in experimental model cells (it is unfortunate that "model" has this ambiguity) such as HEK293, Xenopus egg, or isolated as single channels in excised membrane patches have had great success [88]. Voltage clamp errors are not a problem and the effects of superpositions of nearly similar or different multiple channel types can be eliminated. On the down side differences between native and experimental model cell intracellular and/or extracellular constituents, or the absence of these constituents in excised patches might make the native channel function differently than the isolated one. The results of the single channel studies are enormously useful because as it becomes known which channel genes are expressed (microarrays) at which densities in cells (staining methods or single channel population surveys) it will be possible to compare increasingly realistic models to nervous systems (see Figure 2 in [67]). What are optimal ways of deducing the kinetics of channels in single channel excised patch recordings?

Traditionally Markov modeling proceeds by choosing the number of states (nodes) and the transitions (connections) between them at the outset. The transition rates are then calculated or derived from experimental recordings (see [88] or more recently [20, 21, 41, 103]. Providing as many single channel recordings from each type of channel as possible would be invaluable for investigators and for developing automated Markov model construction tools. An open CI question: Is it feasible for optimization routines to search through different Markov model numbers of states and connections as a first step in a procedure that subsequently optimizes the transition rates, to find the simplest optimal model?

14.2.5 From today's EPDBs to tomorrow's

Investigators are currently using private EPDBs. We estimate, for example, there are about 110 papers to date in the Journal of Neurophysiology which report using their own EPDB locally to store neuronal recordings (145 papers total result from a free text search of "database neuron record" with about a 75% (actual EPDB) rate estimated from the first 37 papers). Investigators who are already using databases for their neuronal data might be able to use tools for making data publicly available like Nemesis [78] or Dan Gardner's group's method [30], or the commercial product Axiope [36] to upload datasets, after these tools evolve to interoperate with proposed EPDBs, or each other.

A common objection to EPDBs is that they would store large amounts of data that no one would use. We offer a different path by suggesting that EPDBs start out with just the samples of data that were suggested in this paper as immediately useful. Today's EP data is not too large.

If we estimated the average amount of electrophysiology data recorded for each paper at 10 GB then 100 papers worth is only one terabyte showing that the existing storage requirements are not as demanding as, for example, the fMRI data center faces. An open question is whether investigators will find this data useful for different reasons than for which it was created (e.g. useful for reasons not thought of yet); several groups are optimistic about the possibilities [16, 32].

14.2.6 Attributes of EPDBs

Every conceivable type of data in EPDBs is enumerated in the widely accepted Common Data Model [31]. In this framework we mention a couple of metadata preferences. The raw data could be stored along with (linked to) processing protocol instructions such as: we removed the artifacts from time steps t1 to t2 by zeroing, we filtered with a 4kHz filter, etc. Raw data is preferable because it permits the study of artifacts, noise, and the testing of filters. The explicit documentation of how the data was processed may be useful to other electrophysiologists. Standard processing methods could also be saved so that data could just point to the processing method rather than repeating it for each applicable dataset.

Measured channel densities stored in EPDBs would be highly valued by modelers. The published experimental conductance densities statistics (for example [38, 64, 91, 92] and references in [69, 70]) is infrequent and crucial for understanding synaptic integration and for making biologically realistic models. Channel densities provide modelers with the range of values that are realized in the biological system, further constraining the model.

14.2.7 Additional usefulness of EPDBs

The ability to quantitatively and/or graphically compare new models to experimental data that had been previously published and uploaded to EPDBs would be useful.

Comparing models to EP data have been done since at least HH, lending support to the model; see for example Figure 3 in [45] for a modified HH Na channel. Investigators must currently request traces from experimentalists, or retrieve experimental data with "data thief" or equivalent software [40, 51]. It would be nice to avoid that time and effort. See for example [18], where both experimental data and traces from calculations from experimental data were Data Thief'ed from 4 publications.

14.2.8 Conclusions

EPDBs could provide invaluable sources for comparisons between, and the development of new spike sorting tools and also single cell and ion channel/receptor electrophysiology methods and modeling. The authors hope that work in EPDBs will provide an exceptional example to this assessment from [5]: "Despite excitement about the Semantic Web, most of the world's data are locked in large data stores and are not published as an open Web of inter-referring resources."

We hope that EPDBs will become fertile members of the growing online population of neuroinformatics databases, fitting naturally among the connectivity, neuronal morphology, and modeling databases to support the study of the electrical functioning of the nervous system and excitable membranes.

14.3 Using NeuroML and neuroConstruct to build neuronal network models for multiple simulators[3,4]

14.3.1 Introduction

Computational models based on detailed neuroanatomical and electrophysiological data have been used for many years to help our understanding of the function of the nervous system. Unfortunately there has not been very widespread use of such models in the experimental neuroscience community. Even between computational neuroscientists, there are issues of compatibility of published models, which have been created using a variety of simulators and programming languages. Here we discuss new standards for specifying such models and a graphical application to facilitate the development of complex network models on multiple simulators.

---

[3] Contribution by P. Gleeson, S. Crook, V. Steuber, R.A. Silver.
[4] This work has been funded by the MRC and the Wellcome Trust.

14.3.2 NeuroML (`http://www.neuroml.org`)

The Neural Open Markup Language project, NeuroML [22, 35], is an international, collaborative initiative to create standards for the description and interchange of models of neuronal systems. The need for standards that allow for greater software interoperability is driving the current NeuroML standards project, which focuses on the key objects that need to be exchanged among existing applications and tries to anticipate those needed by future neuroscience applications.

The current standards are arranged in Levels (Figure 14.1), with each subsequent Level increasing the scope of the specification. Level 1 of the standards provides both a framework for describing the metadata associated with any neuronal model (e.g. authorship, generic properties, comments, citations, etc.) and allows specifications of neuroanatomical data, e.g. the branching structure of neurons, histological features, etc. Morphological data from various sources, e.g. Neurolucida reconstructions, can be converted into this format, termed MorphML [22, 80] for reuse in compartmental modeling simulators, etc.

Level 2 allows the specification of models of conductance based multicompartmental neurons. Inhomogeneous distributions of membrane conductances, subcellular mechanisms and passive cellular properties can be described for cells based on MorphML. Models of voltage and ligand gated ion channels and synaptic mechanisms can be described with ChannelML. Level 3 of the specification is aimed at network models. Populations of neurons in 3D can be defined by providing an explicit list of all neuronal locations, or by providing an implicit enumeration (e.g. a grid arrangement or a random arrangement). Similarly, connectivity can be specified using an explicit list of connections or implicitly by giving an algorithm for defining connectivity rules, cell location specificity of synaptic types, etc.

The advantage of using XML for the descriptions is that files can be checked for completeness against a published standard, i.e. any missing fields in a model description can be automatically detected. Another advantage is that XML files can easily be transformed into other formats. There are currently mappings to the GENESIS [11] and NEURON [46] simulators.

The latest version of the NeuroML specifications can be found online at `http://www.morphml.org:8080/NeuroMLValidator` along with example files at each of the Levels described. There is also the possibility of validating NeuroML files to ensure their compliance to the current version of the standards. The NeuroML project is working closely with a number of application developers to ensure wider acceptance of the standard.

14.3.3 neuroConstruct

One application which uses the NeuroML standards to facilitate model development is neuroConstruct [34]. This is a platform independent software tool

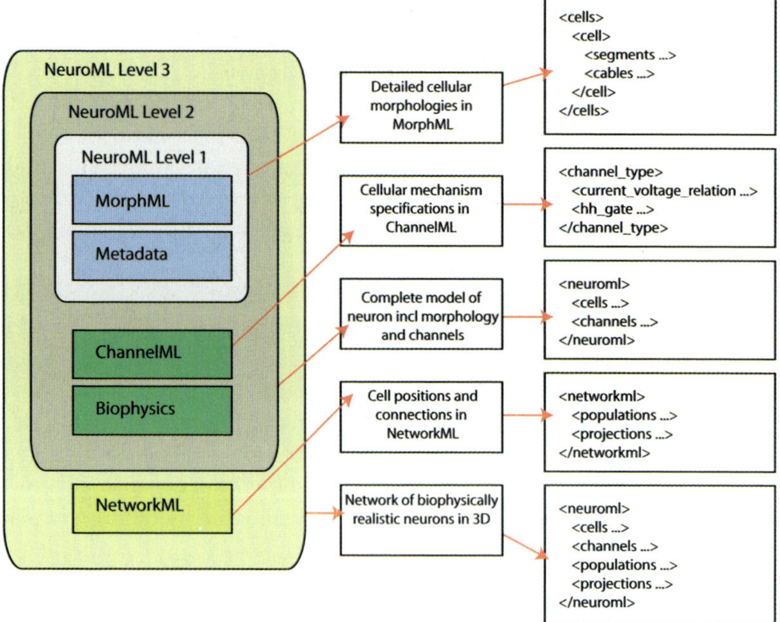

Fig. 14.1. The 3 Levels in the current NeuroML specifications.

for constructing, visualizing and analyzing conductance-based neural network models with properties that closely match the 3D neuronal morphology and connectivity of different brain regions (Figure 14.2). A user friendly GUI allows models to be built, modified and run without the need for specialist programming knowledge, providing increased accessibility to both experimentalists and theoreticians studying network function. neuroConstruct generates script files for NEURON or GENESIS which carry out the numerical integration.

Networks of cells with complex interconnectivity can be created with neuroConstruct, simulations run with one or both of the supported simulators and the network behavior analyzed with a number of inbuilt tools in neuroConstruct. neuroConstruct is freely available from `http://www.neuroConstruct.org`.

14.3.4 Conclusions

The creation of more detailed models of neuronal function will require the interaction of a range of investigators from many backgrounds. Having a common framework to describe experimental findings and theoretical models about function will greatly aid these collaborations. Appropriate tools for creating models and testing ideas are also needed in this process. The ongoing work on NeuroML and neuroConstruct can help towards these goals.

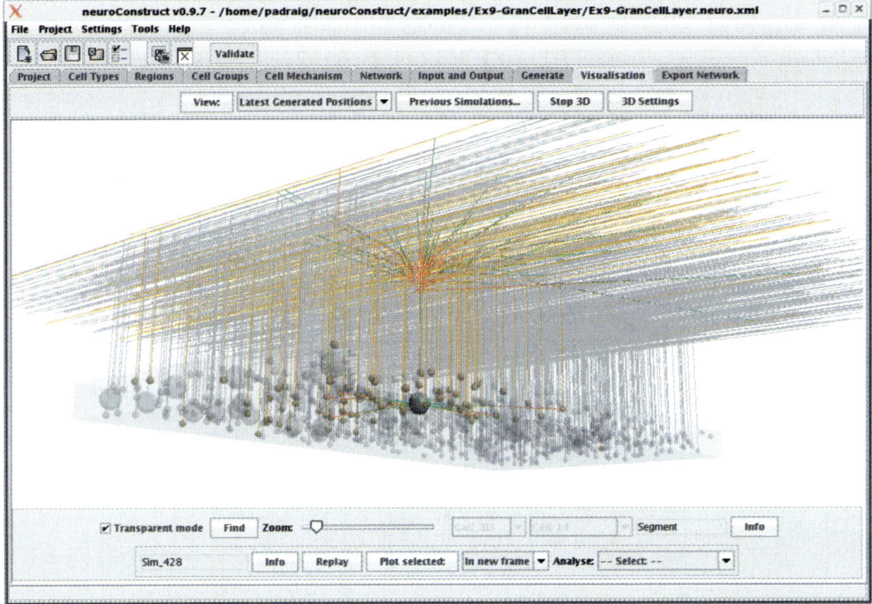

Fig. 14.2. Screenshot of a network model of the granule cell layer of the cerebellum created in neuroConstruct.

14.4 Time saving technique for developing and maintaining user interfaces[5]

Developing computer analyses and simulations for experimental data in neuroscience often requires interactive access for most of the variables used in the computer applications. Most of the time, this interaction is accomplished by developing a user interface that needs to be maintained and updated as the application evolves. This section presents a simple approach that avoids the development of a user interface, yet allows the user full functionality in manipulating the application. Although this project is targeted for the Windows operating system, the same strategy can be used with other operating systems.

The gist of the approach is to allow the application to modify internal variables by receiving real-time commands through the messaging system provided by the operating system. To achieve this goal the following tasks need to be completed: (1) inside the application, build a map between the name of the variables that represent biophysical measures and the actual variables that represent them, (2) implement a function that receives commands from other applications (3) build a small interpreter that can take commands of the form 'Membrane_voltage -80' meaning update the Membrane_voltage variable

---

[5] Contribution by H. Voicu.

with the value -80, and (4) build a small editor that can send commands to other applications through the messaging system of the operating system. The complete project can be downloaded from: `www.voicu.us/software.zip`.

The first task is the most straightforward and can be easily implemented in C++ as shown below:

```
float EL_Ca3_pyr =
  make_float("ca3_pyr_cells.var",&EL_Ca3_pyr,"EL_Ca3_pyr",-60);

float make_float(char *fn,float *p, char *s, float val) {
if (u_i_float_var_counter >= NUM_MAX_OF_UI_FLOAT_VARS)
  return(val);

u_i_var_float_ptr_array[u_i_float_var_counter] = p;
strcpy(u_i_var_float_str_array[u_i_float_var_counter],s);
strcpy(u_i_var_float_file_name_array[u_i_float_var_counter],fn);
u_i_float_var_counter++; return(val);
}
```

Let us assume that `EL_Ca3_pyr` represents the resting potential of pyramidal cells in the CA3 subfield of the hippocampus. This variable is initialized with -60. The function `make_float` builds a map between the location in memory where the value of `EL_Ca3_pyr` is stored and the string "`EL_Ca3_pyr`." The string "`ca3_pyr_cells.var`" defines the name of the file in which `EL_Ca3_pyr` will be listed with its initial value when the program starts. This is handled by a separate function `generate_list_of_variables` which is included in the source code.

Since the messages we plan to send are short we can use the actual content of the pointers `WPARAM wp`, `LPARAM lp` of the function SendMessage to contain the information. This feature makes the application compatible with the GWD editor which we use for sending commands. The function that receives the command must have two important features. It must be able to concatenate partial messages and preserve their order. Preserving temporal order is particularly important since the repeated update of a variable can make messages corresponding to different commands arrive about the same time. To discriminate between partial messages that belong to different commands, each partial message is prefixed by a byte representing its identity. The function `OnMyMessage` takes the partial message contained in the 4 byte long variables wp and lp and concatenates it to the current command. When the full command is received, it is processed using the `Process_Message_From_User` function.

```
LRESULT CMainFrame::OnMyMessage(WPARAM wp, LPARAM lp) {
rm_multiplexer = (unsigned char)(wp & 0xff); wp >>= 8;
if (bool_receiving_message_from_user[rm_multiplexer]) {
for (int i1 =0; i1<3; i1++, rec_msg_cntr[rm_multiplexer]++, wp >>= 8) {
rec_message[rm_multiplexer][rec_msg_cntr[rm_multiplexer]] = wp & 0xff;
```

```
if (rec_message[rm_multiplexer][rec_msg_cntr[rm_multiplexer]]==0)
  goto exec_comm;
}
for (int i1 =0; i1<4; i1++, rec_msg_cntr[rm_multiplexer]++, lp >>= 8) {
rec_message[rm_multiplexer][rec_msg_cntr[rm_multiplexer]] = lp & 0xff;
if (rec_message[rm_multiplexer][rec_msg_cntr[rm_multiplexer]]==0)
  goto exec_comm;
}} else {
rec_msg_cntr[rm_multiplexer] = 0;

for (int i1 = 0 ; i1<3 ; i1++, rec_msg_cntr[rm_multiplexer]++, wp >>= 8) {
rec_message[rm_multiplexer][rec_msg_cntr[rm_multiplexer]] = wp & 0xff;
if (rec_message[rm_multiplexer][rec_msg_cntr[rm_multiplexer]]==0)
  goto exec_comm;
}
for (int i1 =0 ;i1<4; i1++, rec_msg_cntr[rm_multiplexer]++, lp >>= 8) {
rec_message[rm_multiplexer][rec_msg_cntr[rm_multiplexer]] = lp & 0xff;
if (rec_message[rm_multiplexer][rec_msg_cntr[rm_multiplexer]]==0)
  goto exec_comm;
}
bool_receiving_message_from_user[rm_multiplexer] = 1;
}
return 0;

exec_comm:
bool_receiving_message_from_user[rm_multiplexer] = 0;
Process_Message_From_User(rec_message[rm_multiplexer]);
return 0;
}
```

The `Process_Message_From_User` function is a standard interpreter for the commands received by the application. It uses the mapping from variable names to variable content build by the `make_float` function to update variables the same way a user interface does. The same function can be used to interpret commands stored in a file provided in the command line.

The most involved task in this project is the development of a supporting application that can send commands to the analysis or simulation application. Although rather lengthy, this task can be accomplished in a straightforward way. The alternative is to use an application that already has this capability, like the GWD text editor. Another important feature of the supporting application is the ability of increasing and decreasing the values of the variables declared with `make_float`. This ability can be easily implemented in the GWD editor with the help of macros. The source code shows how the value of a variable is increased/decreased by a value determined by the position of the cursor with respect to the decimal point and sent to the application.

A typical "*.var" file looks like this:

```
mfc_hippocampus

gL_Ca3_pyr 0.004000
gL_Ca3_pyr 0.004000
min_max 0.001000 0.009000
```

The first line contains the name of the application where the commands are sent. Each variable appears in two consecutive lines. The macros in the GWD program assume that the second line stores the default value. If the two consecutive lines are followed by a line containing the 'min_max' keyword than the macros do not change the variable below or above the values specified on that line.

There are two important advantages of this technique. First, the introduction of a new variable does not require adding code to the application except the declaration itself. Second, a large number of variables can be organized in different "*.var" files avoiding clutter and the development of additional GUI windows.

14.5 PANDORA's Toolbox: database-supported analysis and visualization software for electrophysiological data from simulated or recorded neurons with Matlab[6]

As the amount of data for EPDBs increase, its organization and labeling becomes more difficult. It is critical to reduce this difficulty since the analysis of the results depends on the accurate labeling of the data. For recordings, the difficulty comes from associating recorded data traces with multiple trials, different recording approaches (e.g., intracellular vs. extracellular), stimulation protocol parameters, the time of recording, the animal's condition, the drugs used, and so on [3, 17]. For simulations, the difficulty comes from the large number of possible simulation parameter combinations of ion channel densities, channel kinetic parameters, concentration and flux rates. The difficulty remains high irrespective of the actual model used, be it a neuron model containing multiple types of Hodgkin-Huxley ion channels [6, 96, 102] or a neuron model with Markov channels [86] and detailed morphological reconstructions, or a model of molecular pathways and processes [95]. Although simpler versions of these models are employed in network simulations, the number of parameters are still large to account for variances in neuron distributions, variable connection weights, modulators, time constants, and delays [50, 58, 97].

Although accurate labeling of data is critical, often custom formats and tools are used for data storage and analysis. These formats range from keeping

---

[6] Contribution by C. Günay.

data in human-readable text files to proprietary or ad-hoc binary formats. It is rare that a proper database management system is employed. Using database management systems has the advantage of automatically labeling the data. This "metadata," which is created when the experimental data is inserted into a database, remains with the data during different analysis steps and it may reach the final plots. Even though for small datasets, the consistency of the experimental parameters can be maintained by manual procedures, having automated systems that keep track of data becomes invaluable for larger datasets. Furthermore, a database system provides formal ways to manage, label and query datasets, and it maintains relationships between dataset elements.

The question is, then, to find the most optimal type of database system for storing and analyzing electrophysiological data. Storing experiment or simulation parameters in a database is simpler compared with storing the raw outputs of the experiments (e.g., voltage traces). Preserving the raw recorded data is essential for analyzing and understanding the results of an experiment. Especially with large datasets, with several hundred gigabytes (billion bytes), it becomes difficult to store, search and process the raw data quickly and efficiently for answering specific questions. But questions can be answered much faster if features that pertain to the possible questions in hand are extracted and placed in a more compact database format. Then, once interesting entries are found from the features and parameters in the database, the raw data can be consulted again for validation, visualization and further analysis. This database of parameters and extracted features can be subject to several types of numerical and statistical analyses to produce higher-level results.

The Neural Query System (NQS) [62] provided such a database system for analyzing results of neural simulations. NQS is a tool integrated into the Neuron simulator [46] to manage simulations and record their features in a database for further analysis. Here, we introduce PANDORA's toolbox[7] that provides this type of database support for both simulated and recorded data. It currently offers offline analysis within the Matlab environment for intracellular neural recordings and simulations in current-clamp mode, stimulated with a current-injection protocol. PANDORA provides functions to extract important features from electrophysiology data such as spike times, spike shape information, and other special measurements such as rate changes; after the database is constructed, second tier numerical and statistical analyses can be performed; and both raw data and other intermediate results can be visualized.

PANDORA was designed with flexibility in mind and we present it as a tool available for other electrophysiology projects. PANDORA takes a simplified approach providing a native Matlab database that has the advantage of being independent of external applications (although it can communicate

---

[7] PANDORA stands for "Plotting and Analysis for Neural Database-Oriented Research Applications."

with them) and thus requires no additional programming in a database language. However, it inherits limitations of the Matlab environment in terms of being not as well optimized for speed and memory usage. In particular, a database table to be queried must completely fit in the computer's memory. PANDORA is distributed with an Academic Free License and can be freely downloaded from `http://senselab.med.yale.edu/SimToolDB`.

14.6 Chronux: A Platform for Analyzing Neural Signals[8],[9]

14.6.1 Introduction

Neuroscientists are increasingly gathering large time series data sets in the form of multichannel electrophysiological recordings, EEG, MEG, fMRI and optical image time series. The availability of such data has brought with it new challenges for analysis, and has created a pressing need for the development of software tools for storing and analyzing neural signals. In fact, while sophisticated methods for analyzing multichannel time series have been developed over the past several decades in statistics and signal processing, the lack of a unified, user-friendly, platform that implements these methods is a critical bottleneck in mining large neuroscientific datasets.

Chronux is an open source software platform that aims to fill this lacuna by providing a comprehensive software platform for the analysis of neural signals. It is a collaborative research effort currently based at Cold Spring Harbor Laboratory that draws on a number of previous research projects [8–10, 27, 60, 71, 77, 98]. The current version of Chronux includes a Matlab toolbox for signal processing of neural time series data, several specialized mini-packages for spike sorting, local regression, audio segmentation and other tasks. The eventual aim is to provide domain specific user interfaces (UIs) for each experimental modality, along with corresponding data management tools. In particular, we expect Chronux to grow to support analysis of time series data from most of the standard data acquisition modalities in use in neuroscience. We also expect it to grow in the types of analyses it implements.

14.6.2 Website and Installation

The Chronux website at `http://chronux.org/` is the central location for information about the current and all previous releases of Chronux. The home page contains links to pages for downloads, people, recent news, tutorials,

---

[8] Contribution by P. Andrews, H. Bokil, H. Maniar, C. Loader, S. Mehta, D. Kleinfeld, D. Thomson, P.P. Mitra.

[9] This work has been supported by grant R01MH071744 from the NIH to P.P. Mitra.

various files, documentation and our discussion forum. Most of the code is written in the Matlab scripting language, with some exceptions as compiled C code integrated using Matlab mex functions. Chronux has been tested and runs under Matlab releases R13 to the current R2007a under the Windows, Macintosh and Linux operating systems. Extensive online and within-Matlab help is available.

As an open source project released under the GNU Public License GPL v2, we welcome development, code contributions, bug reports, and discussion from the community. To date, Chronux has been downloaded over 2500 times. Questions or comments about can be posted on our discussion forum at `http://chronux.org/forum/` (after account registration). Announcements are made through the Google group chronux-announce.

14.6.3 Examples

This section contains examples of Chronux usage, selected to show how it can handle several common situations in analysis of neural signals. We focus on spectral analysis, and local regression and likelihood since these techniques have a wide range of utility.

Spectral Analysis

The spectral analysis toolbox in Chronux is equipped to process continuous valued data and point process data. The point process data may be a sequence of values, such as times of occurrence of spikes, or a sequence of counts in successive time bins. Chronux provides routines to estimate time-frequency spectrograms and spectral derivatives, as well as measures of association such as cross-spectra and coherences. Univariate and multivariate signals may be analyzed as appropriate. Where possible, Chronux provides confidence intervals on estimated quantities using both asymptotic formulae based on appropriate probabilistic models, as well as nonparametric bands based on the Jackknife method. Finally, Chronux includes various statistical tests such as the two-group tests for the spectrum and coherence, nonstationarity test based on quadratic inverse theory, and the F-test to detect periodic signals in a colored background. The latter is particularly useful for removing 50 or 60 Hz line noise from recorded neural data.

Space constraints preclude covering all of spectral analysis here, but the functions generally have a uniform function calling signature. We illustrate three canonical routines below.

mtspectrumc

As a first example, we show how to estimate the spectrum of a single trial local field potential measured from macaque during a working memory task. Figure 14.3 shows the spectrum estimated by the Chronux function

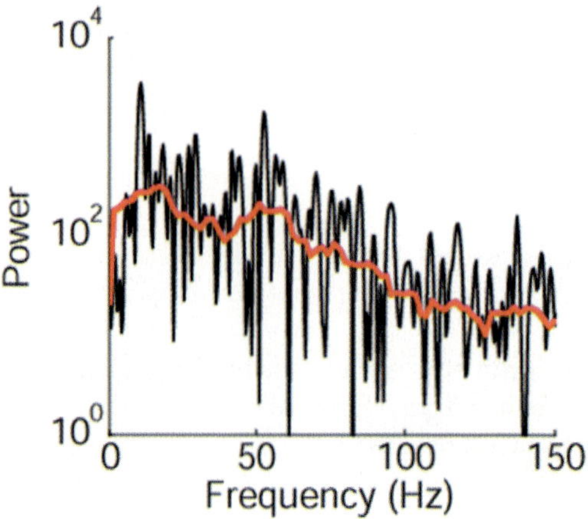

Fig. 14.3. Comparison of a periodogram (black) and multitaper estimate (red) of a single trial local field potential measurement from macaque during a working memory task. This estimate used 9 tapers.

mtspectrumc. For comparison we also display the ordinary periodogram. In this case, mtspectrumc was called with params.tapers=[5 9].

The Matlab calling signature of the mtspectrumc function is as follows:

```
[S,f,Serr] = mtspectrumc( data, params );
```

The first argument is the data matrix in the form of times × trials or channels, while the second argument *params* is a Matlab structure defining the sampling frequency,[10] the time-bandwidth product used to compute the tapers, and the amount of zero padding to use. It also contains flags controlling the averaging over trials and the error computation. In this example, params.tapers was set to be [5 9], thus giving an estimate with a time bandwidth product 5, using 9 tapers (For more details on this argument, see below).

The three variables returned by mtspectrumc are, in order, the estimated spectrum, the frequencies of estimation, and the confidence bands. The spectrum is in general two-dimensional, with the first dimension being the power as a function of frequency and the second dimension being the trial or channel. The second dimension is 1 when the user requests a spectrum that is averaged over the trials or channels. The confidence bands are provided as a lower and upper confidence band at a p value set by the user. As indicated by the last letter c in its name, the routine mtspectrumc is applicable to continuous valued data such as the local field potential or the EEG. The corresponding routines for point processes are mtspectrumpt and mtspectrumpb, applicable

---

[10] The current version of Chronux assumes continuous valued data to be uniformly sampled

to point processes stored as a sequence of times and binned point processes, respectively.

mtspecgramc

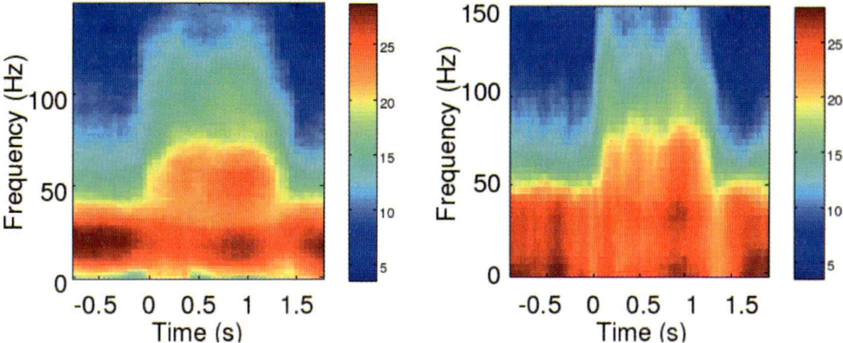

Fig. 14.4. The effect of changing the components of the time-bandwidth product TW. a) T = 0.5s, W = 10Hz. b) T = 0.2s, W = 25Hz. Data from macaque monkey performing a working memory task. Sharp enhancement in high frequency power occurs during the memory period.

The second example is a moving window version of `mtspectrumc` called `mtspecgramc`. This function, and `mtspecgrampt` and `mtspecgrampb`, calculate the multitaper spectrum over a moving window with user adjustable time width and step size. The calling signature of this function is:

```
[S,t,f,Serr] = mtspecgramc( data, movingwin, params );
```

Note that the only differences from the `mtspectrumc` function signature are in the function name, the additional `movingwin` argument and the addition of a return value `t` which contains the centers of the moving windows. The `movingwin` argument is given as `[winsize winstep]` in units consistent with the sampling frequency. The returned spectrum here is in general three dimensional: times × frequency × channel or trial.

The variable params.tapers controls the computation of the tapers used in the multitaper estimate. params.tapers is a two-dimensional vector whose first element, TW, is the time-bandwidth product, where T is the duration and W is the desired bandwidth. The second element of params.tapers is the number of tapers to be used. For a given TW, the latter can be at most $2TW - 1$. Higher order taper functions will not be sufficiently concentrated in frequency and may lead to increased broadband bias if used.

Figure 14.4 shows the effect on the spectrogram of changing the time-bandwidth product. The data again consists of local field potentials recorded from macaque during a working memory task.

coherencycpt

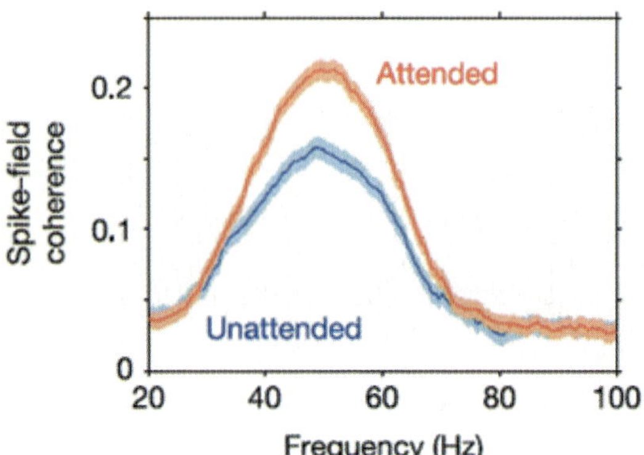

Fig. 14.5. The spike-field coherence recorded from visual cortex of monkey, showing significant differences between the attended and unattended conditions. In addition to the coherence in the two conditions, we also show the 95% confidence bands computed using the Jackknife.

Figure 14.5 [105] shows significant differences in the spike-field coherence recorded from the primary visual cortex of monkeys during an attention modulation task. The coherences were computed using `coherencycpt`. This function is called with two timeseries as arguments: the continuous LFP data and the corresponding spikes which are stored as event times. It returns not only the magnitude and phase of the coherency, but the cross-spectrum and individual spectra from which the coherence is computed. As with the spectra, confidence intervals on the magnitude and phase of the coherency may also be obtained.

Locfit

The Locfit package by Catherine Loader [61] is included in Chronux. Locfit can be used for local regression, local likelihood estimation, local smoothing, density estimation, conditional hazard rate estimation, classification and censored likelihood models. Figure 14.6 is an example of Locfit local smoothing with a cubic polynomial compared to a binned histogram. Locfit enables computation of both local and global confidence bands. In this case, the figure shows 95% local confidence bands around the smoothed rate estimate.

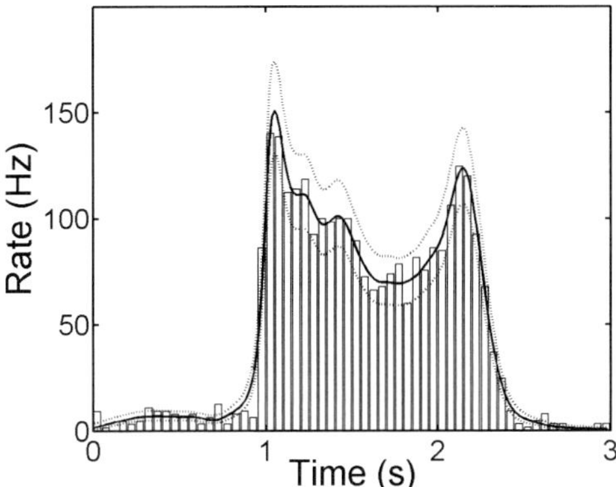

Fig. 14.6. A traditional binned histogram and a Locfit smoothed estimate of the same data set. The Locfit estimate shown uses a nearest-neighbor fraction of 35% when calculating the distribution.

14.7 Repeating synaptic inputs on a single neuron: seeking and finding with Matlab[11]

A single neuron in the mammalian cortex receives synaptic inputs from up to thousands of other neurons. Whole-cell patch-clamp electrophysiological recordings of single neurons reveal many of these synaptic events to the investigator, allowing analyses of the large and active neuronal network that impinges on this single neuron. The goals of the analyses are twofold: to find patterns in this barrage of synaptic inputs, and then determine whether these patterns are produced by design or chance.

While we do not know a priori the patterns to search for, we can still attempt to find patterns. One approach is to search for repeats-that is, sequences of synaptic inputs that repeat later in the recording with significant precision (Figures 14.7, 14.8, 14.9). This is analogous to a study of language by a completely naive observer: a language has a finite vocabulary and set of phrases that can be identified through a search for repeats. This search involves comparing all segments of the recording with itself in an iterative process.

The cross-correlation function is at the heart of this analysis. This function quantifies the temporal similarities of those waveforms and can initially identify whether or not two segments of a long recording might be significantly similar. This initial identification of a repeat is tentative, but the interval at which the repeat is found is stored and examined more carefully in subsequent analyses. Segments that do not pass a minimum threshold are passed over and not analyzed further, saving some time in the subsequent intensive analysis.

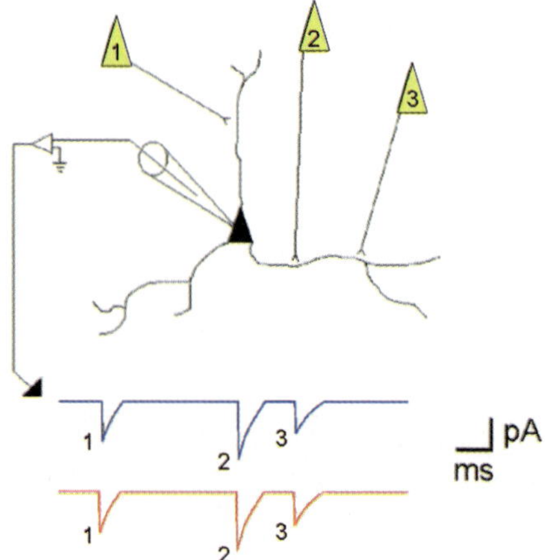

Fig. 14.7. Cartoon of a single neuron recorded within a 4 neuron network. Imagine that the 3 neurons connected to this one recorded neuron fire a sequence of action potentials. This sequence may be reflected in the synaptic currents recorded intracellularly (blue trace). If this same sequence is repeated some time later, the intracellular recording may reflect this repeat of synaptic activity (red trace).

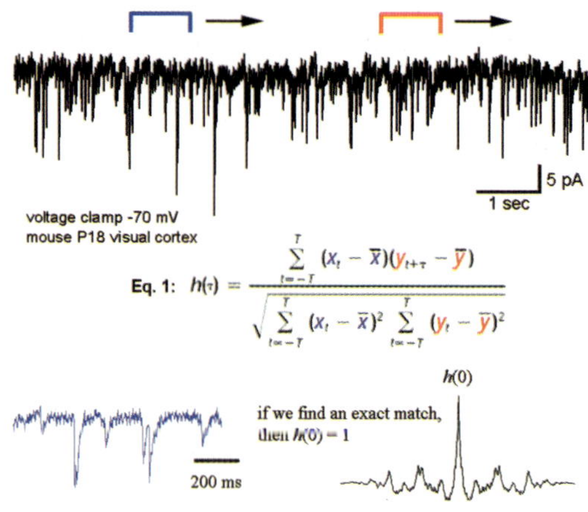

$$\text{Eq. 1: } h(\tau) = \frac{\sum_{t=-T}^{T} (x_t - \bar{x})(y_{t+\tau} - \bar{y})}{\sqrt{\sum_{t=-T}^{T} (x_t - \bar{x})^2 \sum_{t=-T}^{T} (y_t - \bar{y})^2}}$$

Fig. 14.8. Searching for repeats. A continuous 10 second stretch intracellularly recording postsynaptic currents (PSCs) is displayed. The program scans this recording, comparing every one second interval with every other one second interval. Here, the blue and red brackets represent these 1 second scanning windows. These one second segments are compared against each other via a cross-correlation equation (Eq. 1). If there were a perfect repeat of intracellular activity, then the correlation coefficient at the zeroth lag time would be 1.0.

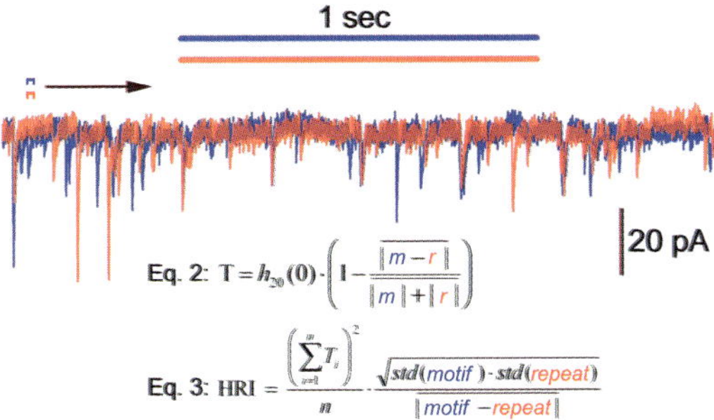

$$\text{Eq. 2: } T = h_{20}(0) \cdot \left(1 - \frac{\overline{\| m - r \|}}{\| m \| + \| r \|} \right)$$

$$\text{Eq. 3: } HRI = \frac{\left(\sum_{i=1}^{w} T_i \right)^2}{n} \cdot \frac{\sqrt{std(motif) \cdot std(repeat)}}{\| motif - repeat \|}$$

Fig. 14.9. Motif-repeat segments that yield a minimum threshold $h(0)$ value are remembered and subsequently analyzed via a high resolution index (HRI). The 1 second segments are aligned according to the best h(0) value, and then they are scanned with a 20 msec time window that computes many cross-correlation values. These correlation values are then adjusted according to the amplitude differences between these short segments (Eq. 2), and values passing a threshold are recorded and used in Eq. 3. The idea of this process is to find very similar PSCs recurring in a precise sequence. Finally, the number of precisely occurring PSCs in a long segment and the precision of those repeats are calculated in the HRI equation (Eq. 3), yielding an index of repeatability.

We used these search programs in analyzing long voltage-clamp intracellular recordings (8 minutes) from slices of mouse visual cortex. These were spontaneous recordings, meaning no stimulation was applied to the slices. Thus, the currents identified in the recordings were presumably the result of synaptic activity, created in large part by the action potential activity of synaptically-coupled neurons. The search algorithms described here were able to find instances of surprising repeatability, as judged by eye (Figure 14.10).

What occurs after the initial identification of putative repeats depends on the hypothesis being tested as well as the recording conditions. If the search is for repeating sequences of postsynaptic currents from a voltage-clamp recording, then the average time course of a single postsynaptic current becomes a critical parameter. Many cross-correlation analyses are performed at these smaller time windows, increasing the sensitivity of the measurement.

Given a long enough recording (several minutes) and many synaptic events, it should be expected that some repeating patterns emerge. The question is then whether the patterns we find are beyond a level that could be expected to occur by chance. In fact, it is undetermined as to whether the specific examples shown in Figure 14.10 are deterministically or randomly generated, although there is strong evidence that non-random patterns do emerge in these recordings ([53], however, see also [72]). The development of surrogate recordings-artificial data that is based on the real-is one way to deal with this issue. These surrogate data can then be compared to the real to see which produces more putative repeats (repeats being judged by the index methods

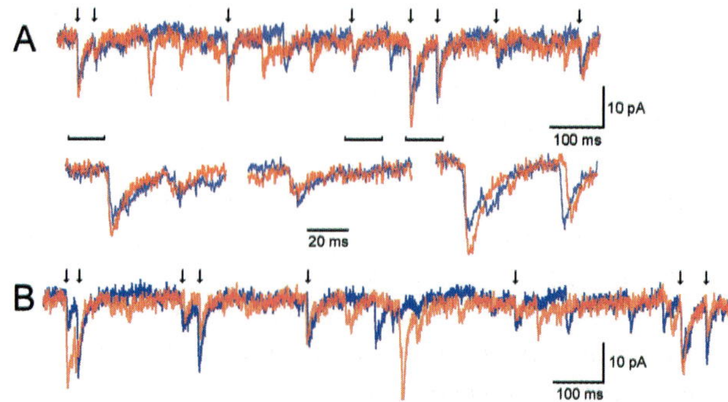

Fig. 14.10. Apparent repeats of synaptic events found. **A:** Two segments from a voltage-clamp recording are displayed, superimposed on each other. The blue trace occurred some time before the red trace, and yet the sequence of synaptic events appear similar. Arrows indicate time points where these synaptic events appear to repeat, and the brackets indicate segments that are temporally expanded below. **B:** Another example of a repeat.

in Figures 14.7 through 14.9). If the real data produces more putative repeats than a large number of generated surrogate data sets, then some evidence is given for a non-random generation of these repeats. The study of this issue is ongoing and involves more than the scope of this chapter. In short, attention must be given to the specific hypothesis tested, as well as the limits and sensitivity of the detector itself (described in Figs 14.7 through 14.9). Perhaps a more convincing method is to perturb the cortex in a biologically relevant manner and see how such manipulations affect or even produce the repeatable patterns (as shown in [63]).

14.8 A method for discovering spatio-temporal spike patterns in multi-unit recordings[12]

14.8.1 Introduction

In previous work we have shown that neurons *in vitro* responded to the repeated injections of somatic current by reliably emitting a few temporal spike patterns differing from each other by only a few spikes [29]. The techniques used to find these patterns were applied to data obtained in vivo from the Lateral Geniculate nucleus in the anesthetized cat [82] and from area MT in the behaving monkey [15]. In these two preparations, the animals were presented with repeated occurrence of the same visual stimulus. Single neurons were recorded and spiking was found to be very reliable when time-locked to the stimulus. In both datasets, however, groups of temporally precise firing

____

[12] Contribution by J.-M. Fellous.

patterns were identified above and beyond those that were reported. These patterns could last a few hundreds of milliseconds (monkey) to several seconds (cat), and were the result of the complex but largely deterministic interactions between stimulus features, network and intrinsic properties [29]. In principle, this clustering technique can be used on any ensemble of spike trains that have a common time scale: either time locked to the same event (repeated stimulus presentations for one recorded neuron, as above), or recorded simultaneously as in multi-unit recordings. In the first case, groups of trials are determined that share common temporal structures relative to stimulus presentation. In the second case, groups of neurons (called here neural assemblies) are identified that show some temporal correlations.

Finding neural assemblies can be thought of a dimensionality reduction problem: out of N neurons recorded (N-dimensions) how many 'work together'? Standard methods for dimensionality reduction such as principal or independent component analysis have been used on populations of neurons [19, 57]. Such dimensionality reduction however does not typically result in the selection of a subset of neurons, but on a linear combination of their activity (population vector). The weighting of these neurons may or may not be biologically interpretable. Another approach is that of peer prediction, where the influence of $N - 1$ recorded neurons onto a single neuron is captured by $N - 1$ weights that are 'learned' [44]. The focus of this approach, like that of many information theoretical approaches [74, 84], is to explain a posteriori the firing pattern of neurons, rather than to study their spatio-temporal dynamics. Another approach is that of template matching where the rastergram of multi neuronal activity is binned to generate a sequence of population vectors (a N-dimensional vector per time bin) [23, 73]. A group of T consecutive population vectors is then chosen (NxT matrix, the 'template') and is systematically matched to the rest of the recording. Other methods use the natural oscillations present in the EEG (theta cycle or sharp waves) to define the length of the template [54]. These methods however require the choice of a bin size and the choice of the number of consecutive population vectors (T, another form of binning). We propose here a general method for the detection of transient neural assemblies based on a binless similarity measure between spike trains.

14.8.2 Methods

To illustrate the effectiveness of the technique, we built a biophysical simulation of a small network of neurons using the NEURON simulator [46]. Neurons were single compartments containing a generic leak current, sodium and potassium currents responsible for action potential generation [37], a generic high voltage calcium current [83], a first order calcium pump [25] and a calcium-activated potassium current to control burst firing [25]. In addition, two generic background synaptic noise currents (excitatory and inhibitory) were added to recreate *in vivo* conditions [26, 28, 75]. Synaptic connections were formed with AMPA synapses.

14.8.3 Results

Figure 14.11A shows the typical recording and modeling configuration considered here. Extracellular recording electrodes are lowered near a group of cells. Neurons in this area are interconnected (thin lines, only a few represented for clarity), but some of them form sparse groups linked by stronger synaptic connections (thick black lines). Two of these putative assemblies are depicted in gray and black. The extracellular recording electrodes have typically access to only a fraction of these assemblies (ellipse, 3 neurons each). To illustrate this configuration further, we built a biophysical simulation of fifteen principal neurons containing 2 different assemblies of 5 neurons each. Figures 14.11B and 14.11C show representative traces of the membrane voltage of two neurons: One that does not belong to an assembly and one that does (top and bottom respectively). No difference can be detected in the statistics of the membrane potential between these neurons (average, standard deviation, coefficient of variation of spiking and firing rate were tested).

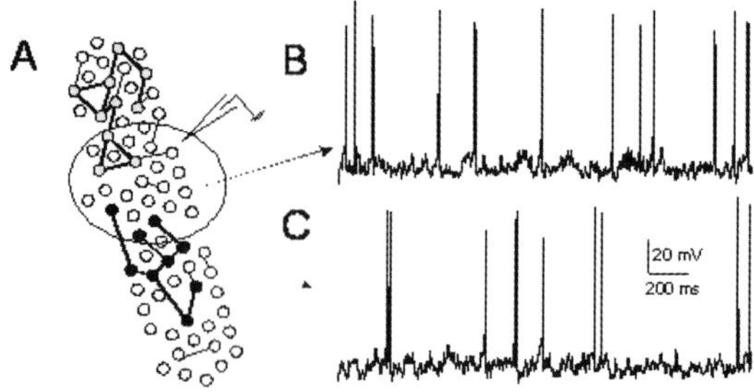

Fig. 14.11. Simulations of neural assemblies. **A:** Schematic representation of a recording configuration in CA3. Panels **B** and **C** show the simulated membrane potential of a cell outside an assembly (**B**) and of a cell within one of the two assembles (**C**, black neuron).

Figure 14.12A shows a rastergram of 2 seconds of spontaneous activity of all 15 neurons. The simulation is tuned so that all neurons have the same mean firing rate (7.5 Hz). Again, no difference can be seen between the spike trains of neurons belonging to an assembly (filled circles labeled on the left of the Y-axis), and those that do not (white circles). Such rastergrams are typically the only neurophysiological information available from multi-unit recordings in the behaving animal. Experimentally, the labels are of course unknown and the goal of this work is to identify the 'hidden' assemblies on the basis of unlabeled rastergrams.

To do so, we first build a similarity matrix computed on the basis of a binless correlation measure used previously [93] (Figure 14.12B). Each square i,j of this 15x15 matrix represents the similarity between the spike trains of neurons i and j (similarities vary between 0 (dissimilar-black) and 1 (identical-white)). A fuzzy clustering method is then applied to this matrix and rows and columns are re-ordered under the constraint that the matrix remains symmetric [29] (Figure 14.12C). Groups of highly similar neurons are now apparent (white regions in the upper left and lower right corners of the matrix). The algorithm identified 3 areas in this matrix (dashed lines). Since each row represents the similarities of one neuron to the 14 others, the matrix reordering can be applied to the spike trains themselves, and the re-ordered rastergram is shown in panel D. The assemblies are correctly identified as evidenced by the grouping of the symbols on the left (assemblies: filled circle, other neurons: white circles). Cluster strengths were 2.4 (black circles), 2.2 (gray circles) and 1.2 (white circles), making the first two clusters significant (above 1.5 [29]). In this simulation, and for illustrative purposes, the two assemblies were implemented in a different manner. The first assembly (gray neurons in Figures 14.11 and 14.12) contained neurons that were correlated because of slightly stronger AMPA intrinsic connections between them. This assembly formed transient assembly-wide correlations (ellipses in Figure 14.12D). The second assembly (black neurons in Figures 14.11 and 14.12) contained neurons that were correlated because they received a common modulation of their mean excitatory background inputs. Unlike with the first assembly, this assembly did not form assembly-wide correlations.

The algorithm successfully identified the assemblies in both cases, showing that it was not sensitive to the source of the correlations and did not require all the neurons within an assembly to be simultaneously correlated. The performance of the clustering algorithm in detecting functionally connected cells, when the connection strength is systematically varied was assessed. The network is analogous to that of Figure 14.11 with gray neurons only. Our simulations (50 independent simulations per connection strength) show that the detection is significantly above chance for connection values that yield probability of postsynaptic firing due to a single presynaptic spike as low as 10%.

The performance of the clustering algorithm was also assessed in the case where the neurons in Figure 14.11A are interconnected by weak synapses (2%) but receive part of their average background inputs from a common source (i.e. black neurons only). The common input was simulated as a modulation of X% of the mean background excitatory synaptic noise, where X is systematically varied. The time course and frequency content of the modulation is obtained from white-noise filtered by AMPA-like alpha-function [65]. Note that in these simulations, the overall firing rate of all cells in the assembly were kept constant and identical to that of the cells outside the assembly, so as to not bias the algorithm into clustering cells by firing rates rather than by temporal correlations. Previous work has however shown that a properly

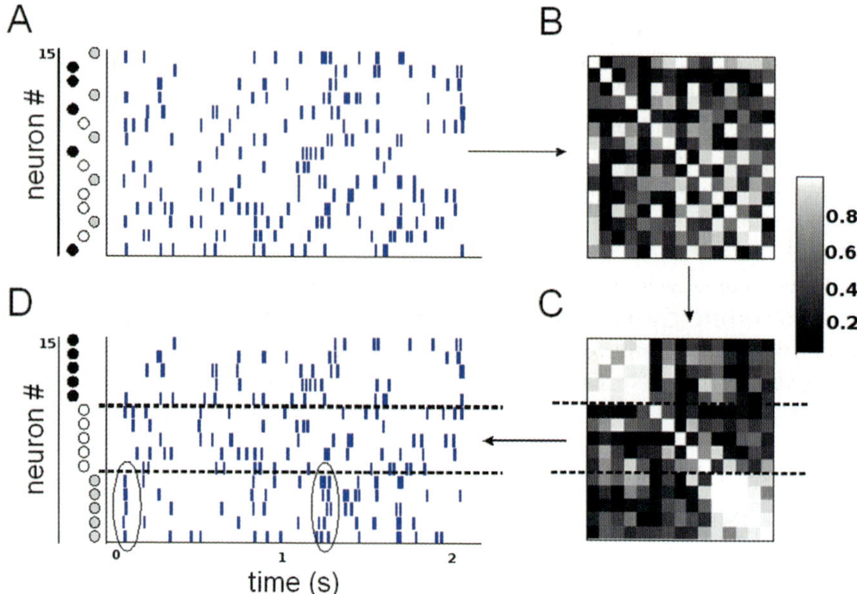

Fig. 14.12. Detection of cell assemblies using Spike Train Clustering. **A**: Original rastergram obtained from a biophysical simulation of 15 neurons. Two assemblies of five neurons are 'hidden' (filled circles). **B**: Similarity matrix based on the whole rastergram. **C**: Reordering of the similarity matrix after fuzzy clustering with 3 clusters [29]. Two significant clusters are apparent in the upper left and lower right corners. Similarity gray scale applies to B and C. **D**: The same reordering in C is applied to the rastergram. The hidden assemblies are completely recovered (grouping of labels).

tuned similarity measure can make the clustering algorithm relatively insensitive to heterogeneous baseline firing rates [29]. Simulations (50 independent simulations per X) show that a modulation of 15% of the background inputs can be effectively detected above chance.

14.8.4 Conclusions

Various methods for multiple spike train analysis have been reviewed elsewhere [12, 14]. In general, the detection of neural assemblies can be accomplished in two steps: 1) Computation of the similarity between the spike trains of the recorded neurons, 2) Use of these similarities to detect eventual groups of neurons that are more similar to each other than they are to the other neurons.

The traditional measures of similarity between spike trains include the cross-correlation coefficient, its generalization with the Joint Peristimulus Time Histograms (where Pearson correlations between two neurons are computed at different time lags) [2, 107] and the cross-intensity function (where estimation of the firing rate of one neuron is made relative to that of another at different time lags) [12]. These methods require some form of binning which involves the a priori choice of a time scale of integration (bin size) and the choice of bin boundaries (with or without overlaps), both of which may introduce artifacts. There is some evidence that computing correlation in the frequency domain (i.e. cross coherence) may help reduce the sensitivity to bin size [76, 99].

Two other concerns arise 1) Correlation coefficients may introduce artificial correlations due to bursting, which is common in many brain areas and 2) correlations are computed on the basis of pairs of neurons, and are unable to capture spike patterns which consist in P neurons firing in one of several possible orders (i.e. in an assembly of P neurons, one neuron may fire before a second at one point in time, or after that neuron later on depending on the firing of the other P-2 neurons) [29]. Because the order of firing between two neurons may not be always the same, the pair wise correlation coefficient could become insignificant.

A few methods have attempted to go beyond pairwise assessments by accounting for correlations between 3 or 4 neurons [1, 24, 52], but those methods are not easily generalizable to 50-100 neurons, which is the typical number of neurons recorded simultaneously. Other methods such as Bayesian estimations [4, 13, 68, 106] or unitary event analysis (statistical detection of spike coincidences that occur above chance) [42, 43] are statistical in nature, and are not suitable for the dynamic assessment of patterns in time. The work we presented here however has that potential [59]. In sum, new methods for detecting spike patterns (i.e. high order correlations) such as that proposed here and ways of comparing their similarities are sorely needed.

14.9 Summary

The authors of this chapter hope that these mini-proceedings will be useful to all those who are already involved in the field of electrophysiology and are looking for currently available data analysis tools and techniques. However, we also hope that Computational Intelligence researchers will find it valuable as their interest will be sparked by the presented discussion of the challenges prevalent in electrophysiology.

The organizers of the workshop would like to thank all the contributors and participants for their invaluable input and involvement in the creation of these mini-proceedings.

References

1. Abeles M, Gat I (2001) Detecting precise firing sequences in experimental data. J Neurosci Methods 107:141–54

2. Aertsen AM, Gerstein GL, Habib MK, Palm G (1989) Dynamics of neuronal firing correlation: modulation of "effective connectivity". J Neurophysiol 61:900–17

3. Baker S, Baseler H, Klein S, Carney T (2006) Localizing sites of activation in primary visual cortex using visual-evoked potentials and functional magnetic resonance imaging. J Clinical Neurophys. 23(5):404–15

4. Barbieri R, Frank LM, Nguyen DP, Quirk MC, Solo V, Wilson MA, Brown E (2004) A Bayesian decoding algorithm for analysis of information encoding in neural ensembles. Conf Proc IEEE Eng Med Biol Soc 6:4483–86

5. Berners-Lee T, Hall W, Hendler J, Shadbolt N, Weitzner DJ (2006) Creating a Science of the Web. Science 313(5788):769–71

6. Bhalla US and Bower JM (1993) Exploring parameter space in detailed single neuron models: Simulations of the mitral and granule cells of the olfactory bulb. J Neurophysiol 69:1948–65

7. Bloom F, et al. (2003) Neuroscience Database Gateway.
Available: `http://ndg.sfn.org`

8. Bokil H, Pesaran B, Andersen RA, Mitra PP (2006) A method for detection and classification of events in neural activity. IEEE Transactions on Biomedical Engineering 53:1678–87

9. Bokil H, Purpura K, Schofflen J-M, Thompson D, Pesaran B, Mitra PP (2006) Comparing spectra and coherences for groups of unequal size. J Neurosci Methods 159:337–45

10. Bokil H, Tchernichovski O, Mitra PP (2006) Dynamic phenotypes: Time series analysis techniques for characterising neuronal and behavioral dynamics. Neuroinformatics Special Issue on Genotype-Phenotype Imaging in Neuroscience 4:119–28

11. Bower JM and Beeman D (1997) The Book of GENESIS: Exploring Realistic Neural Models with the GEneral NEural SImulation System. Springer

12. Brillinger DR (1992) Nerve cell spike train data analysis: A progression of technique. Journal of the American Statistical Association 87:260–71

13. Brown EN, Frank LM, Tang D, Quirk MC, Wilson MA (1998) A statistical paradigm for neural spike train decoding applied to position prediction from ensemble firing patterns of rat hippocampal place cells. J Neurosci 18:7411–25

14. Brown EN, Kass RE, Mitra PP (2004) Multiple neural spike train data analysis: state-of-the-art and future challenges. Nat Neurosci 7:456–61

15. Buracas GT, Zador AM, DeWeese MR, Albright TD (1998) Efficient discrimination of temporal patterns by motion-sensitive neurons in primate visual cortex. Neuron 20:959–69

16. Cannon RC and D'Alessandro G (2006) The ion channel inverse problem: neuroinformatics meets biophysics. PLoS Comput Biol 2(8):e91

17. Carmena JM, Lebedev MA, Henriquez CS, Nicolelis MAL (2005) Stable ensemble performance with single-neuron variability during reaching movements in primates. J Neurosci 25(46):10712–16

18. Chance FS, Nelson SB, Abbott LF (1998) Synaptic Depression and the Temporal Response Characteristics of V1 Cells. J Neurosci 18:4785–99

19. Chapin JK, Nicolelis MA (1999) Principal component analysis of neuronal ensemble activity reveals multidimensional somatosensory representations. J Neurosci Methods 94:121–40

20. Clancy CE and Rudy Y (2002) Na(+) channel mutation that causes both Brugada and long-QT syndrome phenotypes: a simulation study of mechanism. Circulation 105:1208–13
21. Clancy CE and Kass RS (2004) Theoretical investigation of the neuronal Na+ channel SCN1A: abnormal gating and epilepsy. Biophys J 86:2606–14
22. Crook S, Gleeson P, Howell F, Svitak J, Silver RA (2007) MorphML: Level 1 of the NeuroML standards for neuronal morphology data and model specification. Neuroinf. 5(2):96–104
23. Crowe DA, Averbeck BB, Chafee MV, Georgopoulos AP (2005) Dynamics of parietal neural activity during spatial cognitive processing. Neuron 47:885–91
24. Czanner G, Grun S, Iyengar S (2005) Theory of the snowflake plot and its relations to higher-order analysis methods. Neural Comput 17:1456–79
25. Destexhe A, Contreras D, Sejnowski TJ, Steriade M (1994) A model of spindle rhythmicity in the isolated thalamic reticular nucleus. J Neurophysiol 72:803–18
26. Destexhe A, Rudolph M, Fellous JM, Sejnowski TJ (2001) Fluctuating synaptic conductances recreate in vivo-like activity in neocortical neurons. Neuroscience 107:13–24
27. Fee MS, Mitra PP, Kleinfeld D (1996) Automatic sorting of multiple unit neuronal signals in the presence of anisotropic and non-Gaussian variability. J Neurosci Methods 69:175–88
28. Fellous J-M, Rudolph M, Destexhe A, Sejnowski TJ (2003) Variance detection and gain modulation in an in-vitro model of in-vivo activity. Neuroscience 122:811–29
29. Fellous JM, Tiesinga PH, Thomas PJ, Sejnowski TJ (2004) Discovering spike patterns in neuronal responses. J Neurosci 24:2989–3001
30. Gardner D, Abato M, Knuth KH, DeBellis R, Erde SM (2001) Dynamic publication model for neurophysiology databases. Philos Trans R Soc Lond B Biol Sci 356(1412):1229–47
31. Gardner D, Knuth KH, Abato M, Erde SM, White T, DeBellis R, Gardner EP (2001) Common data model for neuroscience data and data model exchange. J Am Med Inform Assoc 8:17–33
32. Gardner D (2004) Neurodatabase.org: networking the microelectrode, Nat Neurosci 7(5):486–87
33. Gardner D (2004) *personal communication*
34. Gleeson P, Steuber V, Silver RA (2007) neuroConstruct: a tool for modeling networks of neurons in 3D space. Neuron 54(2):219–35
35. Goddard NH, Hucka M, Howell F, Cornelis H, Shankar K, Beeman D (2001) Towards NeuroML: model description methods for collaborative modelling in neuroscience. Philos Trans R Soc Lond B Biol Sci 356(1412):1209-1228
36. Goddard NH, Cannon RC, Howell FW (2003) Axiope tools for data management and data sharing. Neuroinformatics 1(3):271–84
37. Golomb D, Amitai Y (1997) Propagating neuronal discharges in neocortical slices: computational and experimental study. J Neurophysiol 78:1199–211
38. Golowasch J, Abbott LF, Marder E (1999) Activity-dependent regulation of potassium currents in an identified neuron of the stomatogastric ganglion of the crab Cancer borealis. J Neurosci 19(20):RC33
39. Golowasch J, Goldman MS, Abbott LF, Marder E (2002) Failure of averaging in the construction of a conductance-based neuron model. J Neurophysiol 87(2):1129–31

40. Gonzalez-Heydrich J, Steingard RJ, Putnam F, Beardslee W, Kohane IS (1999) Using 'off the shelf' computer programs to mine additional insights from published data: diurnal variation in potency of ACTH stimulation of cortisol secretion revealed. Comput Methods Programs Biomed 58(3):227–38

41. Greenstein JL, Wu R, Po S, Tomaselli GF, Winslow RL (2000) Role of the calcium-independent transient outward current I(to1) in shaping action potential morphology and duration. Circ Res 87:1026-1033

42. Grun S, Diesmann M, Aertsen A (2002) Unitary events in multiple single-neuron spiking activity: I. Detection and significance. Neural Comput 14:43–80

43. Grun S, Diesmann M, Aertsen A (2002) Unitary events in multiple single-neuron spiking activity: II. Nonstationary data. Neural Comput 14:81–119

44. Harris KD, Csicsvari J, Hirase H, Dragoi G, Buzsaki G (2003) Organization of cell assemblies in the hippocampus. Nature 424:552–6

45. Herzog RI, Cummins TR, Waxman SG (2001) Persistent TTX-resistant Na+ current affects resting potential and response to depolarization in simulated spinal sensory neurons. J Neurophysiol 86(3):1351–64

46. Hines ML and Carnevale NT (1997) The NEURON simulation environment. Neural Comput 9(6):1179–209

47. Hines ML, Morse T, Migliore M, Carnevale NT, Shepherd GM (2004) ModelDB: A database to support computational neuroscience. J Comput Neurosci 17(1):7–11

48. Hodgkin AL and Huxley AF (1952) A quantitative description of membrane current and its application to conduction and excitation in nerve. J Physiol Lond 117:500-544

49. Hulata E, Segev R, Ben-Jacob E (2002) A method for spike sorting and detection based on wavelet packets and Shannon's mutual information. J Neurosci Methods 117(1):1–12

50. Humphries MD, Stewart RD, Gurney KN (2006) A physiologically plausible model of action selection and oscillatory activity in the basal ganglia. J Neurosci 26(50):12921–42

51. Huyser K and van der Laan J (1992) Data Thief v.1.0.8. Available:http://www.datathief.org/

52. Iglesias J, Villa AE (2007) Effect of stimulus-driven pruning on the detection of spatiotemporal patterns of activity in large neural networks. Biosystems 89:287–93

53. Ikegaya Y, Aaron G, Cossart R, Aronov D, Lampl I, Ferster D, Yuste R (2004) Synfire chains and cortical songs: temporal modules of cortical activity. Science 304(5670):559–64

54. Jackson JC, Johnson A, Redish AD (2006) Hippocampal sharp waves and reactivation during awake states depend on repeated sequential experience. J Neurosci 26:12415–26

55. Kennedy DN and Haselgrove C (2006) The internet analysis tools registry: a public resource for image analysis. Neuroinformatics 4(3):263–70

56. Kuzmick V, Lafferty J, Serfass A, Szperka D, Zale B, Nagvajara P, Johnson J, Moxon K (2001) Novel epileptic seizure detection system using multiple single neuron recordings. Conf Proc IEEE 27th Ann Northeast Bioeng, pp. 7–8

57. Laubach M, Shuler M, Nicolelis MA (1999) Independent component analyses for quantifying neuronal ensemble interactions. J Neurosci Methods 94:141–54

58. Leblois A, Boraud T, Meissner W, Bergman H, Hansel D (2006) Competition between feedback loops underlies normal and pathological dynamics in the basal ganglia. J Neurosci 26(13):3567–83

59. Lipa P, Tatsuno M, Amari S, McNaughton BL, Fellous JM (2006) A novel analysis framework for characterizing ensemble spike patterns using spike train clustering and information geometry. Society for Neuroscience Annual Meeting. Atlanta, GA, pp. 371–76

60. Llinas RR, Ribary U, Jeanmonod D, Kronberg E, Mitra PP (1999) Thalamocortical dysrhythmia: A neurological and neuropsychiatric syndrome characterized by magnetoencephalography. Proceedings of the National Academy Of Sciences of The United States of America 96:15222–27

61. Loader C (1999) Local Regression and Likelihood. Springer

62. Lytton WW (2006) Neural query system - data-mining from within the neuron simulator. Neuroinformatics 4(2):163–75

63. MacLean JN, Watson BO, Aaron GB, Yuste R (2005) Internal dynamics determine the cortical response to thalamic stimulation. Neuron 48:811–23

64. Magee JC (1998) Dendritic hyperpolarization-activated currents modify the integrative properties of hippocampal CA1 pyramidal neurons. J Neurosci 18(19):7613-7624

65. Mainen ZF, Sejnowski TJ (1995) Reliability of spike timing in neocortical neurons. Science 268:1503–6

66. Mamlouk AM, Sharp H, Menne KML, Hofmann UG, Martinetz T (2005) Unsupervised spike sorting with ICA and its evaluation using GENESIS simulations. Neurocomputing, 65-66:275–82

67. Markram H (2006) The blue brain project. Nat Rev Neurosci 7(2):153-160

68. Martignon L, Deco G, Laskey K, Diamond M, Freiwald W, Vaadia E (2000) Neural coding: higher-order temporal patterns in the neurostatistics of cell assemblies. Neural Comput 12:2621–53

69. Migliore M and Shepherd GM (2002) Emerging rules for the distributions of active dendritic conductances. Nat Rev Neurosci 3(5):362–70

70. Migliore M and Shepherd GM (2005) Opinion: an integrated approach to classifying neuronal phenotypes. Nat Rev Neurosci 6(10):810–18

71. Mitra PP and Pesaran B (1999) Analysis of dynamic brain imaging data. Biophysical J 76:691–708

72. Mokeichev A, Okun M, Barak O, Katz Y, Ben-Shahar O, Lampl I (2007) Stochastic emergence of repeating cortical motifs in spontaneous membrane potential fluctuations in vivo. Neuron 53:413–25

73. Nadasdy Z, Hirase H, Czurko A, Csicsvari J, Buzsaki G (1999) Replay and time compression of recurring spike sequences in the hippocampus. J Neurosci 19:9497–507

74. Nirenberg S, Carcieri SM, Jacobs AL, Latham PE (2001) Retinal ganglion cells act largely as independent encoders. Nature 411:698–701

75. Paré D, Shink E, Gaudreau H, Destexhe A, Lang EJ (1998) Impact of spontaneous synaptic activity on the resting properties of cat neocortical pyramidal neurons In vivo. J Neurophysiol 79:1450–60

76. Percival DB, Walden AT (2000) Wavelet Methods for Time Series Analysis. Cambridge University Press

358   C. Günay, T.G. Smolinski, W.W. Lytton, et al.

77. Pesaran B, Pezaris JS, Sahani M, Mitra PP, Andersen RA (2002) Temporal structure in neuronal activity during working memory in macaque parietal cortex. Nature Neuroscience 5:805–11

78. Pittendrigh S and Jacobs G (2003) NeuroSys: a semistructured laboratory database. Neuroinformatics 1(2):167–76

79. Prinz AA, Billimoria CP, Marder E (2003) Alternative to hand-tuning conductance-based models: construction and analysis of databases of model neurons. J Neurophysiol 90:3998–4015

80. Qi W and Crook S (2004) Tools for neuroinformatic data exchange: An XML application for neuronal morphology data. Neurocomp 58-60C:1091–5

81. Rall W (1977) Core Conductor theory and cable properties of neurons. Handbook of Physiology, Sec. 1, The Nervous System, vol. 1, Bethesda, MD: Am Physiol Soc, pp. 39–97

82. Reinagel P, Reid RC (2002) Precise firing events are conserved across neurons. J Neurosci 22:6837–41

83. Reuveni I, Friedman A, Amitai Y, Gutnick MJ (1993) Stepwise repolarization from Ca2+ plateaus in neocortical pyramidal cells: evidence for nonhomogeneous distribution of HVA Ca2+ channels in dendrites. J Neurosci 13:4609–21

84. Rieke F, Warland D, de Ruyter van Steveninck R, Bialek W (1997) Spikes: Exploring the Neural Code. The MIT Press

85. Roth A and Hausser M (2001) Compartmental models of rat cerebellar Purkinje cells based on simultaneous somatic and dendritic patch-clamp recordings. J Physiol 535:445–72

86. Rudy Y, Silva JR (2006) Computational biology in the study of cardiac ion channels and cell electrophysiology. Quart Rev Biophysics 39(1):57–116

87. Rutishauser U, Schuman EM, Mamelak AN (2006) Online detection and sorting of extracellularly recorded action potentials in human medial temporal lobe recordings, in vivo. J Neurosci Methods 154(1-2):204–24

88. Sakmann B and Neher E (1995) Single-Channel Recording (2nd ed). Plenum Press

89. Santhakumar V, Aradi I, Soltesz I (2005) Role of mossy fiber sprouting and mossy cell loss in hyperexcitability: a network model of the dentate gyrus incorporating cell types and axonal topography. J Neurophysiol 93:437–53

90. Santhanam G, Sahani M, Ryu S, Shenoy K (2004) An extensible infrastructure for fully automated spike sorting during online experiments. Conf Proc IEEE Eng Med Biol Soc, pp. 4380–4

91. Schaefer AT, Helmstaedter M, Sakmann B, Korngreen A (2003) Correction of conductance measurements in non-space-clamped structures: 1. Voltage-gated k(+) channels. Biophys J 84:3508–28

92. Schaefer AT, Helmstaedter M, Schmitt AC, Bar-Yehuda D, Almog M, Ben-Porat H, Sakmann B, et al. (2007) Dendritic voltage-gated K+ conductance gradient in pyramidal neurones of neocortical layer 5B from rats. J Physiol 579:737–52

93. Schreiber S, Fellous J-M, Tiesinga PH, Sejnowski TJ (2003) A new correlation-based measure of spike timing reliability. Neurocomputing 52-54:925–31

94. Surkis A, Peskin CS, Tranchina D, Leonard CS (1998) Recovery of cable properties through active and passive modeling of subthreshold membrane responses from laterodorsal tegmental neurons. J Neurophysiol 80(5):2593–607

95. Suzuki N, Takahata M, Sato K (2002) Oscillatory current responses of olfactory receptor neurons to odorants and computer simulation based on a cyclic AMP transduction model. Chem Senses 27:789–801

96. Taylor AL, Hickey TJ, Prinz AA, Marder E (2006) Structure and visualization of high-dimensional conductance spaces. J Neurophysiol 96:891–905

97. Terman D, Rubin JE, Yew AC, Wilson CJ (2002) Activity patterns in a model for the subthalamopallidal network of the basal ganglia. J Neurosci 22(7):2963–76

98. Tchernichovski O, Nottebohm F, Ho CE, Pesaran B, Mitra PP (2000) A procedure for an automated measurement of song similarity. Animal Behaviour 59:1167–76

99. Thomson DJ, Chave AD (1991) Jacknifed error estimates for spectra, coherences, and transfer functions. In: Advances in spectrum analysis and array processing (Haykin S, ed), pp. 58-113. Prentice Hall, Englewood Cliffs, NJ

100. Tóth TI and Crunelli V (2001) Estimation of the activation and kinetic properties of INa and IK from the time course of the action potential. J Neurosci Meth 111(2):111–26

101. Traub RD, Contreras D, Cunningham MO, Murray H, Lebeau FE, Roopun A, Bibbig A, et al. (2005) A single-column thalamocortical network model exhibiting gamma oscillations, sleep spindles and epileptogenic bursts. J Neurophysiol 93(4):2194–232

102. Vanier MC and Bower JM (1999) A comparative survey of automated parameter-search methods for compartmental neural models. J Comput Neurosci 7(2):149–71

103. Venkataramanan L and Sigworth FJ (2002) Applying Hidden Markov Models to the Analysis of Single Ion Channel Activity. Biophys J 82:1930–42

104. Weaver CM and Wearne SL (2006) The role of action potential shape and parameter constraints in optimization of compartment models. Neurocomputing 69:1053–57

105. Womelsdorf T, Fries P, Mitra PP, Desimone R (2006) Gamma-band synchronization in visual cortex predicts speed of change detection. Nature 439:733–36

106. Zhang K, Ginzburg I, McNaughton BL, Sejnowski TJ (1998) Interpreting neuronal population activity by reconstruction: unified framework with application to hippocampal place cells. J Neurophysiol 79:1017–44

107. Zohary E, Shadlen MN, Newsome WT (1994) Correlated neuronal discharge rate and its implications for psychophysical performance. Nature 370:140–43

Part IV

Cognitive Biology

15

Using Broad Cognitive Models to Apply Computational Intelligence to Animal Cognition

Stan Franklin[1] and Michael H. Ferkin[2]

[1] Institute of Intelligent Systems, FedEx Institute of Technology, The University of Memphis, Memphis, TN 38152, USA
franklin@memphis.edu
[2] Department of Biology, The University of Memphis, Memphis, TN 38152, USA
mhferkin@memphis.edu

Summary. The field of animal cognition (comparative cognition, cognitive ethology), the study of cognitive modules and processes in the domain of ecologically relevant animal behaviors, has become mainstream in biology. The field has its own journals, books, organization and conferences. As do other scientists, cognitive ethologists employ conceptual models, mathematical models and sometime computational models. Most of these models, of all three types, are narrow in scope, modeling only one or a few cognitive processes. This position chapter advocates, as an additional strategy, studying animal cognition by means of computational control architectures based on biologically and psychologically inspired, broad, integrative, hybrid models of cognition. The LIDA model is one such model. In particular, the LIDA model fleshes out a theory of animal cognition, and underlies a proposed ontology for its study. Using the LIDA model, animal experiments can be replicated in artificial environments by means of virtual software agents controlled by such architectures. Given sufficiently capable sensors and effectors, such experiments could be replicated in real environments using cognitive robots. Here we explore the possibility of such experiments using a virtual or a robotic vole to replicate, and to predict, the behavior of live voles, thus applying computational intelligence to cognitive ethology.

15.1 Introduction

The analysis of animal behavior cannot be complete without an understanding of *how* behaviors are selected, that is, without the study of animal cognition [1, 14, 17, 66]. The study of cognitive modules and processes in the domain of ecologically relevant animal behaviors (cognitive ethology) has become an exciting research area in biology. The field has its own journals (*e.g.*, *Animal Cognition*), books (*e.g.*, [12]), organization (Comparative Cognition Society) and conferences (*e.g.*, 2006 Comparative Cognition Society Annual Meeting, Melbourne, FL).

S. Franklin and M.H. Ferkin: *Using Broad Cognitive Models to Apply Computational Intelligence to Animal Cognition*, Studies in Computational Intelligence (SCI) **122**, 363–394 (2008)
www.springerlink.com   © Springer-Verlag Berlin Heidelberg 2008

As do other scientists, cognitive ethologists employ conceptual models (*e.g.*, [1]), mathematical models (*e.g.*, [2, 69]) and sometime computational models (*e.g.*, [87]). Most of these models, of any of the three types, are narrow in scope, modeling only one or a few cognitive processes. In contrast to these models, empirical studies of how animal behaviors are selected should be guided by comprehensive theories and integrated conceptual models. While grounded in the underlying neuroscience and consistent with it, these theories and models must be conceptually at a higher level of abstraction, dealing with higher-level entities and processes.

This position chapter advocates studying animal cognition by means of computational control architectures based on biologically and psychologically-inspired, broad, integrative, hybrid models of cognition. Using such a model, experiments with animals could be replicated in artificial environments with virtual software agents controlled by such architectures. Given sufficiently capable sensors and effectors, such experiments could be replicated in real environments using cognitive robots. The LIDA (Learning Intelligent Distribution Agent) model provides just the kind of broad, integrated, comprehensive, biologically and psychologically inspired theory that is needed. In particular, the LIDA model can model animal cognition, and underlies a proposed ontology for its study [57].

15.2 Software agents in robotic simulators as virtual animals

Autonomous agents [58] are systems embedded in, and part of, an environment that sense their environment and act on it, over time, in pursuit of their own agenda. In addition, they must act so as to potentially influence their future sensing, that is, they must be structurally coupled to their environment [76, 77]. Biological autonomous agents include humans, other animals and viruses. As well as computer viruses and some robots, artificial autonomous agents include *software agents*, that is, agents that "live" in computer systems, in databases, or in networks. The "bots" that autonomously explore the internet indexing web pages for Google are examples of software agents. Artificial autonomous agents also include cognitive robots [3, 18, 56].

Robotic simulators are software tools offering often 3D modeling, simulation and animation of any physical system. They are particularly designed as virtual environments for simulations of robots, hence the name. Examples include ARS MAGNA, RoboWorks, Rossum's Playhouse, Khepera Simulator, and very many others. Within such an abstract, virtual world with its own physics, a simulated robot can both sense and act so that it becomes an autonomous software agent. Such a simulated robot typically has a simulated body within the robot simulator.

Modeling such a simulated robot "living" within a robotic simulator after an animal, say a meadow vole, creates an *artificial animal* software agent.

The simulator, thought of as the artificial animal's environment, can be made to contain objects of various sorts, including other agents. Such objects can have simulated weight, rigidity and other realistic physical properties. Such other agents can be made to behave in a relatively realistic manner, as for example do the various agents that occur in video games. An artificial animal (AA) can sense this environment via artificial sights, sounds, odors, touches, tastes, etc., corresponding to the senses available to the animal (robot). AA's artificial effectors can manipulate artificial objects, including itself, in rather realistic ways as compared to a real robot, and in a more or less realistic manner as compared to an animal.

Behavioral experiments with animals typically may involve some sort of structure, such as a runway with two chambers at its end, or a maze. This structure can be simulated within the artificial environment, the robotic simulator. AA's "body" can then be placed appropriately within the simulated structure, an artificial run of the experiment carried out, and data gathered as to how AA responded to the experimental situation in terms of location, action, timing, etc. Repeated artificial runs will allow the virtual replication of the experiment and its dataset. Or, the virtual experiment can be run first. Its data would then predict the results of carrying out the experiment *in vivo*.

As any autonomous agent must, AA has a control structure that interprets its sensory input, selects an action according to its own agenda, and guides its actions on its environment. The computational architecture of any such AA control structure gives rise to a conceptual cognitive model, that is, a theory of how AA and the animal it simulates interprets its sensory data, and chooses and guides its actions. Conversely, any cognitive model can be implemented in a computational architecture that can be used to control AA. Thus, in principle, any scientific hypothesis that arises from the cognitive model can be tested using both a virtual experiment and an *in vivo* experiment.

Using such an experimental paradigm insures that the conceptual cognitive model that gives rise to the computational architecture of AA's control system will be broad and comprehensive. It must involve perception, which makes sense of sensory input. It must contain motivational elements and procedural memory with which to make action selections. Finally, it must include sensory-motor automatisms with which to execute actions. Such comprehensive cognitive models allow for the testing of a broader range of hypotheses than do more constrained models. They also enable more complete, and therefore more satisfactory, explanations of the cognitive processes responsible for the observed behavior. We contend that the adoption of such an experimental paradigm will result in significant advances in the way biological theory guides experimentation in animal behavior.

But there's more to the story. Computational architectures derived from integrated, comprehensive cognitive models are sure to be rife with internal parameters whose values must be tuned (discovered) before the system can perform properly as a control structure for AA. It is well known that a model with sufficiently many free parameters can be tuned so as to reproduce essentially any specific dataset. What is wanted is a tuned set of internal parameters whose values remain constant while a number of disparate datasets are reproduced. Such a tuned parameter set offers reassurance as to the accuracy and usefulness of the model. An inability to find such a tuned parameter set should warn its designers that something is amiss with the model, and that it needs revision. The particular parameters that resist such tuning point researchers to modules and process within the model that are likely to require revision.

But how are such parameters in a computational architecture to be tuned. The problem is essentially a search problem. Given the dataset from one previous *in vivo* experiment that the model should predict and explain, one searches for a set of parameter values that, when implemented, will allow AA to replicate this existing dataset. If found, this search procedure is iterated on the dataset of a second previously performed *in vivo* experiment resulting, hopefully, in a tuned parameter set that will allow the replication of both datasets. Further iteration of this procedure should, if the model is correct, yield a stable set of values for the internal parameters of the computational architecture that should work for replicating a number of different existing *in vivo* experiments. Thus, the replication of existing data sets from previously performed experiments will allow the tuning of internal parameters in the theoretical model. Parameters that resist such tuning over several different data sets indicate flaws in the model that must be repaired. This parameter tuning provides something like a metric for assessing the quality of a cognitive model as a basis for understanding the cognitive processes responsible for the behavior of AA.

In summary, a tuned version of the computational model will allow AA to successfully replicate essentially any simulatable experiment with the animal in question. Successfully accomplishing this goal will provide substantial evidence of the accuracy and usefulness of the conceptual cognitive model. Cognitive hypotheses from the model can then be tested by *in vivo* experiments with real animals to see if their data is predicted by running AA in the same experimental situations. If so, we will have shown the ability of the theoretical model to predict as well as to explain.

The authors propose that their LIDA cognitive model, to be described next, is an appropriate example of a broad, integrated, comprehensive model of the kind we are advocating. We have proposed this model previously as the source of a useful ontology for the study of animal cognition [57].

15.3 The LIDA cognitive model and its architecture

The LIDA model is a conceptual (and partially computational) model covering large portions of human and animal cognition. Based primarily on global workspace theory [4], the model implements and fleshes out a number of psychological and neuropsychological theories including situated cognition [97], perceptual symbol systems [11], working memory [7], memory by affordances [61], long-term working memory [31], and Sloman's [91] cognitive architecture. Viewed abstractly, the LIDA model offers a coherent ontology for animal cognition [57], and provides a framework in the sense of Crick and Koch [21] that can serve to guide experimental research. Viewed computationally, the model suggests computational mechanisms that can underlie and explain neural circuitry.

The LIDA computational architecture, derived from the LIDA cognitive model, employs several modules motivated by computational mechanisms drawn from the "new AI." These include the Copycat Architecture [63], Sparse Distributed Memory [67], the Schema Mechanism [30], the Behavior Net [74], and the Subsumption Architecture [15].

The LIDA model and its ensuing architecture are grounded in the LIDA cognitive cycle. Every autonomous agent [58], be it human, animal, or artificial, must frequently sample (sense) its environment, process (make sense of) this input, and select an appropriate response (action). Every agent's "life" can be viewed as pursuing a continual sequence of these cognitive cycles. Each cycle constitutes a unit of sensing, attending and acting. A cognitive cycle can be thought of as a moment of cognition, a cognitive "moment." Higher-level cognitive processes are composed of many of these cognitive cycles, each a cognitive "atom."

During each cognitive cycle [6, 55] the LIDA agent, be it animal or artificial, first makes sense of its current situation as best as it can. It then decides what portion of this situation is most in need of attention. Broadcasting this portion enables the agent to finally choose an appropriate action and execute it. Please note, that consciousness in the LIDA model refers to functional consciousness, which is the functional role of the mechanism as specified by Baars [4] global workspace theory. The LIDA model takes no position on the issue of phenomenal consciousness in animals.

The cycle (Fig. 15.1) begins with sensory stimuli from the agent's environment, both an external and an internal environment. Low-level feature detectors in sensory memory begin the process of making sense of the incoming stimuli. These low-level features are passed to perceptual memory where higher-level features, objects, categories, relations, situations, etc. are recognized. These recognized entities, comprising the percept, are passed to the workspace, where a model of the agent's current situation is continually being assembled and updated. The percept serves as a cue to two forms of episodic memory, transitive and declarative. The response to the cue consists of local associations, that is, remembered events from these two memories that were associated with elements of the cue. In addition to the current percept, the workspace contains recently previous percepts and the structures assembled from them that haven't yet decayed away. The model of the agent's current situation is assembled from the percept, the associations and the remaining previous models. This assembling process will typically require looking back to perceptual memory and even to sensory memory, to enable the understanding of relations and situations. This assembled new model constitutes the agent's

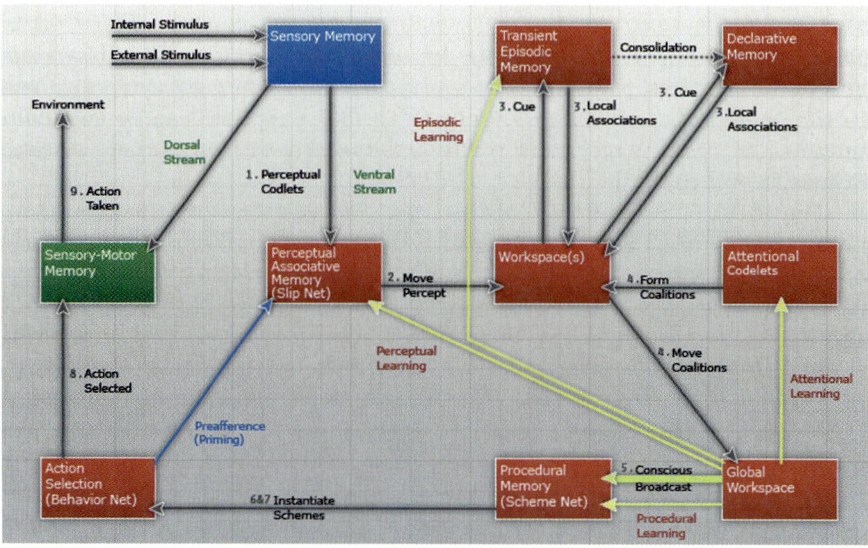

Fig. 15.1. The LIDA Cognitive Cycle.

understanding of its current situation within its world. It has made sense of the incoming stimuli.

For an agent "living" in a complex, dynamically changing environment, this current model may well be much too much for the agent to deal with at once. It needs to decide what portion of the model should be attended to. Which are the most relevant, important, urgent or insistent structures within the model? Portions of the model compete for attention. These competing portions take the form of coalitions of structures from the model. Such coalitions are formed by attention codelets, small, special purpose processors, each of which has some particular type of structure it wants to bring to consciousness. One such coalition wins the competition. The agent has decided on what to attend.

But, the purpose of all this processing is to help the agent decide what to do next. To this end, the winning coalition passes to the global workspace, the namesake of global workspace theory, from which its contents are broadcast globally. Though the contents of this conscious broadcast are available globally, the primary recipient is procedural memory, which stores templates of possible actions including their contexts and possible results. It also stores an activation value for each such template that attempts to measure the likelihood of an action taken within its context producing the expected result. Templates whose contexts intersect sufficiently with the contents of the conscious broadcast instantiate copies of themselves with their variables specified to the current situation. These instantiations are passed to the action selection mechanism, which chooses a single action from these instantiations and those remaining from previous cycles. The chosen action then goes to sensory-motor memory, where it picks up the appropriate algorithm (sensory-motor automatism) by which it is then executed. The action taken affects the environment, and the cycle is complete.

There are neural correlates for each of the modules and processes included in the LIDA cognitive cycle. For each such module or process, there is experimental data supporting these correlations.

The LIDA model hypothesizes that in all animals, including humans, cognitive processing is via a continuing iteration of such cognitive cycles. These cycles occur asynchronously, with each cognitive cycle taking roughly 200 ms in humans and closely related primates. The cycles cascade, that is, several cycles may have different processes running simultaneously in parallel. This cascading must, however respect the serial order of consciousness in order to maintain the stable, coherent image of the world with which consciousness endows us [54, 79]. This cascading, together with the asynchrony, allows a rate of cycling in humans of five to ten cycles per second. A cognitive "moment" is quite short! There is considerable empirical evidence from neuroscience suggestive of such cognitive cycling in humans and closely related primates [75, 89, 96, 99]. None of this evidence is conclusive.

Global workspace theory postulates that learning requires only attention [4]. In the LIDA model this implies that learning must occur with each cognitive cycle. More specifically, learning occurs with the conscious broadcast from the global workspace during each cycle. Learning in the LIDA model follows the established artificial intelligence principle of "generate and test" [65, 100]. New representations are learned in a profligate manner during each cognitive cycle (the generation). Those that are not sufficiently reinforced during subsequent cycles (the test) decay away. Three modes of learning, perceptual, episodic and procedural, employing distinct mechanisms [55, 81] have been designed and are in various stages of implementation. A fourth mode of learning is attentional learning, which has been contemplated but not designed.

Perceptual learning enables an agent to recognize features, objects, categories relations, situations, etc. It seems to be ubiquitous in animals. Episodic learning refers to the memory of events, the what, the where and the when [10, 94]. In the LIDA model such learning is stored in transient episodic memory [20, 55] and in the longer-term declarative memory [55]. At least episodic-like memory, that is episodic memory with no assumption of consciousness, has been demonstrated in many animal species [29] including meadow voles [47]. Procedural learning refers to the learning of new tasks and the improvement of old tasks. In the LIDA model such learning is accomplished in procedural memory [24]. Such procedural learning is widely observed in animal species (*e.g.*, [50]).

Every autonomous agent must be equipped with primitive motivators, sometimes called drives, that motivate its selection of actions. In humans, in animals, and in the LIDA model, these drives are implemented by feelings [59]. Such feelings implicitly give rise to values that serve to motivate action selection. Feelings also act as modulators to learning.

The LIDA theoretical model traverses several levels of biological complexity within the overall rubric of animal cognition. At the highest level it models entire organisms by means of software agents such as a virtual vole. At one step lower, it models various higher-level cognitive processes such as deliberation [51], volition [51], metacognition [101], automization, and non-routine problem solving [24]. Yet another step lower one finds cognitive modules and processes that operate within a single cognitive cycle, that is within a few hundred milliseconds. These lower-level processes include perception [53], various forms of memory [55], attention [6], learning [23], and action selection [82]. At yet a lower level, the nodes and links from LIDA's perceptual memory, implemented via a slipnet [63], provide the common representational currency throughout the model a la Barsalou's [11] perceptual symbol system. Taking a dynamical systems point of view, each such node may be thought of as representing a basin of attraction in the state space of some underlying cell assembly [90]. By spanning these various levels of theoretical complexity, the LIDA model can be expected to contribute to our understanding of several levels of the dynamics of living systems.

15.4 The natural history of the meadow vole, Microtus pennsylvanicus

Meadow voles are small secretive rodents that inhabit ephemeral grasslands in the northern and eastern portions of the United States and Canada. Much is known about their life history. Meadow voles also display striking seasonal differences in behavior. That is, they do most of their breeding during the spring and summer, when the photoperiod or day length is relatively long, about 14 hours of light per 24-hour period. At this time of year, female meadow voles become sexually receptive to males, producing odors that are attractive to males as well as displaying behaviors directed towards males [41–44, 48].

During the breeding season, female meadow voles are also territorial [73]. They defend their nests and territories often by behaving aggressively towards intruders. However, fighting is costly and not frequent [41]. Female meadow voles use other means to defend their territory. Specifically, females scent mark along the borders of their territories and near their nests, and over-mark the scent marks of male and female conspecifics that they encounter. By scent marking and over-marking, female meadow voles are able to delineate boundaries of their territories and also announce their residency in an area.

Male meadow voles are not territorial. Instead, they wander through large home ranges that encompass the territories of one or more females. Males often do not display overt behaviors against male conspecifics. They seldom fight with other males and they do not target the scent marks of other males and over-mark them [41, 42, 48]. However, males scent mark and over-mark in areas containing the marks of female meadow voles [42, 43]. This does not mean that male voles do not compete with one another, they do. However, male-male competition is more subtle; males compete with one another in two ways. First, males recall the location and reproductive condition of females that they encounter during their daily wanderings. That is, they display a memory for what, when, and where [47]. Second, once they locate females that are willing to mate with them, males will assess the risk and intensity of sperm competition. They do so, by determining whether the female has recently encountered other males. Male meadow voles investigate the area near the female, attempting to determine if other males have left their scent marks nearby. If so, the male, when he mates with the female, will increase his sperm investment 116% relative to his investment if he does not encounter fresh scent marks of other males nearby [25, 26, 28].

Although many males may mate with females, the variance in reproductive success among males is highly skewed, so that only a relatively small number of males actually sire offspring [13, 88]. What it is that makes these males more successful is not known. However, studies suggest that the more successful males 1) produce odors that are more attractive than those of other males to females [41], 2) have higher titers of prolactin and gonadal steroids relative to those of other males, which makes the former more attractive and interesting than the latter to females [71], 3) display more behaviors directed at attracting and showing their interest in females [42, 43, 45, 49], 4) have more copulatory interactions with females [25, 27], 5) are better fed than less successful males [84, 85], and 6) are older and more experienced than other male meadow voles [40].

As we mentioned above, meadow voles are seasonal breeders. They generally do not breed during the late fall and winter, when the day length is short and the daily photoperiod is less than 10 hours of light per 24-hour period. During the non-breeding season, females meadow voles relax their territorial borders, produce odors that are no longer attractive to males, but are attractive to females, display behaviors that are directed more at females than at males, and form communal nests with neighboring females and their last litters. Aggressive behavior between females is reduced and is replaced with affiliative and amicable acts [41, 73]. At this time of year, males generally produce odors that are no longer attractive to females. Few males direct behaviors towards females as potential mates. Males appear to be solitary during the winter [73]. During the winter, scent marking and over-marking behavior is no longer directed at opposite-sex conspecifics and with self-grooming behavior serve a role in maintaining the cohesiveness of members of that communal nest [39, 71].

15.5 Meadow voles and cognition - Some case studies

In this section, we summarize the results of some experiments on voles that imply a strong cognitive component to their behavior. For example, meadow voles can distinguish between unfamiliar and familiar conspecifics, littermates and non-littermates, and between sexually receptive and sexually quiescent opposite-sex conspecifics. Meadow voles respond preferentially to the odors of littermates relative to non-littermates by spending more time investigating the odors of the former as compared to those of the latter [33, 46]. Adult female voles behave amicably towards familiar females but not towards unfamiliar females, whereas adult male voles behave agonistically towards familiar males but not unfamiliar males [32]. Male voles over-mark the scent marks of females in heightened sexual receptivity, during postpartum estrous, as compared to those of females that in other states of sexual receptivity [42, 43].

Depending on the social context, the perceptual memory of voles may last several hours to several days [47, 49]. Perceptual memory can be fleeting or long term. For instance, a new person met briefly at a party may not be recognized a few weeks later, while a friend from childhood who hasn't been seen for decades may be recognized in spite of the changes brought by age. Perceptual and episodic memory (what, when, and where) depend to some extent, and in different ways, on association. In perceptual memory an object is associated with its features, a category with its members. Recall from episodic memory is accomplished in animals (and in at least some artificial agents) by means of associations with a cue. Improvement of performance during procedural learning is accomplished in animals by associating particular actions with desired results. Thus association plays different roles in the various memory systems and their various forms of learning, and can be expected to require distinct mechanisms.

First, we asked the question, is it possible for voles to have a sense of number? To address this question, we determined whether voles discriminate between two different scent-marking individuals and identify the individual whose scent marks was on top more often than the other individual [49]. We tested whether voles show a preference for the individual whose scent marks was on top most often. If so, the simplest explanation was that voles can make a relative size judgment, such as distinguishing an area containing more of one individual's over-marks as compared to less of another individual's over-marks. We found that voles respond preferentially to the donor that provided the greater number of over-marks as compared to the donor that provided the fewer number of over-marks. Thus, we concluded that voles might display the capacity for relative numerousness. Interestingly, female voles were better able than male voles in distinguishing between small differences in the relative number of over-marks by the two scent donors.

Next, we conducted a series of experiments to determine whether reproductive condition of female meadow voles affects their scent marking behavior as well as the scent marking behavior of male conspecifics [43]. We did so because, during the breeding season, the reproductive condition of female mammals changes. Females may or may not be sexually receptive. In experiment 1, females in postpartum estrus deposited more scent marks than females that were neither pregnant nor lactating, reference females or ovariectomized females (OVX females). In experiment 2, male voles scent marked more and deposited more over-marks in areas marked by postpartum estrus females than by reference and OVX females. In experiment 3, postpartum estrus females deposited more scent marks and over-marks in areas marked by males than did females in the other reproductive states. The results of these experiments showed that male and female voles may vary the number, type, and location of scent marks they deposit in areas scented by particular conspecifics.

We also tested the hypothesis that male meadow voles posses the capacity to recall the what, where, and when of a single past event associated with mate selection in two experiments [47]. Briefly, male voles were allowed to explore an apparatus that contained two chambers. One chamber contained a day-20 pregnant female (24 hours prepartum). The other chamber contained a reference female. Twenty-four hours after the exposure, the males were placed in the same apparatus, which was empty and clean. At this time, the pregnant female would have entered postpartum estrus, a period of heightened sexual receptivity. Males initially chose and spent significantly more time investigating the chamber that originally housed the pregnant female (now a postpartum estrus female) than the chamber that originally housed the reference female. Male voles also explored an apparatus containing a chamber with a postpartum estrus female and one chamber containing a reference female. Twenty-four hours later, males were placed into an empty and clean apparatus. The males did not display an initial choice and they spent similar amounts of time investigating the chamber that originally housed the postpartum estrus female (now a lactating female) and the chamber that originally housed the reference female. The results of these and additional experiments suggest that male voles may have the capacity to recall the what, where, and when of a single past event, which may allow males to remember the location of females who would currently be in heightened states of sexual receptivity.

We also examined the effects of winning and losing on over-marking behavior of mammals, a behavior associated with intrasexual aggression and competition [34]. We tested the hypothesis that meadow voles adjust their over-marking behavior according to aggressive interactions they had experienced with a same-sex conspecific. The hypothesis was partially supported. That is, female voles that won their encounter over-marked a greater proportion of their opponent's over-marks than did females that either lost their encounter or females that were evenly matched in their encounter. Females that lost their encounter and females that were evenly matched over-marked a similar proportion of their opponent's over-marks. Male voles, however, independent of whether they won, lost, or were evenly matched, over-marked a similar proportion of their opponent's scent marks. The present findings suggest over-marking may not play a major role in male-male competition, but likely plays a large role in female-female competition among meadow voles.

We also determined to what degree meadow voles display self-cognizance and use self-referent phenotype matching for self recognition (Ferkin *et al. unpubl. data*). We tested animals using habituation/dishabituation tasks in which they were exposed to their own current urine scent marks and those of either 1) unfamiliar same-sex conspecifics, 2) same-sex siblings, 3) their past selves (post gonadectomy with no steroid-hormone replacement), 4) their past selves (post gonadectomy with steroid-hormone replacement), and 5) their past selves (intact gonads). Briefly, we discovered that voles behaved as if the scent marks of their past and present selves were the same if the reproductive condition of the voles was not changed and from the same donor. If, however, the voles were gonadectomized and their reproductive condition changed, they behaved as if the scent marks of their past and present selves were the different donors.

Finally, we examined sperm competition in male meadow voles [25, 26, 28], Sperm competition occurs when a female copulates with two or more males and the sperm of those males compete within the female's reproductive tract to fertilize her eggs. The frequent occurrence of sperm competition has forced males of many species to develop different strategies to overcome the sperm of competing males. A prevalent strategy is for males to increase their sperm investment (total number of sperm allocated by a male to a particular female) after detecting a risk of sperm competition. It has been shown that the proportion of sperm that one male contributes to the sperm pool of a female is correlated with the proportion of offspring sired by that male. Therefore, by increasing his sperm investment a male may bias a potential sperm competition in his favor.

We showed that male meadow voles increase their sperm investment when they mate in the presence of another male's odors. Such an increase in sperm investment does not occur by augmenting the frequency of ejaculations, but by increasing the amount of sperm in a similar number of ejaculations. We also found that sperm investment of males exposed to the scent marks of five male conspecifics was intermediate between that of males exposed to the scent marks of one male and that of males exposed to no scent marks of conspecific males [26]. We have recently discovered that males do not increase their sperm investment if the donors of the scent marks are males that are in poorer condition than the male subject, but do so if the male donors are in similar or better condition than the subject male (Vaughn *et al. unpubl. data*). Thus, males can distinguish between different male donors and adjust their sperm investment accordingly. How they do so and what cognitive processes are involved in regulating the physiological response of the vas deferens in the male's testes is under investigation [28].

15.6 Hypotheses

A previous ontology provides a conceptual framework within which to conduct empirical research and fashion hypotheses [57]. Formulating hypotheses is one of the functions of mathematical, computational, and conceptual models. Thus, it's reasonable to formulate potentially testable hypotheses for the LIDA model. By doing so, we hope to encourage empirical testing of our hypotheses. Here we present a few selected testable hypotheses that may be tested with the current LIDA technology.

1. **The Cognitive Cycle**: The very existence of the cognitive cycle in various species, along with its timing (asynchronously cascading at a rate of roughly 5-10 hz) is a major hypotheses. Neuroscientists have provided suggestive evidence for this hypothesis [60, 62, 70].

2. **Perceptual Memory**: A perceptual memory, distinct from semantic memory but storing some of the same contents, exists in humans [55, 81], and in many, perhaps most, animal species.

3. **Transient Episodic-Like Memory**: Humans have a content-addressable, associative, transient episodic memory with a decay rate measured in hours [20]. While perceptual memory seems to be almost ubiquitous across animal species, we hypothesize that this transient episodic memory is evolutionary younger, and occurs in many fewer species [55]. We refer here to episodic-like memory instead of to episodic memory, as in humans, to avoid the controversy over phenomenal consciousness in animals, about which the LIDA model takes no position [47]. Further reference to episodic memory in non-human animals should be read as episodic-like.

4. **Consolidation**: A corollary to the previous hypothesis says that events can only be encoded (consolidated) in long-term declarative memory via transient episodic memory. This issue of memory consolidation is still controversial among both psychologists and neuroscientists (*e.g.* [72]). However, the LIDA model advocates such consolidation.

5. **Consciousness**: Functional consciousness is implemented computationally by way of a broadcast of contents from a global workspace, which receives input from the senses and from memory [4, 5].

6. **Conscious Learning**: Significant learning takes place via the interaction of functional consciousness with the various memory systems (*e.g.* [8, 92]). The effect size of subliminal learning is quite small compared to conscious learning. Note that significant implicit learning can occur by way of unconscious inferences based on conscious patterns of input [86]. All memory systems rely on attention for their updating, either in the course of a single cycle or over multiple cycles [55].

7. **Voluntary and Automatic Memory Retrievals**: Associations from transient episodic and declarative memory are retrieved automatically and unconsciously during each cognitive cycle. Voluntary retrieval from these memory systems may occur over multiple cycles using volitional goals.

8. **Deliberative, volitional decision making**: Such functionally conscious decisions that deliberatively choose between alternatives are, following Global Workspace Theory (Chapter 9 in [4]), are hypothesized in the LIDA model [51] to follow William James' ideomotor theory [64]. Thus a decision is reached in favor of a proposed alternative when no objection to

it is raised. Volitional decision making is inherently a multi-cyclic, higher-order cognitive process.

15.7 Connecting the LIDA Model and the behavior of a meadow vole

In what follows we will describe each of the steps in LIDA's cognitive cycle, stated in *italicized text* as if applying to a human, while also carrying along their application in the mind of a hypothetical male vole.

Imagine a male vole has turned a corner, and encountered scent marks from different conspecifics [37]. Some of these scent marks are old and some are fresh, some are overlapping and some are not. This male vole detects these marks, identifies the donors that deposited the marks, and spends more time investigating the most numerous and the freshest marks [35, 42, 43, 48, 49]. The male vole distinguishes between the different scent donors and responds preferentially to the donors that are of most interest to him. The most interesting donor may likely be a sexually receptive female with whom he would attempt to copulate [25]. The mechanism that the male voles used to discriminate between the different scent donors would likely have involved perceptual learning [57] Keep in mind that the cognitive cycle to be described takes, in total, only a fifth of a second or so to complete.

Here are the nine steps of the LIDA cognitive cycle together with an example interpretation in the mind of our assumed male vole.

1. **Perception**. Sensory stimuli, external or internal, are received and interpreted by perception producing meaning. Note that this step is preconscious.

   In its perceptual memory the male vole categorizes the scent marks as being from males or females (a category), as known (an individual), and as sexually receptive (a feature) [37, 38]. During this step our vole scans its perceptual memory and makes associations between scent marks and scent donors, assessing the identity, sex, and reproductive condition of the scent donors [35, 42, 43, 49].

   This perceptual memory system identifies pertinent feeling/emotions along with objects, categories and their relations.

   In the male vole, feeling nodes for interest and for sexual arousal are somewhat activated. If this is a sexually receptive female, for example, all of these activated nodes are over threshold and become part of the percept.

2. **Percept to Preconscious Buffer**. The percept, including some of the data plus the meaning, is stored in preconscious buffers of LIDA's working memory. In humans, these buffers may involve visuo-spatial, phonological,

and other kinds of information. Feelings/emotions are part of the preconscious percept.

For the male vole, the percept has identified the freshest scent marks coming from a female in postpartum estrus, a highly sexually receptive female. These females readily mate when they encounter males. However, females are only receptive to males for 12 hours after they deliver pups [42].

3. **Local Associations**. Using the incoming percept and the residual contents of the preconscious buffers (content from precious cycles not yet decayed away), including emotional content, as cues, local associations are automatically retrieved from transient episodic memory (TEM) and from declarative memory.

   The contents of the preconscious buffers, together with the retrieved local associations from TEM and declarative memory, roughly correspond to Ericsson and Kintsch's [31] long-term working memory and to Baddeley's episodic buffer [9]. These local associations include records of the agent's past feelings/emotions, and actions, in associated situations.

   Assuming that our male vole possesses declarative memory, the retrieved local associations may include the memory of a previous sexual encounter with this particular female and his reaction to her, a memory for what, when, and where [47] For example, our male vole may have a memory of this female, when she was not in postpartum estrus, but simply pregnant and not sexually receptive [37, 38], which allows our male vole to anticipate that this female will only be in postpartum estrus for a few hours, and then she becomes not interested in mating. Although such expectation may come from either perceptual memory or semantic memory, anticipating the what (a female is highly sexually receptive for a relatively narrow window), the when (a female may no longer be highly sexually receptive), and the where (the location of that female relative to other female voles in the area), suggest that such processing may involve an episodic-like memory [47].

4. **Competition for Attention**. Coalitions of perceptual and memory structures in the workspace compete to bring relevant, important, urgent, or insistent situations to consciousness. (Consciousness here is required only in the functional sense as defined in global workspace theory and as defined by its role in the middle steps of this cognitive cycle. Phenomenal (subjective) consciousness is not assumed.) The competition may also include such coalitions from a recently previous cognitive cycle. Present and past feelings/emotions influence this competition for consciousness. Strong affective content strengthens a coalition's chances of being attended to [52].

   In the male vole, one coalition that is on the lookout for sexual opportunities will carry the other vole's identity, her reproductive status and readiness to mate, some details of the previous encounter, and the feelings associated with the current percept and the previous encounter. This coalition will compete with other such coalitions for "consciousness," but may not win the competition. Suppose our male's first encounter with that female's odor indicated that she has also attracted the attention of a predator, (fresh weasel scent marks are present), which has also become part of the percept, along with a strong fear. In this case, another coalition on the lookout for danger may well win the competition, and the male vole may not respond by seeking out this female.

5. **Broadcast of Conscious Contents**. A coalition carrying content gains access to the global workspace. Then, its contents are broadcast throughout the system.

In humans, this broadcast is hypothesized to correspond to phenomenal consciousness. No such assumption is made here. The conscious broadcast contains the entire content of consciousness including the affective portions.

Now imagine that the male vole did not detect a predator's odor and that the coalition about the female vole was attended to, that is, it came to his "consciousness."

Several types of learning occur. The contents of perceptual memory are updated in light of the current contents of consciousness, including feelings/emotions, as well as objects, categories, actions and relations. The stronger the affect, the stronger the encoding is in memory.

In the male vole, possibly along with others, representation in perceptual memory for the particular female vole, for the category of female voles, for readiness to mate, and for sexual interest would each be strengthened.

Transient episodic memory is also updated with the current contents of consciousness, *including feelings/emotions*, as events. *The stronger the affect, the stronger would be the encoding in memory.* (At recurring times not part of a cognitive cycle, the contents of transient episodic memory are *consolidated* into long-term declarative memory.)

If the male vole possesses a transient episodic memory, and studies suggest that he may [47], the event of having again encountered this particular female vole, her condition, and his reaction to her would be encoded, taking information from the "conscious" broadcast.

Procedural memory (recent actions) is updated (reinforced) with the strength of the reinforcement influenced by the strength of the affect.

For the male vole, the prior acts of turning the corner and sniffing the encountered scent marks would be reinforced. In this case, both acts would have been learned and become familiar.

Thus, perceptual, episodic and procedural learning occur with the broadcast in each cycle.

6. **Recruitment of Resources**. Relevant behavior representations respond to the conscious broadcast. These are typically representations whose variables can be bound from information in the conscious broadcast.

   The responding representations may be those that can help to deal with the current situation. Thus consciousness solves the relevancy problem in recruiting internal resources with which to deal with the current situation. The affective content (feelings/emotions), together with the cognitive content, helps to attract relevant behavioral resources.

   For the male vole, possibly among others, behavior representations for turning the head, for turning the body, for sniffing the scent marks and for moving in the direction that the female vole was traveling, may respond to the information in the broadcast.

7. **Setting Goal Context Hierarchy**. The recruited behavior representations use the contents of consciousness, including feelings/emotions, to instantiate new goal context hierarchies, bind their variables, and increase their activation.

   Goal contexts are potential goals, each consisting of a coalition of behaviors, which, together, could accomplish the goal. Goal context hierarchies can be thought of as high-level, partial plans of actions. It is here that feelings and emotions most directly implement motivations by helping to instantiate and activate goal contexts, and by determining which terminal goal contexts receive activation. Other, environmental, conditions determine which of the earlier goal contexts receive additional activation.

   For the male vole, a goal context hierarchy to seek out the female vole would likely be instantiated in response to information from the broadcast.

8. **Action Chosen**. The action selection mechanism chooses a single behavior, perhaps from a just instantiated goal context or possibly from a previously active goal context.

   This selection is heavily influenced by the various feelings/emotions. The choice is also affected by the current situation, external and internal conditions, by the relationship between the behaviors, and by the residual strengths of various behaviors.

   In the male vole, there may have been a previously instantiated goal context for avoiding the weasel previously sensed. An appropriate behavior in avoiding the predator may be chosen in spite of the presence of the female vole. Alternatively, a beginning step in the goal context for approaching and exploring the female vole may win out.

9. **Action Taken**. The execution of a behavior results in its action being performed, which may have external or internal consequences, or both. This is LIDA taking an action.

*If this particular male that has had few opportunities to copulate with a
female, searching for the female would likely have been selected, resulting in
behavior codelets acting to turn the male in the direction of the female, to
sniff, and to begin his approach. If on the other hand, our vole has frequent
opportunities to mate with females, he may stop his search for this female
when he encounters the odor of a weasel or a male conspecific* [25].

15.8 Sample experiments for tuning a Virtual Vole

The computational LIDA architecture is composed of a number of closely interconnected modules with their associated processes. Their implementation is
outlined in Table 1 below, which specifies the conceptual name of the module,
the name of its implementation in the architecture, the source of inspiration
for the data structure and algorithms employed, and references to detailed
explanations.

Table 15.1. LIDA modules and their implementations

Module	Implementation	Source	References
Perceptual Memory	Slipnet	Copycat Architecture	[56, 63]
Transient Episodic Memory	Sparse Distributed Memory	Sparse Distributed Memory	[22, 67]
Declarative Memory	Sparse Distributed Memory	Sparse Distributed Memory	[22, 67]
Procedural Memory	Scheme Net	Schema Mechanism	[24, 30]
Action Selection	Behavior Net	Behavior Net	[74, 82]

Each of the LIDA modules and their associated processes involve a number
of internal parameters that must be specified before the model can be used
to replicate experimental data. Such specification of parameters, the tuning
of the model, is typically done by trial and error so as to induce the model
to replicate the data from one specific experiment. This provisionally tuned
model is then further tuned to replicate data from both the original experiment and a second experiment. The model is then considered tuned, and
ready to try on other, prospective, experiments.

Some change in the model would be needed if it proves difficult or impossible to successfully tune some parameter. In that case, one must conclude that

the data structure or something in one of its associated algorithms requires adjustment. Thus, the difficulty of tuning the model serves as a sort of implicit metric measuring the correctness of the model.

In the next subsections we describe experiments with voles that might be expected to serve to tune a Virtual Vole, software agent operating within a robotic simulator and simulating a live vole.

15.8.1 Odor preference tests for tuning the virtual vole

In a previous experiment, we quantified the olfactory response of reproductively active voles to the odors of reproductively active same- and opposite-sex conspecifics. The position of the male or female donor was varied on the left- or right-side of the Y-maze to prevent any side bias displayed by the subject [41]. We recorded, continuously for 5 minutes, the amount of time male and female subjects investigated the baskets containing the donor voles. Thus, we showed that voles discriminate between and respond preferentially to opposite-sex conspecifics over same-sex conspecifics [41].

We also performed opposite-sex donor tests in which male subjects were exposed to scent marks of ovariectomized + blank treated females (not sexually receptive) and ovariectomized + estradiol treated females (sexually receptive), and female subjects were exposed to scent marks of gonadectomized + blank treated males (not sexually receptive) and gonadectomized + testosterone treated males (sexually receptive). Each male and female subject was exposed to a unique pair of opposite-sex odor donors. We found that male and female subjects spent more time investigating opposite-sex conspecifics given hormone replacement than opposite-sex conspecifics not given hormone replacement. Thus, voles prefer opposite-sex conspecifics that are sexually receptive to those that are not sexually receptive [36–38].

The Y-maze apparatus used in Ferkin and Seamon [41] can be simulated within a robotic simulator along with the various scent markings. This would allow a virtual vole to act as subject. Knowing ahead of time the desired range of results would allow the tuning of the various parameters in the several modules of the LIDA architecture as implemented in the virtual vole. As mentioned above, replication of these and other such experiments would allow the testing of the implementation of the LIDA model in control of the virtual vole.

15.8.2 Episodic-like memory tests for verifying the tuning of the virtual vole

As the initial tuning of the internal parameters of the virtual vole would have been done using odor preference tests, the question remains of whether these tunings are specific to only those tests. Or, is the tuning of parameters sufficiently general to make the virtual vole a good simulation of live meadow voles in a variety of experimental situations? We suggest testing for the generality of the parameter tuning by replicating other previously performed experiments with live meadow voles. One possibility would be the experiments on episodic-like memory.

Episodic-like memory, the memory for events, allows an animal to recollect the what, the where and the when of what happened. The LIDA model asserts that such episodic-like memory comes in two forms, transient episodic-like memory and declarative memory. Transient episodic-like memory lasts only a relatively short period of time, a few hours or a day in humans. In voles with their much shorter life span, it may be reasonable to assume an even more rapid decay in transient episodic-like memory. Declarative memory is long-term episodic-like memory that in humans may last for decades or a lifetime. This section describes already completed experiments that may be replicated with virtual voles to further test and adjust the tuning of their parameters. They also serve as background for possible future experiments designed to tease out the distinction between transient episodic-like memory and declarative memory in voles, if the latter exists.

Despite the controversy swirling around the ability of animals to recollect specific aspects of past events [19, 95], it is not difficult to imagine that some animals may use information from such past events to secure a mate. An important feature that often characterizes most non-human mammals is that females do not mate with males when they are not in a heightened state of sexually receptivity, such as estrus or postpartum estrus [16]. Thus, for many species of mammals, and particularly the majority of whom in which opposite-sex conspecifics live separately during the breeding season, males should be able to discriminate among females in different states of sexual receptivity. They should be able to identify females that are in a heightened reproductive state, their location, and the amount of time that the females are in this heightened state. Such a capacity would benefit, for example, a male meadow vole, a microtine rodent.

Adult male and female meadow voles live separately during the breeding season. At this time of year, female voles tend to occupy territories that are fixed spatially, but are dispersed widely across the home range of several males [73]. Female voles are induced ovulators and do not undergo estrous cycles [78, 80]. Thus, the reproductive condition and sexual receptivity varies among female voles during the breeding season. That is, female voles may be pregnant, lactating, both pregnant and lactating, neither pregnant nor lactating, or in a period of heightened sexual receptivity during postpartum estrus [68]. Postpartum estrus females are more likely to mate with a male than females that are not pregnant or lactating, or females that are pregnant, lactating or both [27, 42, 43].

Sexual receptivity in female varies and they enter PPE asynchronously. To increase his fitness, male meadow voles should mate with as many females as possible [13], particularly those females that have entered postpartum estrus [27, 42, 43]. Thus, we hypothesize that after a single visit to a female, male voles would later recollect her previous reproductive state (what); her location (where), and how long she would be in that reproductive state (when) [47], thus demonstrating episodic-like memory. The experimental design of this experiment [47] is described below.

All female voles were between 125-135 days of age when used in the tests. Female meadow voles do not undergo estrus cycles [68, 80]. To represent different levels of female receptivity, we used females that were pregnant for 20 days (day 20 pregnant), in postpartum estrus, females that were not pregnant or lactating, termed reference females, and day 2 lactating females. Gestation lasts 21 days in voles, thus day 20 pregnant female voles deliver their litters within 24 hours [68].

Immediately after parturition, these females enter postpartum estrous (PPE), a period of heightened sexual receptivity, which lasts 8-12 hours [27, 42, 43, 68]. The postpartum estrus females had delivered pups 4-6 hours prior to testing. Reference females were not currently pregnant or lactating [38]. The reference females had previously delivered a litter about 3-4 weeks before being used in the experiment (see below); these females had lived singly for approximately 21 days before testing began.

In experimental conditions 4 and 5 (see below), we used females that were in their second day of lactation for each condition. Lactation is 14-16 days in duration, and pups are weaned when they are 16-18 days old [68]. The day 2 lactating females were no longer in postpartum estrus and thus were no longer in a heightened state of sexual receptivity [38]. The postpartum estrus females and day 2 lactating females had not lived with their mate for 17 and 18 days, respectively, before the testing began.

It is important to note that postpartum estrus female voles are in a heightened state of reproductive receptivity and readily mate with males [27, 38, 68]. In contrast, reference females, day 20 females, and day 2 lactating females are not in a heightened state of sexual receptivity, but they may mate [25, 26, 38]. In addition, postpartum estrus females produce odors that are more attractive to males relative to those produced by females that are day 20 pregnant, day 2 lactating, or reference females, who produce odors that are similar in their attractiveness to males [38, 42, 43].

All behavioral observations were performed on voles placed in a T-shaped apparatus (Fig. 15.2). We used two opaque Plexiglas cages with wired tops for observation purposes. The large boxes served to house the female donors. There was a transparent divider with small holes between the females' living area and the area that males explored. This divider allowed males to investigate the female's living area without coming into direct contact with that female.

15.8.3 Test for Episodic-Like Memory

We conducted an experiment, with five experimental conditions, in which male subjects were exposed to unique female donors [47]. Each experimental condition contained two phases, an exposure phase and a test phase. In both phases of the five experimental conditions, a male meadow vole from one of the above treatment groups was placed into the starting box located at the base of the T-shaped arena (Fig. 15.2) for 30 seconds before the gate was lifted and the male was allowed to explore the entire apparatus. Each male underwent a single exposure and single test (see below).

Experimental Condition 1 – During the exposure phase, male voles were placed into an apparatus that housed a reference female in one box and a day 20 pregnant female in the other box (Fig. 15.2). During the exposure phase, we recorded continuously for 10 minutes, the total amount of time male voles spent in the arms of the apparatus that housed each female donor (Fig. 15.2). We also noted the position of the home-boxes (left- or right-side of the apparatus) that housed each particular female donor. The position of a particular female's home-box in the left- or right-side of the apparatus was alternated for each male subject during the exposure phase. After the 10-minute exposure, the male was returned to its own cage. Then, we disconnected the two-female home-boxes from the apparatus, and cleaned and disinfected the apparatus.

The test phase took place 0.5 hour after the exposure phase. During the test phase, the male voles were re-introduced into the apparatus that now contained boxes that housed no female donors; the boxes contained only clean wood chip bedding. We recorded continuously for 10 minutes, the total amount of time that male voles spent investigating the arm of the apparatus that previously housed the reference female that they were exposed to and the arm that previously housed the day 20 pregnant female. During the test phase male voles spent similar amounts of time investigating the arm of the apparatus that would have housed the day 20 pregnant female and the arm of the apparatus that would have housed the reference female [47].

Experimental Condition 2 – Male voles were exposed to an arena containing a day 20 pregnant female and a postpartum estrus female. 0.5 hour later, male voles were allowed to investigate an empty arena. We recorded the initial choice of the male vole and the amount of time that he spends in both arms of the arena. During the test phase male voles spent more time investigating the arm of the apparatus that would have housed the postpartum estrus females than the arm of the apparatus that would have housed the day-20 pregnant female [47].

Experimental Condition 3 – Male voles were exposed to an arena containing a day 2 lactating female and a reference female. 0.5 hour later, male voles were allowed to investigate an empty arena. We recorded the initial choice of the male vole and the amount of time that he spends in both arms of the arena. During the test phase male voles spent similar amounts of time investigating the arm of the apparatus that would have housed the day 2 lactating female and the arm of the apparatus that would have housed the reference female [47].

Experimental Condition 4 – Male voles were exposed to an arena containing a day 20 pregnant female and a reference female. 24 hours later, male voles were allowed to investigate an empty arena. The test phase took place 24 hours after the exposure phase. At this time, the day 20 pregnant female had delivered pups and had entered into postpartum estrus. During the test phase, the male voles were re-introduced into the apparatus that now contained boxes that housed no female donors; the boxes contained only clean wood chip bedding (Fig. 15.2). During the test phase, which occurred 24 hours after the exposure phase, males spent more time investigating the arm of the apparatus that would have contained the postpartum estrus female than the arm of the apparatus that would contained the reference female [47].

Experimental Condition 5 – Male voles were exposed to an arena containing a postpartum estrus female and a reference female. Twenty-four hours later, male voles were allowed to investigate an empty arena. We recorded the initial choice of the male vole and the amount of time that he spends in both arms of the arena. During the test phase, which occurred 24 hours after the exposure phase, male voles spent similar amounts of time investigating the arm of the apparatus that would have housed the day 2 lactating female and the arm of the apparatus that would have housed the reference female [47].

The results of these experiments suggest that male voles may have the capacity to recall the what, where, and when of a single past event, which may allow males to remember the location of females who would currently be in heightened states of sexual receptivity. Viewed from the LIDA model, the outcomes of Experimental conditions 1-3 indicate recollection of an event after a time interval of 0.5 hour between the exposure of a subject male vole to female odor and its later testing can be attributed to transient episodic-like memory [47].

In that the life span of a meadow vole is only about four months in the wild and approximately 18 months in captivity [40, 88], we suspect that the bottom end of the time span for testing for long-term episodic-like memory would be 24 hours or less. However, experiments can be repeated using a 48 hour time interval, which would correspond to long-term episodic memory in humans.

15.8.4 Replication of Episodic-Like Memory experiments using LIDA model virtual voles

. These tests would involve placing a virtual vole in a virtual arena that simulates that described for real voles (Fig. 15.2) and following experimental methods for virtual voles identical to those of the episodic-like memory experiments described above. Specifically, we will use virtual voles and a virtual arena to replicate the tests described above for a live vole in experimental conditions 1-5. During the test trials with the virtual voles, we will identify the initial choice of the virtual male voles and the total amount of time that they spend investigating the arm of the apparatus that previously housed the virtual conspecific females. By doing so, we would be able to compare the response of the virtual male vole with those of the live male voles and test the efficacy of the LIDA model for predicting the behavior of voles. Successful replication of these episodic-like memory experiments with a virtual vole would demonstrate the efficacy of one aspect of the LIDA model. Also, replication of these experiments, both *in vivo* and virtual, would allow the LIDA model to distinguish transient episodic memory in voles with its rapid decay rate from declarative (long-term episodic-like) memory, which can last a lifetime (See hypothesis 3 above).

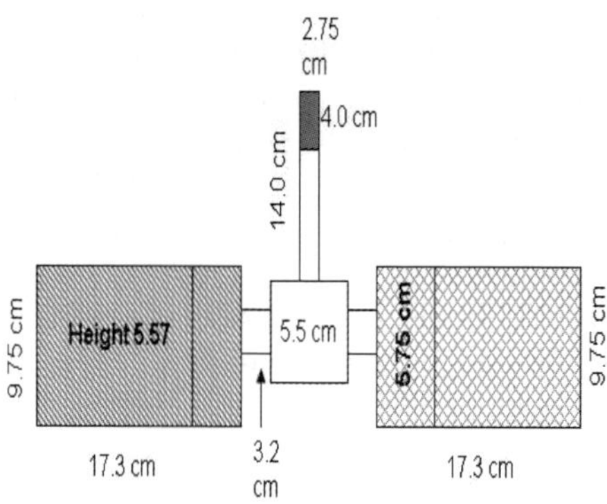

Fig. 15.2. The testing arena for Episodic-like memory in voles.

15.8.5 A possible experiment for testing a LIDA hypothesis

With a properly tuned virtual vole in hand it becomes possible to test the various hypotheses listed above that are derived from the LIDA model. In this section we suggest one such possible experiment designed to test Hypothesis 8. The earlier hypotheses involve processes that are thought to operate within a time frame of a very few hundred milliseconds, making them difficult, though not impossible, to test using live animals. A test of Hypothesis 3 was described in the previous section.

15.8.6 Testing for Volitional Decision Making in Meadow Voles

Hypothesis 8 predicts that some animals are capable of deliberative, volitional decision making. Humans deliberate and make volitional decisions. Do other animals such as meadow voles have this ability? In many animal experiments the subject is faced with a forced choice of response to a stimulus, say push this lever or that. Such experimental situations almost always have involved learning on the part of the subject. In this case, the subject's action selection is likely to have resulted from perceptual recognition and learned action selection, all within a single cognitive cycle, rather than from deliberative decision making. On the contrary, studies of searching for live prey suggest that the jumping spider, *Portia labiata*, may engage in deliberate decision making. In the field studies [98] these spiders were observed to spend a number of tens of minutes out of sensory contact, circling behind and above a prey spider, before lowering itself on a thread and ambushing the prey, which has appeared in a location that was "anticipated" by the spider. Such ambush behavior would seem to require deliberation, and even planning. This behavior on the part of these jumping spiders has also been tested experimentally by Tarsitano [93]. Here we suggest a version of Tarsitano's [93] experiments, adapted to test the hypothesis that meadow voles are capable of making decisions deliberatively. This section will briefly describe such an experiment.

The experimental apparatus consists of a relatively simple maze together with a platform above the maze from which the entire maze can be viewed through a transparent floor. The maze has two disjoint zigzagged arms that interleave with one another in three dimensions in some complex way, with the ends of the two arms separated.

In the exposure phase of the experiment the subject male vole has the run of the platform from one end of which he can see and smell a postpartum estrus female vole positioned at the end of one of the arms of the maze. The subject male vole can sense but not approach the female vole, and can inspect the maze below the platform through its transparent floor. In the test phase of the experiment the subject male vole is positioned at the beginning of the maze where he is faced with a choice of the two entrances of the two arms, and where he is unable to detect the postpartum estrus female vole. Having no procedural learning on which to depend, but only the perceptual learning from his inspection of the maze from his earlier vantage point on the platform, the subject vole, faced with the entrances to the two arms, must carry out a deliberative selection of which of the two arms to explore to encounter the postpartum estrus female. Based on the preferences of male voles, for postpartum estrus females [49], a male would demonstrate deliberative decision making by initially choosing and exploring the arm that will bring him into contact with the postpartum estrus female. Such a response by the male voles would have been the result of prior deliberative planning and a prior volitional decision to seek the postpartum estrus female along the appropriate arm. Such a decision will have likely occurred while the subject vole was exploring the platform, and discovered that one arm of the maze led to the postpartum estrus female and the other arm did not lead to the postpartum estrus female. This choice cannot be successfully made with the sensory information available to the subject vole positioned at the beginning of the maze. The capacity to make the appropriate choice and choose the direct path to the female would provide support for the hypothesis that meadow voles make deliberative decisions.

15.9 Experimenting with a Cognitive Robot

In principle, it should be possible to perform real world experiments using an artificial animal, say an artificial vole, in the form of a cognitive robot controlled by some cognitive architecture based on, for example, the LIDA model. Using such a cognitive robot would retain all the benefits described above for the use of software agent simulations of animals, say virtual voles. In addition, the use of such artificial animals/cognitive robots might be expected to reveal real world issues or difficulties that could be obscured by the use of virtual animals in a simulated environment.

The major problem with designing cognitive robots for such a purpose would seem to be sensing. It is difficult to imagine an artificial vole with the acute sense of smell of a real vole. With the advent of nanotechnology and other new techniques, artificial olfaction is becoming a reality [83]. Replicating experiments using cognitive robots as artificial animals may someday become a reality.

15.10 Conclusion

We conclude that it is in principle possible to employ virtual animals in the form of software agent simulations to benefit biological theory. Controlled by cognitive architectures such as LIDA, such virtual animals allow biologists to test their theories directly by replicating experiments within a virtual environment. To do so requires that the controlling cognitive architecture, like LIDA, be sufficiently broad and comprehensive to serve to control a software agent. Thus broad, comprehensive theories of animal cognition should prove themselves of value to biologists.

References

1. Allen C (1997) Animal cognition and animal minds. In: Machamer P, Carrier M (eds) Philosophy and the Sciences of the Mind. Pittsburgh University Press and the Universittsverlag Konstanz, Pittsburgh, PA, pp. 227–243
2. Alsop B (1998) Receiver operating characteristics from non-human animals: Some implications and directions for research with humans. Psychonomic Bulletin & Review 5:239–252
3. Asada M, MacDorman KF, Ishiguro H, Kuniyoshi Y (2001) Cognitive developmental robotics as a new paradigm for the design of humanoid robots. Robotics and Autonomous Systems 37:185–193
4. Baars BJ (1988) A Cognitive Theory of Consciousness. Cambridge University Press, Cambridge
5. Baars BJ (2002) The conscious access hypothesis: Origins and recent evidence. Trends in Cognitive Science 6:47–52

6. Baars BJ, Franklin S (2003) How conscious experience and working memory interact. Trends in Cognitive Science 7:166–172
7. Baddeley AD, Hitch GJ (1974) Working memory. In: Bower GA (ed), The Psychology of Learning and Motivation. Academic Press, New York, NY, pp. 47–89
8. Baddeley AD (1993) Working memory and conscious awareness. In: Collins A, Gathercole S, Conway MA, Morris P (ed), Theories of memory, Erlbaum, Howe, pp. 11–28
9. Baddeley AD (2000) The episodic buffer: A new component of working memory? Trends in Cognitive Science 4:417–423
10. Baddeley A, Conway M, Aggleton J (2001) Episodic Memory. Oxford University Press, Oxford, UK
11. Barsalou LW (1999) Perceptual symbol systems. Behavioral and Brain Sciences 22:577–609
12. Bekoff M, Allen C, Burghardt GM (2002) The Cognitive Animal. MIT Press, Cambridge, MA
13. Boonstra R, Xia X, Pavone L (1993) Mating system of the meadow vole, Microtus pennsylvanicus. Behavioral Ecology 4:83–89
14. Boysen ST, Himes GT (1999) Current issues and emerging theories in animal cognition. Ann. Rev. Psych. 50:683–705
15. Brooks RA (1991) How to build complete creatures rather than isolated cognitive simulators. In: VanLehn K (ed), Architectures for Intelligence. Lawrence Erlbaum Associates, Hillsdale, NJ, pp. 225–239
16. Bronson FH (1989) Mammalian Reproductive Biology. University of Chicago Press, Chicago, IL
17. Byrne RW, Bates LA (2006) Why are animals cognitive? Current Biology 16:R445–R448
18. Clark A, Grush R (1999) Towards a cognitive robotics. Adaptive Behavior 7:5–16
19. Clayton NS, Griffiths, DP (1998). Episodic-like memory during cache recovery by scrub jays. Nature 395:272–274
20. Conway MA (2002) Sensory-perceptual episodic memory and its context: Autobiographical memory. In: Baddeley A, Conway M, Aggleton J (eds), Episodic Memory. Oxford University Press, Oxford, UK, pp. 53–70
21. Crick F, Koch C (2003) A framework for consciousness. Nature Neuroscience 6:119–126
22. D'Mello SK, Ramamurthy U, Franklin S (2005) Encoding and Retrieval Efficiency of Episodic Data in a Modified Sparse Distributed Memory System. Proceedings of the 27th Annual Meeting of the Cognitive Science Society. Stresa, Italy
23. D'Mello SK, Franklin S, Ramamurthy U, Baars BJ (2006) A Cognitive Science Based Machine Learning Architecture. AAAI 2006 Spring Symposium Series Sponsor: American Association for Artificial Intelligence. Stanford University, Palo Alto, California, USA
24. D'Mello SK, Ramamurthy U, Negatu A, Franklin S (2006) A Procedural Learning Mechanism for Novel Skill Acquisition. In: Kovacs T, Marshall JAR (eds) Proceeding of Adaptation in Artificial and Biological Systems, AISB'06, Vol 1., Society for the Study of Artificial Intelligence and the Simulation of Behaviour, Bristol, England, pp. 184–185

25. delBarco-Trillo J, Ferkin, MH (2004) Male mammals respond to a risk of sperm competition conveyed by odours of conspecific males. Nature 431:446–449

26. delBarco-Trillo J, Ferkin MH (2006) Male meadow voles respond differently to risk and intensity of sperm competition. Behavioral Ecology 17:581–585

27. delBarco-Trillo J, Ferkin MH (2007) Female meadow voles, Microtus pennsylvanicus, experience a reduction in copulatory behavior during postpartum estrus. Ethology 113:466–473

28. delBarco-Trillo J, Ferkin MH (2007) Increased sperm numbers in the vas deferens of meadow voles, Microtus pennsylvanicus, in response to odors of conspecific males. Behavioral Ecology and Sociobiology 61:1759-1764

29. Dere E, Kart-Teke E, Huston JP, De Souza Silva MA (2006) The case for episodic memory in animals. Neuroscience and Biobehavioral Reviews 30:1206–1224

30. Drescher GL (1991) Made-Up Minds: A Constructivist Approach to Artificial Intelligence. MIT Press, Cambridge, MA

31. Ericsson KA, Kintsch W (1995) Long-term working memory. Psychological Review 102:211–245

32. Ferkin MH (1988) The effect of familiarity on social interactions in meadow voles, Microtus pennsylvanicus: A laboratory and field study. Animal Behaviour 36:1816–1822

33. Ferkin MH (1989) Adult-weanling recognition among captive meadow voles (Microtus pennsylvanicus). Behaviour 118:114–124

34. Ferkin MH (2007) Effects of previous interactions and sex on over-marking in meadow voles. Behaviour. 144:1297–1313

35. Ferkin MH, Dunsavage J, Johnston RE (1999) Meadow voles can discriminate between the top and bottom scent of an over-mark and prefer the top scent. Journal of Comparative Psychology 113:43–51

36. Ferkin MH, Johnston RE (1993) Roles of gonadal hormones on controlling sex-specific odors in meadow voles (Microtus pennsylvanicus). Hormones and Behavior 27:523–538

37. Ferkin MH, Johnston RE (1995) Meadow voles, Microtus pennsylvanicus, use multiple sources of scent for sexual recognition. Animal Behaviour 49:37–44

38. Ferkin MH, Johnston RE (1995) Effects of pregnancy, lactation, and postpartum oestrous on odour signals and the attraction to odours in female meadow voles, Microtus pennsylvanicus. Animal Behaviour. 49:1211–1217

39. Ferkin MH, Leonard ST (2005) Self-grooming by rodents in social and sexual contexts. Acta Zool. Sinica. 51:772–779

40. Ferkin MH, Leonard ST (2008) Age of the subject and scent donor affects the amount of time that voles self-groom when they are exposed to odors of opposite-sex conspecifics. In: Beynon R, Hurst J, Roberts C, Wyatt T (eds) Chemical Signals in Vertebrates 11. Springer Press pp. 281–289

41. Ferkin MH, Seamon JO (1987) Odor preferences and social behavior in meadow voles, Microtus pennsylvanicus: Seasonal differences. Canadian Journal of Zoology 65:2931–2937

42. Ferkin MH, Lee DN, Leonard ST (2004) The reproductive state of female voles affects their scent marking behavior and the responses of male conspecifics to such marks. Ethology 110:257–272

43. Ferkin MH, Li HZ, Leonard ST (2004) Meadow voles and prairie voles differ in the percentage of conspecific marks that they over-mark. Acta Ethologica 7:1–7
44. Ferkin MH, Mech SG, Paz-y-Mino C (2001) Scent marking in meadow voles and prairie voles: A test of three hypotheses. Behaviour 138:1319–1336
45. Ferkin MH, Sorokin ES, Johnston RE (1996) Self grooming as a sexually dimorphic communicative behaviour in meadow voles, Microtus pennsylvanicus. Animal Behaviour 51:801–810
46. Ferkin MH, Tamarin RH, Pugh SR (1992) Cryptic relatedness and the opportunity for kin recognition in microtine rodents. Oikos 63:328–332
47. Ferkin MH, Combs A, delBarco-Trillo J, Pierce AA, Franklin S (2008) Meadow voles display a capacity for what, where, and when. Animal Cognition 11:147–159
48. Ferkin MH, Leonard ST, Bartos K, Schmick MK (2001) Meadow voles and prairie voles differ in the length of time they prefer the top-scent donor of an over-mark. Ethology 107:1099–1114
49. Ferkin MH, Pierce A A, Sealand RO, delBarco-Trillo J (2005) Meadow voles, Microtus pennsylvanicus, can distinguish more over-marks from fewer over-marks. Animal Cognition 8:82–89
50. Foote AD, Griffin RM, Howitt D, Larsson L, Miller PJO, Hoelzel AR (2006) Killer whales are capable of vocal learning. Biology Letters 2:509–512
51. Franklin S (2000) Deliberation and Voluntary Action in 'Conscious' Software Agents. Neural Network World 10:505–521
52. Franklin S, McCauley L (2004) Feelings and emotions as motivators and learning facilitators. In: Architectures for modeling emotion: Cross-disciplinary foundations, AAAI 2004 Spring Symposium Series, Technical Report SS-04-02:4851. American Association for Artificial Intelligence, Stanford University, Palo Alto, CA
53. Franklin S (2005) A "Consciousness" Based Architecture for a Functioning Mind. In: Davis DN (ed) Visions of Mind. Information Science Publishing, Hershey, PA, pp. 149–175
54. Franklin S (2005) Evolutionary pressures and a stable world for animals and robots: A commentary on merker. Consciousness and Cognition 14:115–118
55. Franklin S, Baars BJ, Ramamurthy U, Ventura M (2005) The role of consciousness in memory. Brains, Minds and Media 1:1–38
56. Franklin S (2005) Cognitive robots: Perceptual associative memory and learning. Proceedings of the 14th Annual International Workshop on Robot and Human Interactive Communication (RO-MAN 2005), pp. 427–433
57. Franklin S, Ferkin MH (2006) An ontology for comparative cognition: A functional approach. Comparative Cognition & Behavior Reviews 1:36–52
58. Franklin S, Graesser AC (1997) Is it an agent, or just a program? A taxonomy for autonomous agents, Intelligent Agents III. Springer Verlag, Berlin, Germany, pp. 21–35
59. Franklin S, Ramamurthy U (2006) Motivations, values and emotions: Three sides of the same coin. Proceedings of the Sixth International Workshop on Epigenetic Robotics, Vol 128. Paris, France: Lund University Cognitive Studies, pp. 41–48
60. Freeman WJ (2003) High resolution eeg brings us another step closer to the ncc? In: ASSC7: Seventh Conference of the Association for the Scientific Study of

Consciousness: Association for the Scientific Study of Consciousness. Memphis, TN, USA

61. Glenberg AM (1997) What memory is for. Behavioral and Brain Sciences 20:1–19

62. Halgren E, Boujon C, Clarke J, Wang C, Chauvel P (2002) Rapid distributed fronto-parieto-occipital processing stages during working memory in humans. Cerebral Cortex 12:710-728

63. Hofstadter DR, Mitchell M (1995) The Copycat Project: A model of mental fluidity and analogy-making. In Holyoak KJ, Barnden JAORoEe (eds), Advances in Connectionist and Neural Computation Theory, Vol. 2: Logical Connections. Ablex, Norwood, NJ, pp. 205–267

64. James W (1890) The Principles of Psychology. Harvard University Press, Cambridge, MA

65. Kaelbling LP, Littman ML, Moore AW (1996) Reinforcement learning: A survey. Journal of Artificial Intelligence Research 4:237–285

66. Kamil AC (1998) On the proper definition of cognitive ethology. In: Balda R, Pepperberg I, Kamil AO (eds), Animal cognition in nature: The convergence of psychology and biology in laboratory and field. Academic Press, New York, NY, pp. 1–28

67. Kanerva P (1988) Sparse Distributed Memory. The MIT Press, Cambridge, MA

68. Keller, BL (1985) Reproductive patterns. In: Tamarin RH (ed) Biology of New World Microtus. 8th edn. American Society of Mammalogists, Special Publication, Lawrence, KS, pp. 725–778

69. Kruschke JK (2001) Toward a unified model of attention in associative learning. Journal of Mathematical Psychology 45:812–863

70. Lehmann D, Strik WK, Henggeler B, Koenig T, Koukkou M (1998) Brain electric microstates and momentary conscious mind states as building blocks of spontaneous thinking: I. Visual imagery and abstract thoughts. Int J Psychophysiology 29(1):1–11

71. Leonard ST, Ferkin MH (2005) Seasonal differences in self-grooming in meadow voles, Microtus pennsylvanicus. Acta Ethologica 8:86–91

72. Lisman JE, Fallon JR (1999) What maintains memories? Science 283:339–340

73. Madison DM (1980) An integrated view of the social biology of meadow voles, Microtus pennsylvanicus. The Biologist 62:20–33

74. Maes P (1989) How to do the right thing. Connection Science 1:291–323

75. Massimini M, Ferrarelli F, Huber R, Esser SK, Singh H, Tononi G (2005) Breakdown of cortical effective connectivity during sleep. Science 309:2228–2232

76. Maturana HR (1975) The organization of the living: A theory of the living organization. International Journal of Man-Machine Studies 7:313–332

77. Maturana, H R, Varela FJ (1980) Autopoiesis and Cognition: The Realization of the Living. Dordrecht, Netherlands

78. Meek LR, Lee TM (1993) Prediction of fertility by mating latency and photoperiod in nulliparous and primiparous meadow voles (Microtus pennsylvanicus). Journal of Reproduction and Fertility 97:353–357

79. Merker B (2005) The liabilities of mobility: A selection pressure for the transition to consciousness in animal evolution. Consciousness and Cognition 14:89–114

80. Milligan SR (1982) Induced ovulation in mammals. Oxford Reviews of Reproduction 4:1–46
81. Nadel L (1992) Multiple memory systems: What and why. J. Cogn. Neurosci. 4:179–188
82. Negatu A, Franklin S (2002) An action selection mechanism for 'conscious' software agents. Cognitive Science Quarterly 2:363–386
83. Pearce TC, Schiffman SS, Nagle HT, Gardner JW (2002) Handbook of Machine Olfaction: Electronic Nose Technology. Wiley-VCH, Weinheim, Germany
84. Pierce AA, Ferkin MH (2005) Re-feeding and restoration of odor attractivity, odor preference, and sexual receptivity in food-deprived female meadow voles. Physiology and Behavior 84:553–561
85. Pierce AA, Ferkin MH, Williams TK (2005) Food-deprivation-induced changes in sexual behavior of meadow voles, Microtus pennsylvanicus. Animal Behaviour 70:339–348
86. Reber AS, Walkenfeld FF, Hernstadt R (1991) Implicit and explicit learning: Individual differences and IQ. Journal of Experimental Psychology: Learning, Memory, and Cognition 17:888–896
87. Saksida LM (1999) Effects of similarity and experience on discrimination learning: A non associative connectionist model of perceptual learning. Journal of Experimental Psychology: Animal Behavior Processes 25:308–323
88. Sheridan M, Tamarin RH (1988) Space use, longevity, and reproductive success in meadow voles. Behavioral Ecology and Sociobiology 22:85–90
89. Sigman M, Dehaene S (2006) Dynamics of the central bottleneck: Dual-task and task uncertainty. PLoS Biol. 4(7)1227–1238
90. Skarda C, Freeman WJ (1987) How brains make chaos in order to make sense of the world. Behavioral and Brain Sciences 10:161–195
91. Sloman A (1999) What sort of architecture is required for a human-like agent? In: Wooldridge M, Rao AS (eds) Foundations of Rational Agency. Kluwer Academic Publishers, Dordrecht, Netherlands, pp. 35–52
92. Standing L (1973) Learning 10,000 pictures. Quaterly Journal of Experimental Psychology 25:207–222
93. Tarsitano M (2006) Route selection by a jumping spider (Portia labiata) during the locomotory phase of a detour. Animal Behavior 72:1437–1442
94. Tulving E (1983) Elements of episodic memory. Clarendon Press, Oxford, UK
95. Tulving E (2005) Episodic memory and autonoesis: uniquely human? In: Terrace HS, Metcalfe J (eds) The Missing Link in Cognition, Oxford University Press, New York, NY, pp. 3–56
96. Uchida N, Kepecs A, Mainen ZF (2006) Seeing at a glance, smelling in a whiff: Rapid forms of perceptual decision making. Nature Reviews Neuroscience 7:485–491
97. Varela FJ, Thompson E, Rosch E (1991) The Embodied Mind. MIT Press, Cambridge, MA
98. Wilcox S, Jackson R (2002) Jumping spider tricksters: Deceit, predation, and cognition. In: Bekoff M, Allen C, Burghardt GM (eds) The Cognitive Animal. MIT Press, Cambridge, MA, pp. 27–33
99. Willis J, Todorov A (2006) First impressions: Making up your mind after a 100-ms exposure to a face. Psychological Science 17:592–599
100. Winston PH (1992) Artificial Intelligence, 3rd ed., Addison Wesley, Boston, MA

101. Zhang Z, Dasgupta D, Franklin S (1998) Metacognition in software agents using classifier systems, Proceedings of the Fifteenth National Conference on Artificial Intelligence. MIT Press, Madison, Wisconsin, pp. 83–88

Epistemic Constraints on Autonomous Symbolic Representation in Natural and Artificial Agents

David Windridge and Josef Kittler

School of Electronics and Physical Sciences, University of Surrey, Guildford, Surrey, GU2 7XH, United Kingdom
D.Windridge@surrey.ac.uk

Summary. We set out to address, in the form of a survey, the fundamental constraints upon self-updating representation in cognitive agents of natural and artificial origin. The foundational epistemic problem encountered by such agents is that of distinguishing errors of representation from inappropriateness of the representational framework. Resolving this conceptual difficulty involves ensuring the empirical falsifiability of both the representational hypotheses and the entities so represented, while at the same time retaining their epistemic distinguishability.

We shall thus argue that perception-action frameworks provide an appropriate basis for the development of an empirically meaningful criterion for validating perceptual categories. In this scenario, hypotheses about the agent's world are defined in terms of environmental affordances (characterised in terms of the agent's active capabilities). Agents with the capability to hierarchically-abstract this framework to a level consonant with performing syntactic manipulations and making deductive conjectures are consequently able to form an implicitly symbolic representation of the environment within which new, higher-level, modes of environment manipulation are implied (e.g. tool-use). This abstraction process is inherently open-ended, admitting a wide-range of possible representational hypotheses - only the form of the lowest-level of the hierarchy need be constrained a priori (being the minimally sufficient condition necessary for retention of the ability to falsify high-level hypotheses).

In biological agents capable of autonomous cognitive-updating, we argue that the grounding of such a priori 'bootstrap' representational hypotheses is ensured via the process of natural selection.

16.1 Introduction

The aim of the following is to examine the epistemic[1] constraints on self-updating cognition applicable to both artificial and biological agents. In

---

[1]

ep·i·ste·mic:
   1. Of, relating to, or involving knowledge; cognitive.

D. Windridge and J. Kittler: *Epistemic Constraints on Autonomous Symbolic Representation in Natural and Artificial Agents*, Studies in Computational Intelligence (SCI) **122**, 395–422 (2008)
www.springerlink.com

particular, we consider the problem of how the autonomous updating of an embodied agent's perceptual framework in response to the perceived requirements of the environment can occur in a logically-consistent fashion, such the ability to validate the agent's representation of the environment is maintained throughout.

Thus, a cognitive agent employing a representational framework, R, must, upon examination of a set of observations $\{o\}$ relating to the agent's environment, be capable of undergoing spontaneous transition to an updated representational framework R' in which the environment observations are transformed into an alternative set of observables $\{o'\}$ (of possibility differing cardinality) that are deemed to be more 'representative' of the environment via some appropriate criterion of representativity. The question then immediately arises of what form this criterion should take, given that the only access that the agent has to the environment in order to determine the representativity of representations is via those very same representations. In additional to this foundational issue, a further difficulty attaches to the fact that individual representations of the environment are themselves necessarily conjectural, such that even *within* an appropriate representational framework, R, there is a question as to which particular observation, o, is most applicable to the current situation (i.e. classical perceptual uncertainty).

Systems for cognitive updating hence exhibit the potential for ambiguity between perceptual representation and perceived objects unless a means can be found to ensure that the two domains of inference can be empirically related while at the same time maintaining their epistemological distinction. (An autonomous cognitive agent must simultaneously employ some fixed perceptual reference in order to validate environment hypotheses, and a fixed environment representation to validate a particular perceptual framework). We shall hence argue that the notion of cognitive updating is ill-founded unless there exists a framework in which representational hypotheses can be empirically falsified via exploratory activity in the same way as the world representations described in terms of these hypotheses.

By virtue of having adapted to changing environments, sufficiently-evolved natural organisms (those that are complex enough to be considered cognitive) have an implicitly updated framework for environmental representation in which these difficulties are overcome. In such organisms, representational grounding is thus, to a large extent, ensured by natural selection; representa-

---

(The American Heritage©Dictionary of the English Language, 4th Ed. Houghton Mifflin Company, 2004).

2. Of, or relating to, epistemology

(WordNet 1.7.1. Princeton University, 2001).

[From the Greek *epistēmē*, knowledge]

e-pis-te-mol-o-gy:

1. The branch of philosophy that studies the nature of knowledge,
its presuppositions and foundations, and its extent and validity.

(The American Heritage©Dictionary of the English Language, 4th Ed. Houghton Mifflin Company, 2004).

tions that do not meaningfully and efficiently represent the survival prerogatives of the agent in the context of its environment increase the likelihood of its extinction and genetic removal from the heredity of future generations. However, in so far as representations are *learnable*, biological organisms must employ an alternative mechanism for ensuring that the way in which the world is represented remains consistent with their survival imperatives. In doing so they must hence also address the problem of perceptual meaningfulness that lies at the heart of attempts to create cognitively autonomous artificial agents. We term the activity of mechanisms capable of achieving this *cognitive bootstrapping*. The concept of cognitive bootstrapping is thus analogous to (and indeed, to the extent that word-concepts are cognitive representations, *exemplified by*) the practice of semantic learning that we employ as infants, in which we must first obtain a sufficient (bootstrap) sub-set of words and word-meanings in order to be able to formulate falsifiable questions concerning meaning of new words, and thereby achieve the ability to expand our vocabulary indefinitely.

To this end, we survey a range of artificial cognitive mechanisms that attempt to address the issue of representational updating, concluding that only embodied perception-action learners capable of hierarchically-abstracting this relationship in such a way as to be manipulable in relational/symbolic terms are able to meet the indicated epistemic requirements. (Perception-action learning agents may be characterized as those for which 'action precedes perception'; that is, agents for which inferred higher-level percept states are considered meaningful only insofar as they relate to the agent's actions). Such artificial systems limit higher-level symbolic learning to that which is immediately relevant to the agent, defining the external world in terms of an increasingly complex set of motor capabilities, with the objects of the world consequently being represented in terms of their *affordances*. Translated into a multi-agent language-learning context, this means that agents engaged in evolving a collective communicative structure can hence only derive a meaningful syntax in relation to a semantics grounded in their respective active (and collective interactive) capabilities in the environment.

This embodied approach to autonomous cognition thus addresses a number of difficulties associated with classical artificial intelligence (in which intelligence is primarily regarded only in terms of the manipulation of symbols of fixed referential content), in particular those of symbol grounding and logical framing. Hence, in asserting that autonomous cognition is meaningful only with regard to embodied agents with limited action capabilities, the study of artificial cognitive systems is brought within the domain of evolutionary systems and adaptive robotics. As such, we believe these developments are of considerable potential interest to biological researchers.

The structure of the chapter will therefore be as follows. We commence, in section 16.2, with a discussion of the epistemological constraints on symbolic representations via an examination of the necessary *a priori* aspects of cognition that must be retained throughout any putative updating of the perceptual framework in order for such updates to be considered empirically meaningful. We consider how such *a priori* representations arise within naturally evolved systems. We then, in section 16.3, introduce the notion of embodiment within the context of artificial cognitive systems, and indicate, with examples, how this approach has the potential to address the symbol grounding and framing problems associated with classical artificial intelligence via notions such as affordance. In section 16.4, we address the nature of evolving representational structures in embodied *communicative* agents. We indicate, in section 16.5, how *perception-action learning* can be employed to hierarchically infer a grounded symbol set in order to create fully autonomous artificial cognitive systems capable of dynamically updating their perceptual framework in relation to the requirements of their perceived environment. We then, in section 16.6, discuss the issues raised in determining the epistemic constraints on the symbolic abstraction of perception-action architectures applicable to both natural and artificial agents. We conclude by giving a concise summary of the requirements of a cognitive system if it is to be capable of bootstrapping symbolic representations in such a way as to meet these constraints.

16.2 Open-Ended Symbolic Representation: A Philosophical Perspective

A *Priori* Constraints on Cognitive Representation

The argument of the survey revolves around a central paradox: how can a cognitive agent capable of changing its perceptual framework (that is, its way of seeing the world) ever *validate* one particular set of perceptual representations over another? The concept of validation would appear (at least in humans) to depend on the *perception* of the inadequacy of one perceived entity in relation to another: however it is not obvious that a perceptual framework could ever itself be an object for perception. The problem is certainly not soluble in terms of either the Cartesian or Classical Empiricist [1] schools of philosophy, since the first claims cognitive agents cannot absolutely validate the existence of anything beyond their own perception (itself built on a framework of pure reason), and the second does not recognize the possibility of the perceptual mediation of the objective world (objects present themselves as they are 'in themselves' directly to cognition).

Kant [2], however, provides an alternative conceptual framework, asserting that cognition, as a matter of *a priori* necessity, *refers* to entities existing beyond of an agent's sensory domain. Percepts hence serve to mediate between agent and object, being crucial to their distinction as ontologically separate entities. Objects are thus never perceived by cognitive agents as they are *in themselves* (being required to conform to the *a priori* requirements of perception): however, neither are they simply reducible to percepts. Instead, external object concepts are accessible to the cognition agent as *ordering concepts* imposed on intuitions (singular, low-level sensory percepts). Objects, as we understand them, are thus not themselves singular *percepts*: they are (in Kant's terminology) *synthetic unities.*

Thus, despite being necessary *a priori*, object concepts are of an inherently hypothetical nature, existing beyond the immediate certainty of the sensory impressions, serving instead as hypothesized *linkages* between those impressions. Immediate sensory impressions thus refer directly to the external world *a priori* in a way that can not be subject to empirical testing (being rather its *condition*). The conditions underlying perception are thus neither logically true nor false; rather they must simply be *assumed* to be true in order for cognition to take place at all. It is hence this *a priori* limitation on the possible updating of cognitive representation that will serve both to sets bounds upon, as well as to ensure the empirical grounding of, the concept of cognitive bootstrapping.

Implicit within this understanding of cognition is hence the idea that sensory intuitions can be linked together via *actions*, actions inherently having the capability to *test* object hypotheses, falsifying those that do not have the relationship between sensory impressions and actions implicit in the object hypothesis. Thus, we might need to *walk around* an object in order to establish whether the entity progressively revealed to the senses conforms to our conception of the object. Actions thus serve to test the *consistency* of an observed sequence of unfolding sensory impressions with respect to the underlying object hypothesis (which, at an appropriately generalized level, is itself necessary to give unity to the immediate sensory impressions). Object concepts thus implicitly serve as *singular* expressions of the functional mapping between individual sensory impressions and agent actions, which, (since they are not simply the equivalent of these functions) are inherently *compressive* in nature (thus, to give an idealized example, specifying an object within a view-independent 3-D coordinate-space is far more compact than setting out the exhaustive set of possible 2-D planar views on that object).

Cognitive Bootstrapping Within Kantian *A Priori* Constraints

It might first appear that the strong Kantian emphasis on *a priori*-limited sensory representation leaves little room for the perceptual updating required of cognitive bootstrapping. However, this is not the case; a significant question arises with regard to object concepts achieving a high-level of empirical confirmation. Since high-level object *hypotheses* link lower-level percepts together in a conjectural unity, these could, in principle (when sufficiently empirically confirmed) serve as the basis for further synthetically-unified object-concepts. In this sense, the original object-concept has become equivalent to a perception, albeit at a higher hierarchical level. Thus, we might, for instance, regard a very familiar object seen from only one perspective as a sensory-impression in its own right, and not in fact as an object-hypothesis that might be falsified by experience. This object might then form the basis of a new object hypothesis (for instance, by using it as a reference point for navigation), such that the new object hypothesis assumes the old one as a pre-assumed (though not fully *a priori*) basis for cognition.

It is hence the argument of this chapter that within the Kantian object-validation framework it is possible for an autonomous cognitive agent to update and validate its own *perceptual* categories (which is to say, engage in cognitive bootstrapping), but *only* by proceeding via a bottom-up approach built on the assumption of the *a priori* referentiality of the lowest level of the agent's perceptual hierarchy. Correspondingly, we require the *a priori* consistency and relevance of the lowest-level of the agent's *motor* space (so that, for instance, a cognitive entity cannot meaningfully query the topology of its motor-space independently of that of its perceptual-space).

The possibility of empirical validation of high-level synthetic percept/action unities thus rests on an *a priori* sensorimotor foundation. It is then these higher-level synthetic unities that enable perceptual hypothesis validation experiments of the following form: (where A_x are high-level actions and O_n high-level observational states, or perhaps stochastic distributions over states.)

If perceptual hypothesis H_1 is true then, for all definable perceptual transitions $O_m \rightarrow O_n$ such that $O_m \neq O_n; O_m, O_n \in \{O\}$, there exists a unique $A_x \in \{A\}$ such that $\{\cup A_x\} \Leftrightarrow \{A\}$

Which retains an empirical contrast with *object* hypothesis validation experiments of the form:

If object hypothesis H_1' is true then performing action A_x will result in observation O_m.

The former perceptual hypothesis validation experiment hence attempts to determine whether the proposed high-level perceptual framework represents the proposed high-level actions in the most efficient (i.e. least redundant) manner possible. In order that a space exists in which to perform this test of perceptual compression, there must be an underlying *a priori* space of actions and perceptions available to the agent which are not themselves subject to hypothetical uncertainty. Thus, in general, while an autonomous cognitive agent may be free to reinterpret the world in the sense of being able to make an arbitrary high-level choice of perceptual hypothesis, H_n, by which the world is to be interpreted, it is not free to choose an alternative set of action primitives, $\{A'\}$, or an alternative set of sensory primitives, $\{O'\}$, upon which the higher-level $\{A\}$ and $\{O\}$ are based (e.g. $\{A\}$ and $\{O\}$ might be legitimately defined in terms of arbitrary functions of n-ary action/perception concatenations: $\{A|A \rightarrow n^{A'}\}$ and $\{O|O \rightarrow n^{O'}\}$). $\{A'\}$ and $\{O'\}$ are hence the terms upon which the perceptual validation criterion is implicitly constructed (and without which perceptual reinterpretation is completely unconstrained)[2].

---

[2] Obvious candidates for the sets $\{A_n'\}$ and $\{O_n'\}$ in human cognition are, respectively, the motor complex and the space of visually and kinesthetically-determined

The relationship between $[\{A'\}, \{O'\}]$ and $[\{A\}, \{O\}]$ is clearly recursive. It is therefore the object of the following to propose that if an agent's *a priori* perception-action can be hierarchically extrapolated in this manner, it will be possible to arrive at at a sufficiently abstracted perception-action relation such that there exists a concept of the *symbolic representation* of the world. We hence now look at the the subject of *hermeneutics*, the branch of philosophy that deals with the interpretation of symbols, and, ultimately, the mechanism of cognitive *understanding*.

The Hermeneutic Circle and Cognitive Bootstrapping

Hermeneutics emerged initially as the branch of philosophy that deals with the interpretation of texts, only later acquiring its interpretation as the branch of philosophy that concerns the mechanism of human *understanding*. Central to the latter school of hermeneutics is Dilthey's notion of the 'hermeneutic circle' by which symbols can acquire an *objective* meaning. Thus, in order, say, to arrive at a dictionary of word meanings for a corpus of ancient texts, one simply proposes any *a priori*-plausible initial set of symbol meanings (for instance, a core set of words in an ancient text of known meaning with modern-day meanings attributed to the remainder), and then carries-out a reading of the entire corpus of work on this basis in order to arrive at an overall interpretation. This collective understanding is then utilized to *reinterpret* the component texts in the context of the whole. These reinterpreted component texts are again utilized to arrive at a new extrapolated interpretation of the corpus, and so on (errors in the initial set can also be corrected to a strictly limited extent in this manner).

Hence, generalizing this idea into the domain of active agents, the hermeneutic circle involves, firstly, an iterated movement from the outward manifestation of actions to an assumption about their inner, symbolically-determined motivation, and, secondly, proceeding from this assumption back again to a predictive conjecture about the outward manifestation of agent behavior, in process of circular empirical refinement. It is hence tacitly understood (though not explained) by Dilthey that this reiteration will achieve a degree of convergence on a final, stable set of symbol meanings (convergence to stability being the only possible criteria of finality). In this latter hermeneutic context, the attribution of *meaning* to symbolic terms is thus dependent upon the *embodiment* of the symbol-manipulating agent within the objective world; meaning cannot be conferred simply by the manipulation of symbolic entities (without descending into semantic tautology, such as when attempting to derive the meaning of *every* word in a language using only dictionary definitions).

---

body-relative positions of prioperception. Candidates for the inferred $\{A_n\}$ and $\{O_n\}$ might be e.g. the intentional act of cutting and the perceptual grouping of knife-like objects as constituting a distinct class of agent-utilisable entities.

For Heidegger [3] this tendency to regard cognitive meaning as being supervened upon by the action possibilities of an agent reached its apotheosis. He proposed, in his ontological hermeneutics, that one's *sensations* are completely defined by one's *acts* and one's *possibilities*. Heidegger thus envisaged our immediate sensation as being based on *instrumentality* (Vorhanden), in which, for instance, the perception of a pen would be *fully* determined by our possibilities of using it, in particular our possibility of using it to *write*, with further social and contextual signification resting on what we *may choose* to write. Thus, the *entirety* of our physical being is employed in the perception of the pen, rendering the notion of an abstract mental plane of representations underlying our perception entirely redundant. This notion also extended to the derivation of the objects of knowledge (Zuhanden) from the praxical knowledge of action. Objective knowledge is now an abstraction from practical knowledge, and not its precursor. In asserting that knowledge is *intentional*, there is hence a complete rejection of the notion that knowledge is *representational*; this is merely an artifact of dualistic Cartesian thought that falsely separates the body from cognition. Only when intended actions fail to account for our actual percepts, do we, 'stand back' from our perceptions and form a concept of objective existent independent from our selves; in the usual run of things objects are *transparent* to cognitive agents - we only perceive our own active potentialities in the world unless these fail to be realized as expected. There is hence in Heidegger's conception of objective knowledge, an implicit *ontological hermeneutic circle*.

However, notwithstanding these arguments, the subject of *artificial cognition* has traditionally been founded on Cartesian assumptions (namely, that cognition is essentially the rational manipulation of symbols that cannot have objective meaning). The difficulty associated with non-embodied symbol representation has consequently created considerable philosophical argument.

Hermeneutics and the Possibility of Artificial Cognition

One of the more persistent critics of the idea of artificial cognition is Dreyfus [4], who argues that the Representational Theory of Mind (in which the mind performs permutations of representations of the outside world) fails to take account of the contextuality, relevance and holism of perception. Discrete, atomic symbolic computation cannot account for the immediacy of the the cognitive situation. He suggests that only embodiment can provide a semantics of ordinary meaning, which left to symbolic computation alone would collapse into merely empty syntactic considerations. Moreover, this syntax, even if it existed, could never be available to cognition without involving problems of infinite regress. Thus, there can be no 'algorithm' underlying cognition which we could isolate and implement; only the situated, symbol-manipulating agent with an actual, sensible connection to the world can be truly cogent. The world, in effect, provides the 'being' behind the insubstantial formal categorizations of mind. Artificial cognition might thus exist, but not in any systematically pre-formalizable way.

Suber [5] makes the argument that if *mind* can be expected to emerge from computation alone, then we should reasonably expect that *semantics* can emerge from syntax alone. However the Löwenheim-Skolem theorem of the branch of mathematics known as model theory demonstrates that even syntactic specifications with an infinite cardinality are incapable of uniquely determining a concrete, existing model. A very large degree of semantic ambiguity would therefore appear to be associated with any finitely formalizable set of syntactic rules, with the corresponding difficulty that this implies for the grounding of any putative 'laws of cognition' without a corresponding embodiment.

A similar view is given by Winograd in [6] who argues that the fallacy of cognitive objectivism (the view that cognition can be tangibly formalized) is caused by overly formal logical structure of early attempts at simulated cognition (for instance his own SHRDLU algorithm, which is capable of passing the Turing test for intelligent behavior provided queries are restricted to the very limited but complete ontology of it's internally represented world). Winograd argues that formal completeness of the logical system in which an agent is embodied is never available to that agent as a demonstrable fact (this would, in effect, constitute a Gödel proposition [7]). Instead the *embodied* agent can only allocate finite and partial resources to comprehending the world[3]. This

---

[3] Following historical difficulties associated with Russell's paradox (i.e. when the logical consequences of set-self membership are explicitly considered), we have become used to questioning the admissibility of a finite system such as the human brain encompassing a complete self-representation within itself (particularly a demonstrable self-representation). However, partial or temporally-retrograde self-models would appear to be permissible, so that it is possible for a human-being to use a linguistic token 'I' meaningfully and accurately, or, on a computational level, to build mobile robots capable of building accurate models of their *position* in space, if not of their full internal state-space. Complete and immediate self-models are ruled-out completely, though (see for instance [8] for a discussion of the limits to self-observation under finite, Markovian and infinite state-space assumptions, and [9] under quantum-physical assumptions). The use of partial self-models is thus, in essence, to adopt the hierarchical solution of Russell to his paradox via the *theory of types*, in which sentential reference can only be made to individual entities on the lowest level of the hierarchy, with sentential references to *sentences* about individuals being made only on the level immediately above this, and so on. Complete, but temporally retrograde self-models, on the other hand, occur at each iteration of the Universal Turing Machine that implicitly attempts to emulate itself within the Halting problem.

These issues give further impetus to the notion of artificial cognitive bootstrapping: since the underlying mechanism of human cognition can not be *knowingly* expressed in a finite and formally complete manner by any human being, it can not therefore be directly implemented using conventional methods of computational engineering. A human-equivalent artificial cognitive capability can thus

naturally leads him to abandon the notion of formally closed ontologies in any world description given by an agent; world descriptions have only to be (and indeed can only be) locally, and not globally, valid. Thus an artificially constructed cognitive agent is feasible in practice, but must necessarily be of an open-ended design (he was later [10] to reject the possibility that any physically existing device for formal symbol manipulation can have intrinsic meaning outside of that given to it by a subjective, situated agent; hence a computer program a performs a 'task' with 'goals' only if we so designate it).

We thus conclude that it is possible, in principle, for natural and artificial systems to overcome the paradoxes associated with open-ended representation in classical cognition and implement an embodied hermeneutic circle for attaching meaning to spontaneously-generated symbols. We turn now to the *science* of autonomous symbol generation in natural and artificial agents.

16.3 The Epistemology of Symbol Generation Within Embodied Agents

From the perspective of cognitive science it is possible to give a rather different argument for the form that cognitive bootstrapping must take from that given above, but to arrive at exactly the same conclusion. In this case, the argument is framed in terms of the problem of *grounding* symbols employed by autonomous artificial agents. An autonomous cognitive agent is, by definition, one capable of adapting to its in environment in behavioral and representational terms that go beyond those implied by its initial set of 'bootstrap' symbolic assumptions, in order to find representations more suited to the particular environment in which the agent finds itself. Doing so necessitates the use of mechanisms of generalization, inference and decision making in order to modify the initial perceptual symbol set in the light of novel forms of sensory data (and also the mechanisms of differentiation and analysis to validate modifications).

Any representation that is capable of abstract generalization is implicitly governed by protocols such as those of predicate logic. As such, the generalized entities must observe strictly formalized laws of interrelationship, and consequently, in abstracting the symbol set away from the original set of innate percept-behavioral pairings, there is a danger of them becoming detached from any intrinsic meaning in relation to the agent's environment. A related difficulty, known as the *frame problem* [11], also arises in such generalized formal domains; it is by no means clear which particular set of logical

---

only be achieved via an evolving, self-updating design approach. This is, in effect, to transpose the negative conclusion of the Hilbert programme (the attempt, in the 1920s to construct, in advance, a formal axiomatization of *all* mathematics) from a mathematic context to that of cognitive science, where the *laws of cognition* are hence the quantity that is incapable of a provably - i.e. *knowingly*-complete analytic formulation.

consequences (given the infinite number of possibilities) that the generalized reasoning system should concern itself with.

There is hence a problem of symbol relevance and 'grounding' unless additional mechanisms can be put in place to form a bridge between the formal requirements of logical inference applicable to symbols, and the constraint of the relevance of this symbol set to the agent within the context of both its goals and the intrinsic nature of the environment in which these goals are to be fulfilled. In terms of the philosophy of cognition, this necessitates a move from a Quinean [12] to a Wittgensteinian [13] frame of reference, in which symbol *meaning* is intrinsically contextual, and environment-dependent, rather than being a matter of arbitrary ontological assumption.

For cognitive agents in the animal kingdom the grounding of symbols is enforced by the mechanism of Darwinian natural selection; representations that do not meaningfully and efficiently represent the survival prerogatives of the agent in the context of its environment increase the likelihood of its extinction and genetic removal from the heredity of future generations [14]. This mechanism, however, is not readily available to artificial cognitive agents other than in the context of self-replicating agents within a simulated environment (see Sipper's An Introduction to Artificial Life [15] for an overview of this sub-field). For artificial cognitive agents embodied within the real world (that is to say, *robots*), the form that this symbol grounding framework must take is, by an increasing consensus ([16], [17], [18]), one of hierarchical stages of abstraction that proceed from the 'bottom-up'. At the lowest level is thus the immediate relationship between percept and action; a change in what is perceived is primarily brought about by actions in the agent's motor-space. This hence limits visual learning to what is immediately relevant to the agent, and significantly reduces the quantity of data from which the agent must construct its symbol domain by virtue of the many-to-one mapping that exists between the pre-symbolic visual space and the intrinsic motor space [19]. It is consequently apparent that classical A.I. approaches to artificial cognition were of only limited success in that they attempted to build high-level environmental representations *prior* to considering agent actions within this model, rather than allowing this representation to evolve via hierarchical abstraction of the *a priori* percept-action relation [20]. Representative priorities were thus specified in advance by the system-builder and not by the agent, meaning that an *autonomous* agent would have had to build its goals and higher-level representations in terms of the assumed representational modes, with all the redundancy that this implied. Furthermore, novel modes of representation were frequently ruled out in advance by this pre-specification of scene-description.

The issue of representation is thus of the first importance to cognitive science. A central historical concern of the field has consequently been to determine whether mental acts can be interpreted as the action of a large collection of individual computational elements (neuronal models, derived from physiological knowledge of the human, mammalian and reptilian brains), or whether they are to be interpreted at a higher level in terms of representations or schema. These two schools are respectively labeled the *connectionist* and the *symbolic*. This distinction of approach is perhaps best reflected in their respective attitudes towards *simulation* of the human mind, both within the field of cognitive science as well as in the correlated engineering discipline of machine learning. Simulation of mental states is thus carried out either via emulation of large numbers of individual neurons, in which case we expect mental properties to arise as *emergent properties*, or else the simulation is executed at the schematic or representational level, in which case the actual underlying computational mechanics are of no inherent significance. In the former case, simulation is independent only of the underlying computational *substrate* (a logical unit can equally well be enacted by a radio-valve as a transistor), in the latter case simulation is independent of the particular computational implementation of the representational algorithm.

A central problem for symbolic interpretations of cognitive psychology is thus to capture the fact the mental formalisms must be simultaneously both *computational* and *representational*; that is mental symbols must be manipulable by logical rules and also capable of referring to aspects of the world. Newell and Simon [21] were the first both to posit and to propose a solution to this problem from the perspective of cognitive psychology, centering on the concept of *physical symbol systems*. Here, physical relations (proximities, causalities and so on) provide the referential basis for symbol structures expressed within the brain.

Environmental adaptation (through Darwinian natural selection) is consequently the assumed agency constraining the formal symbol structure to mimic the physical environment (or at least those aspects of it that are relevant to the survival of the symbolic agent) within biological agents expressing Newell and Simon's ideas. This aspect of the symbolic account was further brought out by Pinker and Bloom in the context of language evolution [22], who argued that 'grammar is a complex mechanism tailored to the transmission of [physically representable] propositional structures through a serial interface', the serial interface being the vocal communication channel. Biologically-based accounts of symbolic causality thus agree that the *representativity* of mental symbols is characterized by their capacity to ensure the continuing existence of the symbol-manipulating agent (or at least its genetically-contiguous progeny). Thus, while the symbolic manipulation system may be completely formal, the representativity of the symbols in the symbolic account is contingent and environmentally determined.

In this wider biological context, the particular symbolic model proposed by Newell and Simon can then be considered explicitly one of cognitive bootstrapping in the sense that world-model updates are achieved via genetic variations through mutation or sexual reproduction (equating to the hypothesis updating stage of cognitive bootstrapping), and are empirically checked for their referencing ability in terms of the agent's attempts to survive within the environment (equating to the hypothesis verification stage). The initial bootstrap symbol set is thus arrived at contingently, but the iterative convergence of the symbol reference system rapidly removes all traces of its random origin until an appropriate representation is arrived at (if only asymptotically).

The above model assumes a relatively constant environment in relation to which the organism in question evolves. Conversely, where environments are not constant, and are changing at a faster rate than genetic adaptation can allow for, we would expect to find that the innate symbols acquire an inappropriate reference (such as, for instance, amongst humans, where animal threat assessments are calibrated to our hunter-gatherer past, rather than our urban/agrarian present; notably, the human instantiation of the primate's innate fear of the larger carnivores). It is therefore necessary, if Newell and Simon's notion of physical symbol systems is to be extended to symbolic inference mechanisms capable of autonomously updating themselves, that the Darwinian mechanism of bootstrapping be replaced by a more rapidly-updating technique that nonetheless retains the former mechanism's groundedness in the environmental survival imperatives of the cognitive agent: this shall be the subject of later discussion. We note for the present, however, that the innate, naturally-selected physical symbol set serves effectively as an initial perceptual meaning hypothesis for cognitive bootstrapping.

In contrast to the formal mechanics of the Symbolic approach, Connectionist accounts seek to comprehend agent meaning attribution in terms of the aggregate information processing abilities of arrays of neuronal units, in intentional replication of mammalian or reptilian brain physiology. Cognitive properties can thus arise emergently, without explicit formal structure. An example of this is Complementary Reinforcement Back-Propagation (CRBP) training within artificial neural networks [23], which is proposed as way of achieving self-volitional behavior in robots through neuronal constraints alone. Marshall *et al.* thus conjecture that self-directed learning behavior comes about as the result of competing tensions, such as that between the compulsion to model existing perceptual states effectively and the compulsion to seek out novel states. The 'homeostasis' thus achieved allows the network to bootstrap increasingly complex behavior patterns. CRBP directly models this behavior by, in addition to allowing back-propagation to reinforce internal goals in the conventional manner, also allowing the *complement* of the goal state to serve as negative behavior reinforcement during back-propagation. The tension between these contrary goal imperatives is hence directly modeled within the neural network structure, forcing the agent to test cognitive models by deliberately seeking areas in which they break down, and thus to refine them.

A key milestone of the Connectionist approach was thus the demonstration of the Boolean-logic completeness of such neuronal aggregates via the *multi-layer* perceptron (MLP) model. However, the MLP model lacks Turing-completeness due to the absence of memory associated with individual neurons (as opposed to the neuronal *network* as a whole, which does exhibit memory capability). It was hence determined by Franklin and Garzon [24] that the standard McCulloch-Pitts net augmented with expandable memory *is*

Turing-complete and hence capable of arbitrary formal-language manipulation. The Symbolic and Connectionist approaches had, for the first time, thus achieved a demonstrable equivalence. Gärdenfors [17] later constructed a propositional language system based on the theory of functional dynamics applied to (purely abstract) information states. A neural network that undergoes learning generalization of the Hebbian kind in response to new information is thus shown to perform an *inductive inference* of the kind recognized in formal logic. Hence the symbolic/connectionist equivalence is not simply an *interpretation* of the the underlying neural connectionist model; it has actual referential capability.

At a more general remove, another approach to unifying the symbolic and connectionist accounts, involving a common model for both artificial neural-network classification functions as well as formal symbolic constructs such as verbal grammar, is to view brain cognition as a form of *compression.* This approach, first suggested by Wolf [25], sees the essence of cognitive agency within the world as being the ability to represent the varied mass of sensory information in a compact (and thus, generalized) form. Hence, grammatical rules may be regarded as compressed expressions of *language* possibility, and *classification* may be seen as a compression of sense-data. The *object concept* itself can be derived by the redundancy or commonality between stereoscopic, or multi-angular images (compare this with the Kantian notion of the object concept as a unifier of perspectives).

In animal cognition, the mechanism motivating this compression is Darwinian natural-selection; biological agents employing better generalizers (which is to say, better compressors) use fewer neurons to find food by encoding successful hunting strategies in the most general manner possible. Since such agents inherently require less food to sustain their smaller neuronal budgets, there ensues a 'virtuous circle' in which they stand a greater change of surviving and reproducing than their less efficiently-compressing relatives. Progressive generations thus increasingly enhance the likelihood of agents with ever more economized cognitive capacities (which is to say efficient sensory compression mechanisms). Moreover, when the environmental requirements are not static (as, for instance, in the context of hominid evolution), the selection pressure is towards ever more generalized *representative* capabilities (which is to say towards mechanisms of ever more efficient compression of *non-specific* data). This is hence a fully open-ended cognitive bootstrapping mechanism - the continuous need of the species to which the agent belongs to compress general, previously unexperienced sensory data amounts to a process of perceptual *hypothesis formation,* since the generalizability of the compression must be tested by feeding the hypothesis back into environment to establish its usefulness to the agent (in a process of hypothesis verification). The agent's percept categories hence become self-founding in a process akin to the *hermeneutic circle.* We now look more closely at the specific form that the perception-action relation must take in embodied agents.

The Embodied Perception-Action Relation in Cognitive Biology

The notion that the form of our conscious perception of the external world is dictated by, or further, *defined within the terms* of the actions that we may perform within it, is common both to phenomenology (as indicated in the Philosophy section), and also to several long-standing schools of cognitive science. (Dewey had argued as early as 1896 [26] that perception, thought and action must be considered as part of the same stratum). A paradigmatic example of action-based perception in cognitive science is given in the study of environmental *affordance*, a term first coined by James Gibson [27], and specified in [28] as having the following properties:

- 1. An affordance exists relative to the action capabilities of a particular agent.
- 2. The existence of an affordance is independent of the agent's ability to perceive it.
- 3. An affordance does not change as the needs and goals of the agent change.

Affordances, being the action possibilities of the agent's environment, are thus objective in the sense of being invariant to arbitrary shifts in interpretation. However, a complementarity is implicated between perceiver and perceived: the criterion of accuracy for perceptual representation now depends on the agent's ability to represent *its own* active possibilities, i.e its self-model. Related schematizations of embodied cognition include Lakoff's [29] argument that *reason* is itself patterned by the spatial awareness of agency. Glenberg [30] similarly argues that conceptualization is constrained by the structure of the environment, our bodies, and our memory capacity. On the applied side of cognitive science are the searches for neural correlates of embodied cognition, for instance Berlucchi and Aglioti's [31] argument that the imitation of movements within neonates is indicative of an implicit neural body-structure model from which later neural body-structure models are determined. This model provides a reference frame that further extends to the neural determination of *inanimate* object models. The mechanism of object understanding is thus a cognitive bootstrap to the extent that it requires, firstly, an initial set of *a priori* assumptions (the implicit model) in terms of which the world model is first defined and, secondly, a constructive engagement between the world and agent's world-representation in order to refine this model. This work, and others like it, thus serve to validate Piaget's [32] notion that higher cognitive functions have their roots in lower-level biological mechanisms.

A similar idea is expressed by Millikan [14] with regard to language and intentionality, arguing that *function* can only be attributed to an entity within a biological context. She hence proposes a biological solution to the Kripke-Wittgenstein paradox, which relates to the apparent impossibility (at least in Kripke's reading of Wittgenstein) of establishing absolute conceptual or perceptual identity between communicating agents, since an unbounded notion such as the concept of 'addition' could never be proven to be the same for both agents. For example, one agent's rule of addition might be the 'correct' one; $\forall x, y \; z := x + y$, whereas the other agent's rule might be some near approximation such as; $\forall x, y \; x < 5 \times 10^9, \; y < 5 \times 10^9 \; z := x + y$; else $z := 5$. In any reasonably finite scenario these agents would falsely form the impression that they both had the same understanding of the addition concept. Millikan's resolution of the paradox is to propose that natural selection serves to remove the latter formulation of the addition rule on the grounds of its inefficiency; it does the same essential referring as the former rule with regard to reasonably small numbers (such as those the agents typically experience in their biological lifetime), but uses more computation to do so. Hence aggregate natural selection will favor the smallest generalization consistent with the biologically necessary referents (thus providing a basis for Occam's Razor).

Millikan's work thus overcomes the classical *problem of reference*, where the relation between percept and object appears to be arbitrary (we might, for instance, ask why we regard the perceptual class *animal* as a singular entity, rather than as a collection of organic sub-objects or as a subpart of a species-collective). Millikan argues that the particular form the percept takes in relation to the object and the agent-object interaction has an inherent survival value for the agent (we have traditionally hunted animals for food, and so regard an individually huntable unit as a single perceptual entity). Percept models that do not efficiently model the survival-relative aspects of the object in relation to the agent's action possibilities simply cease to exist on an evolutionary time-scale.

16.4 Linguistic Signification and Embodied Agency

Perhaps the most obvious manifestation of the autonomous learning of symbolic representations occurs in human language. In attempting communication with another cognitive entity, agents must necessarily find a representation of the *commonalities* of their experience prior to allocating exchangeable linguistic tokens capable of standing in for these representations. That is, we must abstract from our immediate perceptions in order to find that aspect of them that is accessible to a real or putative second entity embodying a similar perceptive capability. As we have seen, the possibility of the abstraction of aspects of our perception/action experience into the third person is, for Kant, already implicit in our perception of the world. Perceptions are inherently *experienced* as having a certain unifying constancy under the transformations associated with agent actions; that is, we perceive *objects* from *perspectives*, rather than pure sensory impressions. The abstraction of our experience required for communication is thus *implicit at the outset*. However, this rigid, predetermined ontological structure might not initially appear to allow for the possibility of *learning* a language, or for the spontaneous evolution of an appropriate language between cognitive entities attempting to describe their cognitive world at an appropriate level of detail. How is it then possible, in a communicative context, for cognitive entities to establish a common symbolic representation of the world that goes beyond what is necessitated *a priori*?

Implicit in this idea is the formulation of a symbolic representation of the agent itself. Rohrer [33], for instance, suggests that linguistics should properly be regarded as a sub-science of cognitive science, proposing that the basis for language is the projection of one's own agency model into the perceptual domain; that is, a *de-relativizing* of experience in order to establish a common frame of reference. Perry [34], Bermudez [35] and Metzinger [36] also agree that cognitive self-awareness (as manifested by the linguistic token 'I') *requires* all communicating parties to have internal representations of both the world and of the various inter-communicating agents; in no other circumstances can one explicitly attribute *perceptions* to *oneself.* (Viezzer argues in [37] that the symbol grounding problem can only really be solved by modeling both the agent's world [at the perception/action level] *and* the agent's modeling of the world in order to permitting *genuine* representational updating by the agent). Pinker [38] argues that language derives from an initial cognitive orientation attributable to an active agent (so that, for example, the fundamental noun/verb split mimics the perception/action division), which then develops along more complex lines via a semantic bootstrapping mechanism.

Spontaneous Language Formation in Embodied Agents

The study of spontaneous language formation in simulated agents gains its philosophical imperative in consequence of the *symbol grounding problem* first enunciated by Harnad [39]. Harnad's thesis demands a semantic interpretation of formal symbol systems that transcends the (merely syntactic) interrelationships available to the symbolic manipulation system in question. The problem Harnad identifies is analogous to the learning of non-native languages in humans; this is much more meaningful when attempted *in situ* amongst other speakers of the language than when learned from a dictionary. Harnad consequently proposes two forms of symbolic grounding in particular; 'iconic representations', which are effectively equivalent to class perceptual medians, and 'categorical representations', which consist of both learned and *a priori* feature *invariants.* Steels gives perhaps the paradigmatic demonstration of semantic grounding in the formation of language in the 'Talking Heads Experiment' [40], the motivation for which is to demonstrate that 'communication through language is the main driving force in bootstrapping the representational capacities of intelligent agents'. Language and meaning are consequently coeval in this scenario; symbolic syntax arises *at the same time* as semantics.

The talking heads experiment hence consists in a pair of robotic agents each equipped with a video camera and a set of predetermined low-level feature descriptors that can be arbitrarily mapped to internally-generated words. One agent is initially designated the 'speaker', and the other the 'hearer'. The agents occupy an environment in which planar objects of various colors are distributed at random (for instance red squares, blue triangles etc). The designated speaker then chooses one item at random from this common context and attempts to describe it using its own internal lexicon (which it *cannot* simply assume is shared by the hearer). The hearer must then guess the correct item and point at it, failure to do so requiring the hearer to update its internal lexicon by generating a new word definition that successfully *disambiguates* the indicated item. The role of hearer and listener are then exchanged over a series of language games in order that an *objective* world description be finally obtained by both agents (as opposed to the identical, but speaker-subjective world description that would arise if the roles of speaker and hearer were fixed). Word definitions are thus characterized in terms of combinations of *a priori* feature descriptors of a visual nature; for instance, color, horizontal object positions, vertical object positions, etc. For example, consider an experimental context in which two objects A and B, a red triangle located at the top of the field of view and a blue square located at the bottom of the field of view, are the respective objects of interest. These might be disambiguated by word-descriptors of the form: A: vertical−position > 0.5; B: vertical−position < 0.5. Or, equivalently, by descriptors of the form: A: red; B: blue

There is hence no unambiguously 'correct' object word-representation in this scenario, and consequently no ground truth perceptual space accessible to the agents. If these two alternative sets of lexical designations were allocated to the speaker and hearer, respectively, it would consequently only be within an *expanded* experimental context that the discrepancy in description would come to light. For instance, only if a third blue object were introduced and located towards the bottom of the field of view, would the speaker be required to learn to distinguish the concept of *color* as a distinct perceptual category (though it always inherently had the latent capacity to do so), in order to distinguish every object employed within the word-game (perhaps correlating with the neonatal synaesthesia hypothesis [41]). Equally, the hearer would need to evolve word descriptions that incorporated spatial considerations only in order to distinguish all three objects within the *extended* scenario. Steels' achievement is consequently in demonstrating that lexical convergence between speaker and hearer does indeed occur. Moreover, provided that there exists a sufficient richness in the range of object scenarios, the talking heads experiment demonstrates that this convergence is *objective* (in the sense that the final word distinctions correspond to our ground truth descriptions in terms of the *a priori* features).

This result is consequently consistent with the hypothesis that 'third-person' cognitive modeling lies at the heart of the symbol/referent relation. The *objectivity* (or subject-independence) of the final convergence of the word designations hence comes about because language conjectures are projected by the speaker back into the environment for validation *on the assumption* of the presence of a hearer with a linguistic and indicative capability similar (in *a priori* terms) to it own; self-modeling of perceptual agency is thus implicit in the experimental scenario. In philosophical terms, the talking heads experiment embodies the Wittgensteinian (cf [13]) view of communicative activity as a 'language game' in which agents invent words and meanings during their interactions, and opposes the Quinean [12] view that sees language as a series of inductive abstractions of perceptual correlations between word and object.

16.5 Approaches to the Spontaneous Generation of Symbolic Representations in Embodied Artificial Agents

The engineering field in which embodied cognitive bootstrapping receives its most tangible expression is thus robotics; the study of programmable machine systems. When this programmability extends to the notion of self-programmability, we are concerned with the particular field subset known as *autonomous robotics*. When the goal is further to *construct* a sensory model of (presumably previously unexperienced) environments, we are then implicitly in the realm of artificial embodied cognition or *cognitive robotics*. Recent advances in the computer processing power available for real-time computation have allowed robotics to begin to employ *cognitive vision* methods, for which the sensory input consists of mono-, stereo- or multi-scopic camera feeds. Environmental modeling in the cognitive vision regime is hence analogous to that exhibited by the mammalian cognitive vision system (particularly when dealing with with stereo and multiscopic camera feeds, for which a significant computational burden is the three-dimensional reconstruction of the environment from planar projections). Typical low-level cognitive tasks thus include edge detection, object segmentation, motion registration, and so on, with potentially ever higher levels of cognitive abstraction possible beyond the immediate low-level vision tasks.

One particular area of investigation that implicates the notion of cognitive bootstrapping occurs at the interface of visual and haptic perception (e.g. [42], [43]). When a mammalian agent interacts with the environment, it implicitly updates its visual model of the environment by *haptic contact*, using the *a priori* certainly of touch data to reduce the ambiguity present in visual data (particularly the ambiguities of binocular scene reconstruction). Moreover, it appears that the mammalian brain achieves this Bayes-optimally. The cognitive bootstrap in this model is thus the use of visual perception to *motivate* sensorimotor actions such as those involved in grasping for an object *in order to test the validity of those same visual perceptions*. As before, the bootstrapping of an initial, partially representative model and the iterative convergence between percepts and percept-motivated actions hence acts to overcome the logical paradox inherent in a self-validated perceptual system. More generally the concept of the perception-action cycle implicit in these visual-haptic models can by seen as the most tangible basis on which to implement an artificial cognitive bootstrap mechanism. Perceptions are hence seen as environmental *hypotheses* while actions are *hypothesis validation steps*. More specifically, *vision* is to be understood as a *hypothetical* linkage between possible instances of haptic contact (such as in 3D object reconstruction), and *vision-motivated actions* test the validity (or at least consistency) of these models.

The degree to which artificial cognition can be made fully open-ended is thus a matter of architecture; however, it is necessary, or at least, vastly simplifying, to incorporate a number of *a priori* constraints on the cognitive reinterpretation process, the general minimum being the presence of a sensory topology that defines the arena in which the autonomous robot is active as a *space*. However, this spatial representation need not necessarily occur at the lowest level of the vision hierarchy, a point that will become apparent in the following discussion.

Hierarchical Percept-Action Approaches to Cognitive Robotics

Hierarchical approaches to autonomous robotics were first proposed by Brooks in [44], who employed the term *subsumption architecture*. The assumption of such architectures is that agent abilities are arranged in *levels*, with higher-level competences incorporating lower-level competences. For instance, the ability to plan a route presupposes the ability to avoid obstacles. Higher architectural layers hence control the behavior of the lower via the mechanism of inhibition, allowing the possibility of open-ended development of the cognitive agent's responses. Brooks notes that different forms of environment representation are appropriate to the differing levels, and that these levels can be extended indefinitely; however, the possibility of autonomously abstracting these higher hierarchical levels *along with* an appropriate environment representation is not directly considered. For this, we require an *abstractable percept-action* architecture.

Modayil [45] hence proposes a method of bootstrapping progressively higher levels of symbolic representation, up to and including the concept of *objects*, via the clustering of representations from lower levels of the OPAL (Object Perception and Action Learner) architecture. Bootstrap learning thus allows the system to move from egocentric (view-centered) and allocentric (object-centered) sub-symbolic descriptions to symbolic object-based description by ascending a four-fold hierarchy; Individuation, Tracking, Image Description and Categorization. Individuation involves the use of occupancy grids to classify individual sensor readings as either static or dynamic. Clusters of dynamic readings are then tracked over time to provide an object model; stable shape models are then constructed from the consistent aspects of the objects so formed. OPAL is thus capable of autonomously discretizing the sensory environment into a static background, the learning robot, and a set of movable objects via the abstraction of a perception-action architecture.

Granlund [18] provides a still more general architecture for cognitive robotics based on the notion that scene description is *not* required prior to action. Thus, it is argued that the failure of conventional cognitive architectures is due to the categoric abstraction of objects at an intermediate stage between percept formation and action specification. What is lost in this approach are the contextual modifiers necessary for precise specification of agent action; in short, we gain *descriptivity* at the expense of *intentionality*, the latter being relevant only to an embodied agent in a particular context. Granlund hence proposes a bootstrap mechanism for the initial learning of the embodied system based on a perception-action feedback cycle. Here, in the learning phase of the perception-action mapping, action *always* precedes perception. Thus, the potentially exponential complexity of the percept domain is limited by considering only those percepts directly related to actions, which consequently occupy a far smaller state-space. (An idea of the information-theoretic disparity between these two different types of environmental modeling, the agent-specific and the agent-non-specific, is found in [46]). In the absence of explicit scene-representation actions are hence driven by biologically-motivated random exploration impulses (literally random walks in the action state space).

The percept-action mapping can thus be made subject to various optimization procedures that allow compact representation, and implicitly, therefore, *generalization*. The random actions and subsequent compact percept mappings thus amount to an unsupervised training of the architecture. There consequently exists a natural stopping criterion for the random action impulses at the point at which the compact representation of the percept-action mapping no longer undergoes significant change (learning having converged). At this point the random action impulses can ascend to a greater level of abstraction and operate on the higher-level percept-action representations that have been generated by the compact generalization. These higher level action impulses themselves generate further training data at the lower levels, allowing for robust and adaptive learning across the whole of the hierarchical structure so formed.

These compact representations within the hierarchical percept-action structure are *symbols*, corresponding, for instance, to the symbols employed in verbal communication. Such communication might hence be considered a low-bandwidth interaction between agents that allows complex actions to be initiated in one agent by another by virtue of the 'unpacking' of the compact representations that takes place as information travels down the percept-action hierarchy from the highest to the lowest levels. Symbolic communication between such agents is hence *always* grounded. The cognitive architecture thus defined is clearly one of cognitive bootstrapping; the inferred higher-level cognitive hypotheses validate themselves *in terms of* the lower-level hypotheses by virtue of the 'filtering-down' effect wherein action impulses in the high-level abstracted cognitive categories result in progressively more *contextualized* low-level actions. Only at the highest goal-setting level is there thus a requirement for environment representations that are completely logically self-consistent (such as a coarse-grained reconstruction of the three-dimensional volume in which the agent acts): lower hierarchical levels need only be *para-consistent*.

Sun [47], in setting out a foundation for artificial cognitive architectures, similarly argues that human cognition is essentially 'bottom-up' and further, that minimal initial bootstrap models are necessary to avoid over-representational models that may fail to generalize. Stein [48] also argues that goal-based behavior in cognitive robots should be considered, not only at an abstract symbolic level, but also at the *lowest sensorimotor levels.* Hence, in projecting a goal, a robotic agent should utilize *exactly* the same exploratory and learning processes that it uses to interact with the real world, but instead substitute a 'virtual reality' interface at the very lowest level of the sensors and actuators. This virtual reality is precisely the sensory map formed by the currently hypothesized world-model. 'Cognition', for Stein, is hence simply the *imagined sensation and action* implicit in tracing out an action path to a particular goal state in the world-model. Stein's MetaToto hence *self-trains* its higher-level cognitive abilities using only its internal representations. There is perhaps a Darwinian justification for this imaginative self-training; a biological agent that tests its action hypotheses in imagination can rule out potentially unsurvivable actions without endangering itself. Such agents are thus more likely to prevail and reproduce than equivalent unreflective agents. In human terms, this principle may also relate to the phenomenon of sleep paralysis (treated more completely in the discussion and conclusions section).

A framework for autonomous perception-action learning that employs inductive logic programming to establish environment protocols and bootstrap appropriate high-level symbolic representations is given by the author in [49]. For a generic sensor-actuator coupling placed within a specific environment, only certain of the set of possible actions will serve to alter the percept space in a consistent fashion. Hence, after randomized exploration and induction of the rules governing this action legitimacy, the cognitive system sets out to eliminate redundant perceptual predicates in the inferred clauses in order to express a new, higher-level percept-action correspondence in which its actions are *always* successful. Such higher level perception-action representation is always of a more symbolic and abstracted nature than the generic sensor-actuator coupling, ultimately defining an open-ended series of logically-described environmental affordances of a form appropriate to verbal communication.

16.6 Discussion and Conclusions

We have looked at the problem of autonomous symbol generation in a range of natural and artificial agents, and have identified the mechanism of 'cognitive bootstrapping' as a means of accomplishing this in a maximally open-ended and epistemologically-consistent fashion. Cognitive Bootstrapping is hence the iterative mechanism by which cognition can become *self-founding* without falling into Quine's ontological relativism [50], in which *any* world representation can be considered valid. The mechanism thus iterates between interpretation (in which percept categories are applied to the world) and exploration (in which sensory-data that has the potential to clarify the validity of the conjectured percepts is sought). Cognitive bootstrapping hence constitutes a form of the *hermeneutic circle* within a perception-action learning context.

Critically, since the exploratory phase is conducted *in terms* of the existing and potentially invalid percept categories, the initial 'bootstrap' hypothesis must have a degree of *a priori* validity in order to allow progressive convergence on an 'objective' model. Furthermore, there must exist an *a priori* criterion of percept-hypothesis validation/falsification implicit in the bootstrap hypothesis (such as haptic contact in the case of autonomous visual-haptic robotics). These *a priori* percept categories (often taking the form of contact-sensing and motor-space feedback within physically-embodied cognitive entities) are thus not admissible to the perceptual updating procedure, and represent the sole limitations on the extent to which cognition can become *self-determining*. (We may hence legitimately doubt the visual perception of an object but not the fact of our haptic contact with it, or the muscle-articulations involved in reaching out to it).

We thus overcome the paradox inherent in constructing a cognitive agent with unlimited capacity for forming novel percept categories with which to view the world, which must nonetheless be able to *perceive* whether these categories are representative of the world. Overcoming the paradox by bootstrapping requires that we have an initial set of *low-level* percept categories that we *must assume* are 'correct', and then *hierarchically* progress from there to higher-level categories via percept-hypothesis formation and action-based testing. This initial category set, we argue, is the set of Kantian *a priori* cognitive categories capable of providing a framework in which Popperian [51] falsification of percept category hypotheses can be adequately formulated. Without this mechanism a perceiving subject could not distinguish internal perceptual and external object states with any epistemological certainty.

The question then arises as to what constitutes the minimal *a priori* category set required for cognitive bootstrapping in the artificial cognitive domain; the *a priori* cognitive categories underlying the cognitive bootstrap need not be structurally identical with those of humans. For instance, in a cognitive architecture such as Granlund's [18], rather than an *object* category being imposed *a priori*, we have instead the broader-based *a priori* notion of *invariant percept subspaces* from which compact and invariant symbolic entities of increasing hierarchical complexity can be progressively defined, *including* the synthetic category of 'object'.

The context of symbol-hypothesis falsification in this architecture is then the percept-action link coupled with an exploratory imperative (even a simple 'random walk' imperative will suffice). Thus, the architecture presumes that the output of symbol manipulation must always result in an actual or potential *action*, the effectiveness of which the agent must determine from within the percept space (which itself incorporates the higher level symbolic entities). Hence, an action imperative derived at the symbolic level (for instance, the placing of one particular object on top of another) can only be evaluated as having been carried-out successfully by utilizing *both* the higher-level symbolic categories (since the imperative was formulated in these terms) *and* the lowest-level object representation (since this provides the primary link between the symbolic layer and the *a priori* sensory level of which it is an invariant subspace category). The symbol system is thus always semantically grounded; the system can spontaneously form and evaluate the suitability of invariant categories (which are always *hypothesized*), subject only to the constraint that it can not re-evaluate the validity of the *a priori* sensory level, or the invariant subspace categorization mechanism itself.

In terms of biological agents, *a priori* environment representation proceeds via Darwinian natural selection. However, environmental selection pressures on replicating agents in a rapidly changing environment (relative to the evolution rate) will always tend to favor cognitive architectures that generalize to the greatest extent given their initial *a priori* configuration. Such agents must hence evolve via a bootstrap process toward a minimization of the disparity between the biological agent's internal world representation and the species-based survival imperatives imposed by the environment. Human societal (as opposed to genetic) evolution meets this criterion, with survival demands on human communities typically changing on generational, rather than evolutionary, time scales. Here, the means of replication of human behavior and understanding is not gene-based (which would respond only very slowly to environmental pressures) but rather meme-based, that is to say, replicated via linguistic communication, and is hence capable of far more rapid evolution (see [52]). We hence agree with Millikan [14] that the *a priori* representativity of congenital human percepts is granted via natural selection (so that, for instance, if human beings' innate perception of *ingestability* did not, to some degree, correlate with those objects in the environment that met with their nutritional requirements, then the species would not have proved biologically viable in the long term). Any artificial autonomous agent would similarly require a minimal set of guaranteed referential percept categories, but, in the absence of a framework of natural selection, these would have to be imposed by their designers, perhaps motivated along biological lines.

Given that the referentiality of perception must be ensured at the outset, the question then arose of how, within the confines of these Kantian restrictions, open-ended cognitive development is actually to be accomplished. We have seen that, in general, the perceptual optimization strategy adopted by biological and artificial agents is one of perceptual *compression*; the idea being to reduce the total sensory stream into a relatively few significant data. This, however, is still not sufficient, in itself, to determine the appropriateness of a proposed perceptual update - after all, it is always possible to map every percept to a single datum, giving maximal compression at the expense of all environmental information. Thus, any novel perceptual inference must be allied with an action complex *within which this perceptual inference is sustained.* We thus *utilize* a percept mechanism of unknown value in order to interpret the external world in such a way that we can gain sufficient information in order to evaluate the worth of that perception mechanism. If it proves insufficient to the task of gathering enough evidence to validate itself, then it automatically fails that validation. Any new percept categorizations must hence be made *in terms of* well-established or perceptual bootstrap categories, such that these new percept categorizations can in turn be treated as the basis for further categorizations in a hierarchical fashion. We thus always maintain a 'fall-back' mechanism for empirical validation, irrespective of the perceptual framework adopted. A consequence of this is that an autonomous agent with no overall goal other than randomized exploration can form an enormous range of intentional *sub-goals* by virtue of the hierarchicality implicit in bootstrapped cognitive structure.

This notion of hierarchically-grounded intentionality would then correlate with the existence of the 'sleep-paralysis' mechanism in mammals. According to the activation-synthesis theory [53], during rapid eye movement (REM) sleep, *randomized* neuronal stimulation is applied to the pons area of the brain as part of its memory consolidation activity. This randomized activity is interpreted at the perceptual level as *dreaming*. Dreaming is hence experienced as high-level visual and auditory stimuli of the same sort that occur in waking life, albeit with an appropriately randomized narrative component. However, this imagery is not merely abstracted symbolism, being rather *hierarchically grounded* in the percept-action complex of the organism. Mammals thus have an innate tendency to *act out* responses to the dream-stimuli in an intentional and physical manner. It is therefore necessary for the brain stem to actively prevent this motor stimulation from making the final connection from the lowest-level of the grounded hierarchy to the muscles: a failure of this mechanism results in the phenomenon of *sleep-walking*.

A further example of the hierarchical grounding of higher-level visual percepts in low-lever percept-action mappings occurs in the mirror-neurons of the primate premotor cortex [54]: these neurons fire in response *both* to motor actions performed by the primate, as well as to those same motor actions performed by other primates *in the observing primate's visual field*. The high level visual percepts corresponding to the observed action must thus be hierarchically grounded in the intentional lower-level action states.

Conclusion

In conclusion, it is apparent from our analysis of Kant that a 'blank slate' approach to cognitive updating is not feasible. Certain minimal categorical assumptions must inhere in perception in order to define it as such, in distinction to the perceived environment. In terms of cognitive robotics these restrictions mean that agents are not simply free to apply arbitrary generalization techniques to the reinterpretation of raw sensory data in order to bootstrap novel perceptual primitives. By the same token, biological agents (e.g. humans) capable of autonomous cognitive updating must employ a certain degree of naturally-selected representative capability in order to serve as a basis for further updating of their representational framework.

Cognitive agents must hence initially characterize their active environment according to pre-specified imperatives (species-survival in the case of biological agents, but potentially more general imperatives for artificial agents). However, we have demonstrated that the perception-action relation is capable of hierarchical abstraction to the symbolic level, with higher-level representations validated in terms of the high-level actions implicit in them. The only limit on the ability of agents employing this approach to bootstrap new perceptual categorizations is then the retention of the *a priori* structures required to give an empirical validation criterion for both the updated representational frameworks as well as the environmental representations themselves.

Acknowledgment: The work presented here was supported by the the EU, grant COSPAL (IST-2003-004176). However, it does not necessarily represent the opinion of the European Community, and the European Community is not responsible for any use which may be made of its contents.
This chapter represents the biology-related aspect of arguments first outlined by the author in the University of Surrey technical report 'Cognitive Bootstrapping: A Survey of Bootstrap Mechanisms for Emergent Cognition' [55].

References

1. Hume D (1999) An Enquiry concerning Human Understanding. Oxford University Press, Oxford/New York
2. Kant I (1999) Critique of Pure Reason. Cambridge University Press
3. Heidegger M (1996) Being and Time. Blackwell
4. Dreyfus H (1972) What Computers Can't Do. New York: Harper and Row
5. Suber P. Mind and baud rate. E-print of the Phil. Dept., Earlham College, Retr. 13/6/2005 http://www.earlham.edu/ ~peters/writing/baudrate.htm
6. Winograd T (1980) What does it mean to understand language?, Cognitive Science 4(3):209–242. Reprinted in D. Norman (ed.), Perspectives on Cognitive Science, Ablex and Erlbaum Associates, 1981, 231-264
7. Gödel K (1931) Über formal unentscheidbare sätze der principia math. & verwandter systeme, 1 Monatshefte für Mathematik und Physik (38):173–198
8. Breuer T (2000) In: M Carrier GM L Ruetsche (ed.) Science at Century's End: Philosophical Questions on the Progress and Limits of Science. Pittsburgh & Konstanz, University of Pittsburgh Press & Universitätsverlag Konstanz
9. Breuer T (1995) The impossibility of exact state self-measurements, Philosophy of Science 62:197–214
10. Winograd T, Flores F (1986) Understanding Computers and Cognition. Addison-Wesley, Reading, MA
11. McCarthy J, Hayes P (1969) Some philosophical problems from the standpoint of artificial intelligence, Machine Intelligence (4):463–502
12. Quine WVO (1960) Word and Object. NY: John Wiley and Sons, MIT
13. Wittgenstein L (2001) Philosophical investigations : the German text with a revised English translation by Ludwig Wittgenstein. Oxford : Blackwell

14. Millikan RG (1987) Language, Thought, and Other Biological Categories: New Foundations for Realism. The MIT Press; Reprint edition
15. Sipper M (1995) An introduction to artificial life., Explorations in Artificial Life (special issue of AI Expert) 4–8
16. Marr D (1982) Vision: A Computational Approach. Freeman & Co., San Fr.
17. Gärdenfors P (1994) How logic emerges from the dynamics of information, Logic and Information Flow 49–77
18. Granlund G (2003) Organization of Architectures for Cognitive Vision Systems, In: Proceedings of Workshop on Cognitive Vision. Schloss Dagstuhl, Germany
19. Magee D, Needham CJ, Santos P, Cohn AG, Hogg DC (2004) Autonomous learning for a cognitive agent using continuous models and inductive logic programming from audio-visual input, In: Proc. of the AAAI Workshop on Anchoring Symbols to Sensor Data
20. Brooks RA (1991) Intelligence without representation, Artificial Intelligence 47:139–159
21. Newell A, Simon H (1976) The Theory of Human Problem Solving; reprinted in Collins & Smith (eds.), In: Readings in Cognitive Science, section 1.3.
22. Pinker S, Bloom P (1990) Natural language and natural selection, Behavioural and Brain Sciences 13(4):707–784
23. Marshall J, Blank D, Meeden L (2004) An emergent framework for self-motivation in developmental robotics, In: Proc. of the Third International Conference on Development and Learning (ICDL '04). Salk Inst.
24. Franklin S, Garzon M (1991) Neural Computability, In: Omidvar O (ed.) Progress in Neural Networks, vol. 1. Ablex
25. Wolff JG (1987) Cognitive development as optimisation, In: Bolc L (ed.) Computational Models of Learning, 161–205. Springer-Verlag, Heidelberg
26. Dewey J (1896) The reflex arc concept in psychology, The Psychological Review (3):356–370
27. Gibson JJ (1979) The ecological approach to visual perception. Houghton-Mifflin, Boston
28. McGrenere J, Ho W (2000) Affordances: Clarifying and Evolving a Concept, In: Proceedings of Graphics Interface 2000, 179–186. Montreal, Canada
29. Lakoff G, Johnson M (1999) Philosophy in the Flesh : The Embodied Mind and Its Challenge to Western Thought. Harper Collins Publishers
30. Glenberg A (1997) What memory is for, Behavioral and Brain Sciences 20(1):1–55
31. Berlucchi F, Aglioti S (1997) The body in the brain: neural bases of corporeal awareness, Trends in Neuroscience 20(5):60–564
32. Piaget J (1970) Genetic Epistemology. Columbia University Press, New York
33. Rohrer T (2001) Pragmatism, Ideology and Embodiment: William James and the Philosophical Foundations of Cognitive Linguistics, In: Sandriklogou, Dirven (eds.) Language and Ideology: Cognitive Theoretical Approaches, 49–82. Amsterdam: John Benjamins
34. Perry J (1997) Myself and I, In: Stamm M (ed.) Philosophie in Sythetisher Absicht, 83–103. Stuttgart:Klett-Cotta
35. Bermudez JL (2001) Non-conceptual self-consciousness and cognitive science, Synthese (129):129–149
36. Metzinger T (2003) Phenomenal transparency and cognitive self-reference, Phenomenology and the Cognitive Sciences (2):353–393

37. Viezzer M (2001) Dynamic Ontologies or How to Build Agents That Can Change Their Mind. Ph.D. Thesis, University of Birmingham, UK
38. Pinker S (1995) The Language Instinct: The New Science of Language and Mind. Penguin Books Ltd. ISBN: 0140175296
39. Harnad S (1990) The symbol grounding problem, Physica D (42):335–346
40. Steels L (1997) The origins of syntax in visually grounded robotic agents, In: Pollack M (ed.) Proceedings of the 10th IJCAI, Nagoya, 1632–1641. AAAI Press, Menlo-Park Ca.
41. Harrison JE, Baron-Cohen S (1996) Synaesthesia: Classic and Contemporary Readings. Blackwell Publishers
42. Saunders J, Knill DC (2004) Visual feedback control of hand movements, J of Neuroscience 24(13):3223–3234
43. Schlicht EJ, Schrater PR (2003) Bayesian model for reaching and grasping peripheral and occluded targets, Journal of Vision 3(9):261
44. Brooks RA (1986) A robust layered control system for a mobile robot, IEEE Journal of Robotics and Automation 14(23)
45. Modayil J. Bootstrap learning a perceptually grounded object ontology. Retr. 9/5/2005 http://www.cs.utexas.edu/users/modayil/modayil-proposal.pdf
46. Nehaniv CL, Polani D, Dautenhahn K, te Boekhorst R, Canamero L (2002) In: Standish B Abbass (ed.) Artificial Life VIII, 345–349. MIT Press
47. Sun R (2004) Desiderata for cog. architectures, Philosophical Psychology 17(3)
48. Stein LA (1991) Imagination and situated cognition. Tech. Rep. A.I. Memo No. 27, MIT AI Laboratory
49. Windridge D, Kittler J (2007) Open-Ended Inference of Relational Representations in the COSPAL Perception-Action Architecture, In: Proc. of International Conf. on Machine Vision Applications (ICVS 2007). Germany
50. Quine WVO (1977) Ontological Relativity. Columbia
51. Popper K (1959) The Logic of Scientific Discovery. (translation of Logik der Forschung). Hutchinson, London
52. Dawkins R (1989) The Selfish Gene (2nd ed.). OUP
53. Hobson J (1988) The Dreaming Brain. Basic Books, New York
54. Gallese V, Goldman A (1998) Mirror neurons and the simulation theory of mind-reading, Trends in Cognitive Sciences 2(2)
55. Windridge D (2005) Cognitive bootstrapping: A survey of bootstrap mechanisms for emergent cognition. Tech. Rep. VSSP-TR-2/2005, CVSSP, The University of Surrey, Gulldford, Surrey, GU2 7XH, UK

Index

Printing: Krips bv, Meppel, The Netherlands
Binding: Stürtz, Würzburg, Germany